Lecture Notes in Mathematics

Edited by A. Dold and B. Eckmann

1329

M. Alfaro J. S. Dehesa F. J. Marcellan
J. L. Rubio de Francia J. Vinuesa (Eds.)

Orthogonal Polynomials and their Applications

Proceedings of an International Symposium
held in Segovia, Spain, Sept. 22–27, 1986

Springer-Verlag
Berlin Heidelberg New York London Paris Tokyo

Editors

Manuel Alfaro
Departamento de Matemáticas, Universidad de Zaragoza
50009 Zaragoza, Spain

Jesús S. Dehesa
Departamento de Física Moderna, Universidad de Granada
18071 Granada, Spain

Francisco J. Marcellan
Departamento de Matemática Aplicada, E.T.S. de Ingenieros Industriales
28006 Madrid, Spain

José L. Rubio de Francia †
Departamento de Matemáticas, Universidad Autónoma de Madrid
28049 Madrid, Spain

Jaime Vinuesa
Departamento de Matématicas, Universidad de Cantabria
39080 Santander, Spain

Mathematics Subject Classification (1980): Primary: 42C05, 33A65
Secondary: 40A30, 39A10, 40A15, 41A55, 41A60, 60J80, 12D10

ISBN 3-540-19489-4 Springer-Verlag Berlin Heidelberg New York
ISBN 0-387-19489-4 Springer-Verlag New York Berlin Heidelberg

Printed in Germany

Printing and binding: Druckhaus Beltz, Hemsbach/Bergstr.
2146/3140-543210

This volume is dedicated to the memory of Jose Luis Rubio de Francia, Professor of Mathematics of the Autonomous University of Madrid, who died on February 1988. His death is a great loss to his country, to science and education, and to his many friends. We wish to record here our profound sense of loss. We hope that the completed volume is faithful to his vision of a book that not only instructs but also inspires.

INTRODUCTION

The present volume contains the Proceedings of the Second International Symposium on Orthogonal Polynomials and Their Applications held in Segovia, Spain, September 22nd to 27th, 1986.

This Symposium continues the idea born at Bar-le-Duc (France) on the occasion of the 150th anniversary of Edmond Nicolas Laguerre's birthday.

It intends to present a comprehensive view of the main achievements, fundamentals and working problems of the orthogonal polynomials and its relationship with other mathematical fields and applied sciences.

The core of the Symposium was a series of ten invited lectures, each covering some of the more fundamental aspects. In addition, there were 60 half-hour communications, including particular aspects of orthogonal polynomials on the real line and over the complex plane, special functions, orthogonal expansions and numerical methods. A special session devoted to open problems in the field of orthogonal polynomials also took place.

This volume contains nine invited lectures, more than ten seminars and a collection of some open questions.

Many institutions and individuals helped make the Symposium possible. We particularly thank, for their financial support, the Comisión Asesora de Investigación Científica y Técnica (CAICYT), Consejo Superior de Investigaciones Científicas (CSIC), Confederación Española de Centros de Investigación Matemática y Estadística (CECIME), Universidad Politécnica de Madrid, IBM-España S.A., Diputación Provincial de Segovia, Ayuntamiento de Segovia y Caja de Ahorros y Monte de Piedad de Segovia. For the hospitality offered to the participants, we thank the Ayuntamiento and Diputación Provincial de Segovia. Finally it is a pleasure to acknowledge Mónica Garay and Russell di Napoli for their secretarial help. We are also greatly indebted to the series editors of the Lecture Notes in Mathematics for giving us the possiblity to publish these Proceedings in this series.

M. Alfaro
J. S. Dehesa
F. Marcellán
J. L. Rubio
J. Vinuesa

TABLE OF CONTENTS

POPOV, B.: A class of polynomials connected with orthogonal polynomials.

RONVEAUX, A.: Semi-classical weights (-∞,+∞) semi-hermite orthogonal polynomials.

SABLONNIERE, P.: Hahn polynomials as eigenvectors of positive operators.

SANSIGRE, G., FERNANDEZ, L., ALVAREZ, M.: On zeros of general orthogonal polynomials on the unit circle.

SMAILI, N.: The class-T discrete semi-classical polynomials.

SOTTAS, G.: Positivity domain of ultraspherical type quadrature formulas with Jacobi abscissas: Numerical investigations.

SZAFRANIEC, F.H.: Orthogonal polynomials and subnormality of related shift operators.

TEMME, N.M.: Laguerre Polynomials: Asymptotics for Large Degree.

VAN DOORN, E.A.: Orthogonal polynomials, chain sequences and oscillation criteria for second order linear difference equations.

VAN ISEGHEM, J.: Zeros of vector-orthogonal polynomials.

WALLIN, H.: Orthogonal polynomials on fractals.

ZAYED, A.: A generalized inversion formula for the continuous Jacobi Trasform.

LIST OF PARTICIPANTS

ALFARO, M.P.
Dpartamento de Matemáticas
Univ. de Zaragoza. Spain.

ALFARO, M.
Departamento de Matemáticas
Univ. de Zaragoza. Spain.

AL SALAM, W.A.
Department of Mathematics.
University of Alberta
Edmonton, Alberta T6G 2G1
Canada

ALVAREZ, M.
Departamento de Matemáticas
E.T.S. de Ingenieros Industriales
Univ. Politécnica de Madrid
Spain

ANTOLIN, J.
Departamento de Física Teórica
Univ. de Zaragoza. Spain.

ARIAS DE VELASCO, L.
Departamento de Matemáticas
E.T.S. de Ingenieros Industriales
Univ. de Oviedo, Gijón. Spain.

BACHENE, M.
U.E.R. Analyse, Probabilité et
Applications.
Univ. Pierre et Marie Curie.
Paris, France.

BACRY, H.
Centre de Physique Theorique.
Luminy
Marseille, France.

BARREIRO, A.
Departamento de Matemáticas.
E.T.S. de Ingenieros Industriales
Univ. Politécnica de Madrid.
Spain.

BAVINK, H.
Department of Mathematics
and Computer Science.
Technical University Delft.
The Netherlands.

BELMEHDI, S.
Department de Mathematiques
Université de Savoie.
Chambery, France

BEN CHEIKH, Y.
Department de Mathematiques
Faculté de Sciences et Techniques
Monastir. Tunisia

BOUKHEMIS, A.
U.E.R. Analyse, Probabilités et
Applications.
Univ. Pierre et Marie Curie.
Paris, France.

BREZINSKI, C.
U.E.R. d'I.E.E.A.
Univ. des Sciences et Techniques
Lille I.
Villeneuve d'Ascq. France.

CACHAFEIRO, A.
Departamento de Matemáticas.
E.T.S. de Ingenieros Industriales
Univ. de Santiago. Vigo. Spain.

CASASUS, L.
Dpto. de Ecuaciones Funcionales
Univ. de La Laguna. Spain.

DEHESA, J.S.
Dpto. de Física Teórica
Univ. de Granada. Spain.

DJEHAICHE, F.
U.E.R. Analyse, Probabilités et
Applications.
Univ. Pierre et Marie Curie.
Paris, France.

DOUAK, K.
U.E.R. Analyse, Probabilités et
Applications
Univ. Pierre et Marie Curie.
Paris, France.

DRAIDI, N.
Laboratoire de Mathématique
Université d'Orsay. France.

DRAUX, A.
U.E.R. d'I.E.E.A.
Univ. des Sciences et Techniques
Lille I.
Villeneuve d'Ascq. France.

EIERMANN, M.
Institut fur Praktische Matematik
Universitat Karlsruhe. FRG.

FERNANDEZ, L.
Departamento de Matemáticas
E.T.S. de Ingenieros Industriales
Univ. Politécnica de Madrid.
Spain.

GARCIA LAZARO, P.
Departamento de Matemáticas
E.T.S. de Ingenieros Industriales
Univ. Politécnica de Madrid
Spain.

GATTESCHI, L.
Dipartimento di Matematica
Universitá di Torino. Italy.

GAUTSCHI, W.
Department of Computer Science
Purdue University.
West Lafayette,
Indiana. U.S.A.

GERONIMO, J.S.
School of Mathematics.
Georgia Institute of Technoloty.
Atlanta, Georgia, U.S.A.

GILEWICZ, J.
Centre de Physique Teorique.
Luminy.
Marseille, France.

GIORDANO, C.
Dipartimento di Matematica
Universita di Torino. Italy.

GODOY, E.
Departamento de Matemáticas.
E.T.S. de Ingenieros Industriales.
Univ. de Santiago. Vigo, Spain.

GONZALEZ VERA, P.
Dpto. de Ecuaciones Funcionales
Univ. de La Laguna. Spain.

GONZALEZ-CONCEPCION, C.
Dpto. de Ecuaciones Funcionales
Univ. de La Laguna. Spain.

GORI, L.
Dipartimento di Metodi e Modelli
Matematici per le Scienze
Applicate.
Universita degli Studi di Roma,
"La Sapienza". Italy.

GUADALUPE, J.J.
Departamento de Matemáticas
Univ. de Zaragoza. Spain.

GUADALUPE, R.
Departamento de Matemáticas
Facultad de Informática
Univ. Politécnica de Madrid. Spain.

GUERFI, M.
U.E.R. Analyse, Probabilités et
Applications.
Univ. Pierre et Marie Curie.
Paris, France.

HENDRIKSEN, E.
Department of Mathematics.
University of Amsterdam.
Amsterdam 1018 WB. The Netherlands.

ISMAIL, M.E.
Departament of Mathematics.
Arizona State University.
Tempe, Arizona. U.S.A.

JACOBSEN, L.
Institutt for Mathematik.
Universitet Trondheim AVH.
Dragvoll, Norway.

KOORNWINDER, T.H.
Centre for Mathematics and Computer
Science
Amsterdam 1009 AB. The Netherlands.

KRALL, A.M.
Departament of Mathematics.
Pennsylvania State University.
University Park,
Pennsylvania. U.S.A.

KURATSUBO, S.
Department of Mathematics.
Faculty of Science.
Hirosaki University.
Aomori 036, Japan.

LAFORGIA, A.
Dipartimento di Matematica
Universita di Torino, Italy.

LASCOUX, A.
U.E.R. Mathematiques
Université Paris VII, France.

LITTLEJOHN, L.L.
Department of Mathematics.
Utah State University.
Logan, Utah. U.S.A.

LOPEZ, G.
Facultad de Matemáticas y
Cibernética.
Univ. de La Habana, Cuba.

MAGNUS, A.
Institut Mathematique.
Univ. de Louvain-la-Neuve,
Belgium.

MARCELLAN, F.
Departamento de Matemáticas
E.T.S. de Ingenieros Industriales
Univ. Politécnica de Madrid,
Spain.

MARKETT, C.
Lehrstuhl A fur Mathematik.
RWTH Aachen. FRG.

MARONI, P.
U.E.R. Analyse, Probabilités et
Applications.
Univ. Pierre et Marie Curie.
Paris, France.

MASON, J.
Computational Mathematics Group.
Royal Military College of
Science.
Shrivenham, Swindon, Wilts. U.K.

MEHTA, M.L.
Laboratoire de Physique
Theorique.
C.E.N. de Saclay.
Gift-sur-Yvette, France.

MEIJER, H.G.
Department of Mathematics and
Computer Science.
Technical University Delft.
The Netherlands.

MENA, A.C.
Grupo de Matemática Aplicada
Univ. de Porto. Portugal.

MICCHELLI, C.A.
Thomas Watson I.B.M.
Research Center
Yorktown Heights, New York,
U.S.A.

MONTANER, J.
Dpto. de Matemática Aplicada
Univ. de Zaragoza. Spain.

MORAL, L.
Dpto. de Matemática Aplicada
Univ. de Zaragoza. Spain.

NEVAI, P.G.
Department of Mathematics.
Ohio State University
Columbus, Ohio, U.S.A.

NJASTAD, O.
Departament of Mathematics.
Univ. of Trondheim NTH
Trondheim, Norway.

ORSI, A.
Dipartimento di Matematica.
Facolta di Ingegneria.
Politecnico di Torino, Italy.

ORTIZ, J.
Departamento de Matemáticas.
E.T.S. de Ingenieros Industriales.
Univ. de Oviedo, Gijón. Spain.

ORTIZ, E.L.
Department of Mathematics.
Imperial College of Science &
Technology.
Queen,s Gate, London SW7 2BZ. U.K.

ORSTEIN, A.
Department of Mathematics.
Institute of Technology
Technion-Israel
Haifa,Israel.

OSOWSKI, S.
Institute of Electrical Engineering
Technical University.
Warsaw. Poland.

PEREZ ACOSTA, F.
Dpto. de Ecuaciones Funcionales.
Univ. de La Laguna. Spain.

PEREZ GRASA, I.
Departamento de Matemáticas.
Facultad de Ciencias Empresariales.
Univ. de Zaragoza. Spain.

PEREZ RIERA, M.
Departamento de Matemáticas.
Univ. de Zaragoza. Spain.

PIÑAR, M.
Depto. de Matemática Aplicada.
Colegio Universitario de Jaén.
Spain.

PITTALUGA, G.
Dipartimento di Matematica.
Universita di Torino. Italy.

POPOV
Macedonian Academy of Sciences.
Skopje. Yugoslavia.

RAHMANOV, E.A.
Steklov Institute of Mathematics.
USSR Academy of Sciences.
Moscow. U.S.S.R.

REZOLA, M.L.
Departamento de Matemáticas.
Univ. de Zaragoza. Spain.

RIVLIN, T.J.
Thomas Watson IBM Research Center
Yorktown Heights, New York. U.S.A.

RODRIGUES, M.J.
Grupo de Matemática Aplicada
Univ. de Porto. Portugal.

RODRIGUEZ, I.
Departamento de Matemáticas.
E.T.S. de Ingenieros Industriales.
Univ. de Oviedo, Gijón, Spain.

RODRIGUEZ SOCORRO, G.
Esc. Formación Prof. de E.G.S.
"Santa María".
Univ. Autónoma de Madrid, Spain.

RONVEAUX, A.
Departament de Physique.
Facultés Universitaires Notre
Dame de la Paix.
Namur, Belgium.

RUBIO, J.L.
División de Matemáticas.
Universidad Autónoma de
Madrid, Spain.

SABLONNIERE, P.
U.E.R. d'I.E.E.A.
Univ. des Sciences et Techniques
de Lille I.
Villeneuve d'Ascq. France.

SACRIPANTE, L.
Dipartimento di Matematica
Universita di Torino. Italy.

SAFF, E.B.
Department of Mathematics.
Univ. of South Florida.
Tampa, Florida, U.S.A.

SANSIGRE, G.
Departamento de Matemáticas.
E.T.S. de Ingenieros Industriales.
Univ. Politécnica de Madrid. Spain.

SMAILI, N.
U.E.R. Analyse, Probabilités et
Applications.
Univ. Pierre et Marie Curie,
Paris, France.

SOTTAS, G.
Section de Mathematiques.
Univ. de Geneve. Switzerland.

STAHL, H.
Department of Mathematics.
Technical University Berlin.
Berlin, F.R.G.

SZAFRANIEC, E.H.
Instytut Matematyki.
Universytet Jagiellonski.
Krakow, Poland.

TASIS, C.
Departamento de Matemáticas.
E.T.S. de Ingenieros Industriales.
Univ. de Oviedo, Gijón, Spain.

TEMME, N.M.
Centre for Mathematics and
Computer Science.
Amsterdam 1009 AB.
The Netherlands.

TORRANO, E.
Dpto. de Matemáticas para Químicos
Univ. Complutense
Madrid, Spain.

VALENT, G.
Laboratoire de Physique Theorique
et des Hautes Energies.
Université Paris, France.

VAN DOORN, E.A.
Dep. of Applied Mathematics.
Twente University of Tecnology.
Enschede AE 7500. The Netherlands.

VAN ISEGEHM
U.E.R. d'I.E.E.A.
Université des Sciences et
Techniques Lille I.
Villeneuve d'Ascq. France.

VAN ROSSUM, H.
Instituut voor Propedeutische
Wiskunde.
Univ. of Amsterdam. The Netherlands.

VARONA, J.
Departamento de Matemáticas.
Colegio Universitario de La Rioja
Univ. de Zaragoza. Spain.

VIGIL, L.
Departamento de Matemáticas.
Univ. de Zaragoza. Spain.

VINUESA, J.
Dpto. de Teoría de Funciones.
Univ. de Cantabria.
Santander. Spain.

WAADELAND, H.
Institutt for Mathematikk.
Universitet Trondheim AVH.
Dragvoll, Norway.

WALLIN, H.
Department of Mathematics.
University of Umea. Sweden.

WIMP, J.
Department of Mathematics.
Drexel University. Philadelphia
Pensylvania 19104. U.S.A.

ZARZO, A.
Departamento de Matemáticas.
E.T.S. de Ingenieros Industriales.
Univ. Politécnica de Madrid. Spain.

ZAYED, A.I.
Departament of Mathematics.
California Polytechnic State
University.
San Luis Obispo. California, U.S.A.

ERROR ESTIMATE IN PADÉ APPROXIMATION

Claude BREZINSKI

Laboratoire d'Analyse Numérique et d'Optimisation
U.F.R. I.E.E.A.
Université des Sciences et Techniques de Lille Flandres-Artois
59655 - Villeneuve d'Ascq - Cedex
FRANCE

ABSTRACT.

The aim of this paper is to extend Kronrod's procedure for estimating the error in Gaussian quadrature formulas to Padé approximation. Some results on Stieltjes polynomials are also given. Examples show the effectiveness of the method. The procedure is then applied to the ε-algorithm, which is a convergence acceleration method related to Padé approximation. General principles for estimating the error in series approximations and sequence transformations are also brought to light.

INTRODUCTION.

In 1965, A.S. Kronrod published a small book [10] in which he proposed an heuristic method for estimating the error in Gaussian quadratures. It simply consists in comparing the results obtained by two different quadrature formulas. If, for example, one makes use of two Gaussian quadratures with respectively n and n+1 nodes then one has to evaluate the integrand at 2n+1 points. This procedure is not the best since a Gaussian quadrature with n+1 nodes exactly integrates polynomials up to the degree 2n+1 while we are using 2n+1 points for the same accuracy. We are wasting the integrand evaluations. Thus, Kronrod's idea was to use a Gaussian quadrature formula first and then to add it new nodes in an optimal way that is in order to obtain a new formula exact for polynomials of the highest possible degree. The difference between the two results gives an estimate of the error.

Since Padé approximants can be viewed as formal Gaussian quadratures [2], the aim of this paper is to extend Kronrod's idea to Padé approximation.

Let us recall some results. Let f be a formal power series

$$f(t) = \sum_{i=o}^{\infty} c_i t^i \quad , \quad c_i \in \mathbb{C}$$

Let c be the linear functional acting on the space of complex polynomials, defined by

$$c(x^i) = c_i \quad i = 0,1,\ldots$$

We formaly have

$$f(t) = c((1-xt)^{-1}).$$

Let v be an arbitrary polynomial of degree k. We define w as

$$w(t) = c\left(\frac{v(x)-v(t)}{x-t}\right)$$

where c acts on the variable x, t being a parameter. w is a polynomial of degree k-1. w is the polynomial associated to v. We set

$$\tilde{v}(t) = t^k v(t^{-1}) \quad \text{and} \quad \tilde{w}(t) = t^{k-1} w(t^{-1}).$$

The rational fraction $\tilde{w}(t)/\tilde{v}(t)$ is called a Padé-type approximant of f. It is denoted by $(k-1/k)_f(t)$. v is called its generating polynomial. We have

$$f(t) - (k-1/k)_f(t) = \frac{t^k}{\tilde{v}(t)} \, c\left(\frac{v(x)}{1-xt}\right) = O(t^k).$$

Padé-type approximants can be related to polynomial interpolation. Let P be the Hermite interpolation polynomial of $(1-xt)^{-1}$ (as a function of x, t being a parameter) at the zeros of v. We have

$$(k-1/k)_f(t) = c(P(x)).$$

Thus (k-1/k) can be viewed as a formal interpolatory quadrature formula for $(1-xt)^{-1}$ with the zeros of v as nodes. The above error expression is just the classical property of an interpolatory quadrature formula with k nodes to be exact for polynomials of degree at most k-1. Moreover, from that expression, we have

$$f(t)-(k-1/k)_f(t) = \frac{t^k}{\tilde{v}(t)} \{c(v(x)) + c(xv(x))t + \ldots + c(x^{k-1} v(x))t^{k-1} + t^k \, c\left(\frac{x^k v(x)}{1-xt}\right)\}$$

Since v can be arbitrarily chosen, let us take it such that

$$c(x^i v(x)) = 0 \quad i = 0,\ldots,k-1.$$

This is always possible if the Hankel determinant

$$H_k^{(0)} = \begin{vmatrix} c_0 & c_1 & \cdots & c_{k-1} \\ c_1 & c_2 & \cdots & c_k \\ \cdots & \cdots & \cdots & \cdots \\ c_{k-1} & c_k & \cdots & c_{2k-2} \end{vmatrix}$$

is different from zero, condition which will assumed to be true in the sequel. If v satisfies the preceding conditions, it is the orthogonal polynomial of degree k with respect to the functional c. It will be then denoted by P_k, assumed to be monic, and we shall have

$$f(t) - (k-1/k)_f(t) = O(t^{2k}).$$

This means that (k-1/k) is the usual [k-1/k] Padé approximant.
[k-1/k] can be viewed as a formal interpolatory quadrature formula using the zeros of P_k (an orthogonal polynomial) as nodes. Thus [k-1/k] is a formal Gaussian quadrature and the preceding approximation result is the classical property of such a quadrature formula to be exact for polynomials of degree at most 2k-1.

KRONROD'S PROCEDURE.

Since [k-1/k] is a formal Gaussian formula we shall apply Kronrod's idea of adding n new nodes to the k preceding ones (the zeros of P_k). In other words, we shall consider the Padé-type approximant (n+k-1/n+k) with the generating polynomial $v(x) = P_k(x)\ V_n(x)$ where V_n is an arbitrary monic polynomial of degree n.

We set

$$W_n(t) = c(P_k(x)\ \frac{V_n(x) - V_n(t)}{x - t}).$$

The polynomial w asoociated to v is given by

$$w(t) = W_n(t) + Q_k(t)\ V_n(t)$$

where Q_k is the polynomial associated to P_k. Moreover

$$\tilde{v}(t) = t^{n+k}v(t^{-1}) = \tilde{P}_k(t)\tilde{V}_n(t) \text{ with } \tilde{P}_k(t) = t^k P_k(t^{-1}),\ \tilde{V}_n(t) = t^n\ V_n(t^{-1})$$

$$\tilde{w}(t)=t^{n+k-1}w(t^{-1}) = t^k\tilde{W}_n(t)+\tilde{V}_n(t)\tilde{Q}_k(t) \text{ with } \tilde{Q}_k(t) = t^{k-1}Q_k(t^{-1}),\ \tilde{W}_n(t)=t^{n-1}W(t^{-1}).$$

Thus we have proved the

Theorem 1 : *Let* (n+k-1/n+k) *have the generating polynomial* $v(x)=P_k(x)V_n(x)$. *Then*

$$(n+k-1/n+k)_f(t) - [k-1/k]_f(t) = t^k\ \frac{\tilde{W}_n(t)}{\tilde{P}_k(t)\tilde{V}_n(t)}$$

$$f(t)-(n+k-1/n+k)_f(t) = \frac{t^{n+k}}{\tilde{P}_k(t)\tilde{V}_n(t)}\ c(\frac{P_k(x)V_n(x)}{1-xt}).$$

Remarks : These results are still valid (after replacing $[k-1/k]$ by $(k-1/k)$) if P_k is replaced by an arbitrary polynomial of degree k.

$[V_n(x) - V_n(t)]/(x-t)$ is a polynomial of degree $n-1$ in x. Thus, by the orthogonality property of P_k, W_n is identically zero if $n \le k$ and consequently $(n+k-1/n+k)$ is identical to $[k-1/k]$. For the same reason, when $n > k$, W_n is a polynomial of degree n-k-1. Thus the smallest possible value of n is k+1. It is the reason why we shall now consider the case $n = k+1$. V_{k+1} will be chosen such that

$$c(x^i P_k(x) V_{k+1}(x)) = o \quad i = 0,\ldots,k.$$

Such polynomials were introduced by Stieltjes in a letter to Hermite dated November 8, 1894, seven weeks before his death [1]. They are now called Stieltjes polynomials. They have been studied by Monegato in the classical case [12] and by Prévost in the formal one [14]. Since the above relations defining V_{k+1} are equivalent to a triangular system of linear equations whose diagonal elements are all equal to $c(x^k P_k(x))$, then V_{k+1} exists and is unique if and only if $H_k(c_o) \neq o$.

We have the

Theorem 2 : *If the generating polynomial* $v(x) = P_k(x)V_{k+1}(x)$ *of* $(2k/2k+1)$ *is such that* $c(x^i v(x)) = o$ *for* $i = 0,\ldots,k$ *then*

$$f(t) - (2k/2k+1)_f(t) = \frac{t^{3k+2}}{\tilde{v}(t)} \, c\left(\frac{x^{k+1} v(x)}{1-xt}\right)$$

$$(2k/2k+1)_f(t) - [k-1/k]_f(t) = \frac{t^{2k}}{\tilde{V}(t)} \, c(x^k P_k(x)).$$

Proof.

The first result immediately follows from theorem 1 and the above relations defining V_{k+1}, by expanding the error term.

Obviously $W_{k+1}(t) = c(x^k P_k(x))$, $\tilde{W}_{k+1}(t) = t^k c(x^k P_k(x))$ and the result follows theorem 1. □

We shall now give a result which shows why Kronrod's procedure provides a good estimate of the error of $[k-1/k]$.

Let us first begin with some general principles. Assume we are given two approximations (any style) R_1 and R_2 of f such that

$$f(t) - R_1(t) = O(t^{r_1})$$
$$f(t) - R_2(t) = O(t^{r_2}).$$

We have

$$f(t) - R_2(t) - (f(t) - R_1(t)) = R_1(t) - R_2(t)$$

$$= f(t) - R_2(t) + O(t^{r_1})$$

$$= -(f(t) - R_1(t)) + O(t^{r_2}).$$

Thus $R_1(t) - R_2(t)$ is an approximation of order r_2 of the error $f(t) - R_1(t)$ and of order r_1 of the error $f(t) - R_2(t)$. If $r_1 < r_2$, R_2 is a better approximation of f than R_1 but, by way of compensation, $R_1(t) - R_2(t)$ is a better estimation of the error $f(t) - R_1(t)$. Moreover $[R_2(t) - R_1(t)] / [f(t) - R_1(t)] = 1+O(t^{r_2-r_1})$ which shows that, when t approaches zero, the difference $R_2(t) - R_1(t)$ becomes a good estimation of the error $f(t) - R_1(t)$.

On the other hand $[R_1(t) - R_2(t)] / [f(t)-R_2(t)] = O(t^{r_1-r_2})$ which shows that this ratio tends to infinity when t goes to zero and that the difference $R_1(t)-R_2(t)$ cannot be a good estimation of the error $f(t) - R_2(t)$.

Applying these general principles to Kronrod's procedure explains its effectiveness. However let us be more specific or, in other words, let us express completely the relations :

Theorem 3 :

$$(2k/2k+1)_f(t) - [k-1/k]_f(t) = f(t) - [k-1/k]_f(t) - \frac{t^{3k+2}}{\tilde{V}_{k+1}(t)\ \tilde{P}_k(t)} c\left(\frac{x^{k+1}P_k(x)V_{k+1}(x)}{1-xt}\right).$$

$$\frac{f(t) - [k-1/k]_f(t)}{(2k/2k+1)_f(t) - [k-1/k]_f(t)} = 1+ t^{k+2} c\left(\frac{x^{k+1}P_k(x)\ V_{k+1}(x)}{1 - xt}\right)/c(x^k\ P_k(x)).$$

Proof.

Let e be the linear functional defined by

$$e(x^i) = c(x^i\ P_k(x)) = e_i \qquad i = 0,1,\ldots$$

and let $g(t) = \sum_{i=0}^{\infty} e_i t^i$. By definition V_{k+1} is the orthogonal polynomial of degree k+1 with respect to e and W is its associated polynomial. Thus

$$\tilde{W}_{k+1}(t)/\tilde{V}_{k+1}(t) = [k/k+1]_g(t).$$

But, thanks to the orthogonality property of P_k with respect to c, $e_i = o$ for $i = o,\dots,k-1$. We set

$$g(t) = t^k g_k(t).$$

By a well known property of Padé approximants we have

$$[k/k+1]_f(t) = t^k[O/k+1]_{g_k}(t).$$

We previously saw that $\tilde{W}_{k+1}(t) = e_k t^k$. Thus

$$e_k/\tilde{V}_{k+1}(t) = [O/k+1]_{g_k}(t) = g_k(t) - \frac{t^{k+2}}{\tilde{V}_{k+1}(t)} c\left(\frac{x^{k+1}P_k(x)\ V_{k+1}(x)}{1 - xt}\right)$$

and, by theorem 2, we get

$$(2k/2k+1)_f(t) - [k-1/k]_f(t) = \frac{t^{2k}}{\tilde{P}_k(t)} g_k(t) + O(t^{3k+2}) = \frac{e_k t^{2k}}{\tilde{V}_{k+1}(t)\tilde{P}_k(t)}.$$

On the other hand we know that [2]

$$f(t) - [k-1/k]_f(t) = \frac{t^{2k}}{\tilde{P}_k(t)} c\left(\frac{x^k P_k(x)}{1-xt}\right) = \frac{t^{2k}}{\tilde{P}_k(t)} g_k(t)$$

and both results follow. □

<u>Remarks</u> : Although V_{k+1} is an orthogonal polynomial with respect to e it does not satisfy, in general, a three term recurrence relationship [12, 14]. The reason is that the linear functional e depends itself on the index k. It is easy to check that

$$\tilde{w}(t)/\tilde{V}_{k+1}(t) = [2k/k+1]_{f\tilde{P}_k}(t).$$

If f is the expression of a rational fraction whose numerator has degree k-1 and whose denominator has degree k then $H_k^{(o)} \neq 0$ and $H_{k+1}^{(o)} = 0$. Thus $f(t) - [k-1/k]_f(t) = (2k/2k+1)_f(t) - [k-1/k]_f(t) = 0$.

<u>STIELTJES POLYNOMIALS.</u>

In the classical case, that is for polynomials orthogonal on an interval with respect to a weight function, the main practical problem for implementing Kronrod's procedure is to know the zeros of V_{k+1}. The study of these zeros leads to examinate some properties of Stieltjes polynomials. As we shall see in the next section, the extension of Kronrod's idea to Padé approximation does not involve the knowledge of these zeros. However, for the sake of completeness, we shall now extend to the formal case some properties of Stieltjes polynomials.

Since the $\{P_i\}$ form a basis of the set of polynomials, we can write

$$V_{k+1}(x)\ P_k(x) = \sum_{i=o}^{2k+1} t_i P_i(x)$$

where the t_i's are constants depending on the index k. For $j \le k$ we have

$$c(x^j V_{k+1}(x) P_k(x)) = \sum_{i=o}^{2k+1} t_i\ c(x^j P_i(x)) = 0$$

by definition of V_{k+1}. Thus for $j = o$

$$t_o\ c(P_o(x)) + \ldots + t_{2k+1}\ c(P_{2k+1}(x)) = 0 = t_o\ c(P_o(x))$$

by the orthogonality properties of the P_i's. Thus $t_o = o$. Similarly for $j = 1$

$$t_1\ c(xP_1(x)) + \ldots + t_{2k+1}\ c(xP_{2k+1}(x)) = 0 = t_1\ c(x\ P_1(x)),$$

thus $t_1 = 0$ and so on up to $t_k = 0$. Consequently

$$V_{k+1}(x)\ P_k(x) = \sum_{i=k+1}^{2k+1} t_i P_i(x).$$

This is a generalization of a known property of quasi-orthogonal polynomials [8, p. 64].

If $x_1,\ldots,x_k$ are the zeros of P_k, assumed, for the moment, to be distinct, then

$$V_{k+1}(x_i)\ P_k(x_i) = 0 \quad i = 1,\ldots,k.$$

Since $V_{k+1}(x)\ P_k(x)$ is monic we shall have $t_{2k+1}=1$ and thus we get :

$$\begin{array}{ll}
t_{k+1}\ P_{k+1}(x) + \ldots + t_{2k} P_{2k}(x) - V_{k+1}(x)\ P_k(x) & = -\ P_{2k+1}(x) \\
t_{k+1}\ P_{k+1}(x_1) + \ldots + t_{2k} P_{2k}(x_1) & = -\ P_{2k+1}(x_1) \\
\text{---} & \text{-----------} \\
t_{k+1}\ P_{k+1}(x_k) + \ldots + t_{2k} P_{2k}(x_k) & = -\ P_{2k+1}(x_k)
\end{array}$$

or

$$\begin{pmatrix} P_{k+1}(x) & \cdots & P_{2k}(x) & -1 \\ P_{k+1}(x_1) & \cdots & P_{2k}(x_1) & 0 \\ \cdots & \cdots & \cdots & \cdots \\ P_{k+1}(x_k) & \cdots & P_{2k}(x_k) & 0 \end{pmatrix} \begin{pmatrix} t_{k+1} \\ \vdots \\ t_{2k} \\ V_{k+1}(x)P_k(x) \end{pmatrix} = - \begin{pmatrix} P_{2k+1}(x) \\ P_{2k+1}(x_1) \\ \vdots \\ P_{2k+1}(x_k) \end{pmatrix}$$

Thus

$$V_{k+1}(x)P_k(x) = \frac{\begin{vmatrix} P_{k+1}(x) & \cdots & P_{2k+1}(x) \\ P_{k+1}(x_1) & \cdots & P_{2k+1}(x_1) \\ \cdots & \cdots & \cdots \\ P_{k+1}(x_k) & \cdots & P_{2k+1}(x_k) \end{vmatrix}}{\begin{vmatrix} P_{k+1}(x_1) & \cdots & P_{2k}(x_1) \\ \cdots & \cdots & \cdots \\ P_{k+1}(x_k) & \cdots & P_{2k}(x_k) \end{vmatrix}}$$

If now x_i is a zero of multiplicity r_i of P_k then $v^{(j)}(x_i) = o$ for $j = o,\ldots,r_i-1$ and the corresponding rows in the numerator and denominator of the above determinantal expression must be replaced by

$$P_{k+1}^{(j)}(x_i) \;\cdots\; P_{2k(+i)}^{(j)}(x_i) \qquad j = 0,\ldots,r_i-1.$$

This result generalizes a result of Christoffel [16, p. 29]. The proof is an adaptation of that of Szegö and Patterson [13]. See also [11].

From the practical point of view $V_{k+1}(x)\ P_k(x)$ can be computed, for a fixed value of x, by the E-algorithm [3] which is a general extrapolation algorithm computing ratios of determinants similar to the preceding one. In this algorithm one has to set the initial conditions

$$E_o^{(o)} = P_{k+1}(x),\ldots,E_o^{(k)} = P_{2k+1}(x)$$

and

$$g_{o,i}^{(o)} = P_{k+1}(x_i),\ldots,g_{o,i}^{(k)} = P_{2k+1}(x_i) \text{ for } i = 1,\ldots,k$$

and the corresponding modifications for multiple zeros. Then, applying the E-algorithm to these initial conditions, we shall obtain

$$E_k^{(o)} = V_{k+1}(x)\; P_k(x).$$

Instead of normalizing V_{k+1} such that $t_{2k+1} = 1$ we should have impose the condition

$$t_{k+1} + \ldots + t_{2k+1} = 1.$$

We should have obtain a ratio of determinants with exactly the same numerator as above and a denominator where the first row $P_{k+1}(x)\ldots P_{2k+1}(x)$ is replaced by $1\ldots1$. In that case, $V_{k+1}(x)P_k(x)$ could be computed by a general algorithm for recursive projection (among other things) called the CRPA [4].

A determinantal expression for V_{k+1} is given in [14]. Let us just remark that if we write

$$V_{k+1}(x) = \sum_{i=o}^{k+1} b_i P_i(x)$$

where the b_i's depend on k, then $b_{k+1} = 1$ since V_{k+1} and P_{k+1} are both monic ; but, due to the orthogonality property of P_k

$$c(V_{k+1}(x)\; P_k(x)) = \sum_{i=o}^{k+1} b_i\; c(P_i(x)\; P_k(x)) = b_k\; c(P_k^2(x))$$

which shows that $b_k = 0$. Thus V_{k+1} can be written as

$$V_{k+1}(x) = P_{k+1}(x) + \sum_{i=o}^{k-1} b_i P_i(x).$$

In the general case, that is without any assumption on the functional c, it does not seem to be possible to prove other interesting results. For the classical case with an even weight function in [-1, +1], see [6].

IMPLEMENTATION AND APPLICATIONS.

In the case of classical Gaussian quadrature formulas, the implementation of Kronrod's procedure needs the knowledge of the zeros of P_k (the nodes of the Gaussian formula) and those of V_{k+1} (the new added nodes). In the case of Padé approximation the only needed knowledge is that of the polynomials P_k and V_{k+1}. Their zeros are of no use, which leads to a major simplification in the implementation of the method.

The polynomial P_k is first computed by the standard techniques that is either directly by solving a system of linear equations on recursively, even if the Padé table is non-normal [9].

The polynomial V_{k+1} can be then very easily obtained. We set

$$V_{k+1}(x) = a_o + \dots + a_k x^k + x^{k+1}.$$

Writing the orthogonality relations which define V_{k+1} and using the orthogonality of P_k, we get

$$c(v(x)) = a_k e_k + e_{k+1} = 0$$

$$c(xv(x)) = a_{k-1} e_k + a_k e_{k+1} + e_{k+2} = 0$$

$$\text{---}$$

$$c(x^k v(x)) = a_o e_k + a_1 e_{k+1} + \dots + a_k e_{2k} + e_{2k+1} = 0$$

This system immediately provides the a_i's if the e_i's are known. If we set

$$P_k(x) = b_o + \dots + b_k x^k$$

the e_i's are given by

$$e_i = c(x^i P_k(x)) = b_o c_i + \dots + b_k c_{i+k}.$$

Let us give some examples. We shall begin with the exponential function. We have

$$[0/1] = \frac{1}{1-t} \qquad (2/3) = \frac{-11t^2-24t+36}{(1-t)(7t^2-24t+36)}$$

$$[1/2] = \frac{6+2t}{6-4t+t^2} \qquad (4/5) = \frac{379/4t^4+468t^3-630t^2-4200t+15750}{(6-4t+t^2)(-62t^3+420t^2-1575t+2625)}$$

t	$(2/3)-[0/1]$	$e^t-[0/1]$	$(4/5)-[1/2]$	$e^t-[1/2]$
- 3	- 0.236842	- 0.200213	0.512535 10^{-1}	0.497871 10^{-1}
- 2	- 0.214286	- 0.197998	0.244553 10^{-1}	0.242242 10^{-1}
- 1.5	- 0.184615	- 0.176870	0.126534 10^{-1}	0.126038 10^{-1}
- 1.2	- 0.157343	- 0.153351	0.708997 10^{-2}	0.707656 10^{-2}
- 1	- 0.134328	- 0.132121	0.424741 10^{-2}	0.424308 10^{-2}
- 0.8	- 0.107239	- 0.106227	0.217551 10^{-2}	0.217449 10^{-3}
- 0.5	- 0.060301	- 0.060136	0.470093 10^{-3}	0.470053 10^{-3}
- 0.3	- 0.028431	- 0.028412	0.774809 10^{-4}	0.774799 10^{-4}
- 0.1	- 0.004253	- 0.004253	0.122459 10^{-5}	0.122459 10^{-5}

0.1	- 0.005940	- 0.005940	0.157761 10^{-5}	0.157761 10^{-5}
0.3	- 0.078637	- 0.078712	0.165559 10^{-3}	0.165556 10^{-3}
0.5	- 0.349515	- 0.351279	0.166270 10^{-2}	0.166245 10^{-2}
0.8	- 2.70677	- 2.77446	0.162582 10^{-1}	0.162386 10^{-1}
1.1	12.0531	13.0042	0.864646 10^{-1}	0.860166 10^{-1}

The second example concerns the logarihm. We have

$$\frac{1}{t}\ \mathrm{Log}\ (1+t) = 1 - \frac{t}{2} + \frac{t^2}{3} - \frac{t^3}{4} + \dots$$

$$[0/1] = \frac{2}{2+t} \qquad (2/3) = \frac{11t^2 + 60t + 60}{(2+t)(3t^2+30t+30)}$$

$$[1/2] = \frac{6+3t}{6+6t+t^2} \qquad (4/5) = \frac{73t^4+1440t^3+6480t^2+10080t+5040}{15(6+6t+t^2)(t^3+30t^2+84t+56)}$$

t	$(2/3) - [0/1]$	$t^{-1}\mathrm{Log}(1+t)-[0/1]$	$(4/5) - [1/2]$	$t^{-1}\mathrm{Log}(1+t)-[1/2]$
- 0.9	0.678051	0.740246	0.218735	0.218002
- 0.7	0.180001	0.181500	0.169347 10^{-1}	0.169044 10^{-1}
- 0.5	0.529100 10^{-1}	0.529610 10^{-1}	0.167941 10^{-2}	0.167898 10^{-2}
- 0.3	0.124450 10^{-1}	0.124459 10^{-1}	0.105293 10^{-3}	0.105291 10^{-3}
- 0.1	0.973577 10^{-3}	0.973578 10^{-3}	0.720349 10^{-6}	0.720349 10^{-6}
0.1	0.720845 10^{-3}	0.720846 10^{-3}	0.4636470 10^{-6}	0.436470 10^{-6}
0.3	0.498223 10^{-2}	0.498233 10^{-2}	0.228335 10^{-5}	0.228333 10^{-4}
0.5	0.109290 10^{-1}	0.109302 10^{-1}	0.119409 10^{-3}	0.119405 10^{-3}
0.7	0.172938 10^{-1}	0.172996 10^{-1}	0.322896 10^{-3}	0.322866 10^{-3}
0.9	0.234991 10^{-1}	0.235158 10^{-1}	0.640397 10^{-3}	0.640272 10^{-3}
1.5	0.393185 10^{-1}	0.394319 10^{-1}	0.216639 10^{-2}	0.216484 10^{-2}
2	0.490196 10^{-1}	0.493061 10^{-1}	0.385675 10^{-2}	0.385160 10^{-2}
3	0.612245 10^{-1}	0.620981 10^{-1}	0.757325 10^{-2}	0.755267 10^{-2}
5	0.700280 10^{-1}	0.726376 10^{-1}	0.141567 10^{-1}	0.140986 10^{-1}
7	0.703417 10^{-1}	0.748409 10^{-1}	0.188054 10^{-1}	0.187126 10^{-1}

Let us now examinate how the method works for complex values of the variable t. We shall consider again the case of the exponential function. We set $t = x+iy$ with $x^2+y^2 = r^2$. We get :

	x	(2/3) - [0/1]	e^t - [0/1]
r=1	0.8	1.371070 - i 0.302249	1.336818 - i 0.243365
	0.1	0.089950 + i 0.373320	0.101780 + i 0.374194
	0	0.031757 + i 0.336627	0.040302 + i 0.341471
	- 0.6	- 0.113906 + i 0.143426	-0.117639 + i 0.143693
r=0.5	0.3	0.167072 - i 0.090205	0.166379 - i 0.089725
	0.1	0.117975 + i 0.053957	0.118039 + i 0.053453
	0	0.077217 + i 0.079609	0.077583 + i 0.079426
	- 0.1	0.039494 + i 0.087744	0.039790 + i 0.087898
	- 0.3	-0.020187 + i 0.072123	-0.020364 + i 0.072272
r=0.2	0.1	0.017211 - i 0.015734	0.017206 - i 0.015731
	0	0.018526 + i 0.636581 10^{-2}	0.018528 + i 0.636164 10^{-2}
	-0.1	0.420143 10^{-2} + i 0.162548 10^{-1}	0.420198 10^{-2} + i 0.162584 10^{-1}
r=0.1	0.08	-0.101538 10^{-2} - i 0.563004 10^{-2}	-0.101520 10^{-2} - i 0.563000 10^{-2}
	0.03	0.471704 10^{-2} - i 0.226453 10^{-2}	0.471689 10^{-2} - i 0.226457 10^{-2}
	0	0.490512 10^{-2} - i 0.823659 10^{-3}	0.490516 10^{-2} + i 0.823516 10^{-2}
	-0.03	0.341639 10^{-2} + i 0.328104 10^{-2}	0.341653 10^{-2} + i 0.328107 10^{-2}
	-0.08	-0.162156 10^{-2} + i 0.407171 10^{-2}	-0.162169 10^{-2} + i 0.407170 10^{-2}

In the preceding numerical examples we see that the precision of the error estimate of Kronrod's procedure increases with k for a fixed value of t. In fact we have

$$\frac{(2k/2k+1)_f(t) - [k-1/k]_f(t)}{f(t) - [k-1/k]_f(t)} - 1 = - \frac{(2k/2k+1)_f(t) - f(t)}{[k-1/k]_f(t) - f(t)}$$

which shows that this ratio tends to 1 when k goes to infinity if and only if the sequences of Padé type approximants $((2k/2k+1)_f(t))_{k\in\mathbb{N}}$ tends to f(t) faster than the sequence $([k-1/k]_f(t))_{k\in\mathbb{N}}$ for the value of t under consideration.

It arises very often in the practice that one does not compute a single Padé approximant but a sequence of them $([m_k/m_k])_{k\in\mathbb{N}}$ with $m_k+n_k > m_{k-1}+n_{k-1}$. In such a case we have

$$\frac{[m_k/n_k]_f(t) - [m_{k-1}/n_{k-1}]_f(t)}{f(t) - [m_{k-1}/n_{k-1}]_f(t)} = 1 + O(t^{m_k+n_k-m_{k-1}-n_{k-1}})$$

Any such sequence can be recursively computed by using the universal program given in [9] which can be used, as well as the above procedure, for non-normal Padé tables. However since most often in practical cases $m_k+n_k-m_{k-1}-n_{k-1}$ is egal to 1 or 2,

Kronrod's procedure will be more effective. Moreover its implementation is easy and the supplementary work needed is small.

As explained in the above general principles an estimation of the error of $[k-1/k]$ can be obtained with any other approximation R_2 such that $R_2(t) = f(t) + O(t^{3k+2})$. Since the application of Kronrod's procedure needs the knowledge of $c_o, \ldots, c_{3k+1}$, R_2 can be any Padé approximant $[m/n]$ with $m+n = 3k+1$ or any Padé-type approximant (m/n) with $m = 3k+1$. Among these approximants the easiest one to construct is of course

$$R_2(t) = f_{3k+1}(t) = \sum_{i=o}^{3k+1} c_i t^i .$$

Thus the question arises to know when the estimate given by Kronrod's procedure provides a better estimate of the error than this crude one.

We consider the ratio

$$r(t) = \frac{\dfrac{(2k/2k+1)_f(t) - [k-1/k]_f(t)}{f(t) - [k-1/k]_f(t)} - 1}{\dfrac{f_{3k+1}(t) - [k-1/k]_f(t)}{f(t) - [k-1/k]_f(t)} - 1}$$

If $|r(t)| < 1$, $(2k/2k+1)_f(t) - [k-1/k]_f(t)$ is a better estimate of the error $f(t) - [k-1/k]_f(t)$ than $f_{3k+1}(t) - [k-1/k]_f(t)$. We have

$$r(t) = \frac{(2k/2k+1)_f(t) - f(t)}{f_{3k+1}(t) - f(t)}$$

Thus the answer to our question needs the study of the convergence and convergence acceleration behaviour of the Padé-type approximants $(2k/2k+1)$ which is a difficult problem with no general solution. However we have

$$r(t) = \frac{1}{\tilde{P}_k(t)\tilde{V}_{k+1}(t)} \; \frac{c\left(\dfrac{x^{k+1}P_k(x)V_{k+1}(x)}{1-xt}\right)}{c\left(\dfrac{x^{3k+2}}{1-xt}\right)}$$

and

$$\lim_{t\to o} r(t) = \frac{1}{\tilde{P}_k(o)\tilde{V}_{k+1}(o)} \; \frac{c(x^{k+1}P_k(x)V_{k+1}(x))}{c_{3k+2}}$$

which is easy to compute if c_{3k+2} is known. Thus, if $|r(o)| < 1$, Kronrod's procedure provides a better estimate of the error than f_{3k+1}, at least in a neighbourhood of the origin.

Let us consider the previous numerical examples.

For $f(t) = e^t$ and $k = 1$ we get $r(o) = -16/9 = -1.777\ldots$ and

t	(2/3) - [0/1]	e^t - [0/1]	$f_4(t)$ - [0/1]	r(t)
-3	-0.2368421052	-0.2002129316	0.1125 10^{1}	-0.2764 10^{-1}
-2	-0.2142857143	-0.1979980501	0.1388 10^{-16}	-0.8226 10^{-1}
-1.5	-0.1846153846	-0.1768698398	-0.1265625	-0.1540
-1.2	-0.1573426573	-0.1533512426	-0.1361454	-0.2320
-1.	-0.1343283582	-0.1321205588	-0.1250000	-0.3101
-0.8	-0.1072386059	-0.1062265914	-0.1038222	-0.4209
-0.5	-0.6030150754 10^{-1}	-0.6013600695 10^{-1}	-0.5989583 10^{-1}	-0.6891
-0.3	-0.2843152740 10^{-1}	-0.2841254854 10^{-1}	-0.2839326923 10^{-1}	-0.9844
-0.1	-0.4253609660 10^{-2}	-0.4253491055 10^{-2}	-0.4253409091 10^{-2}	-1.4470
-0.01	-0.4917615327 10^{-4}	-0.4917615182 10^{-4}	-0.4917615099 10^{-4}	-1.7407
0.01	-0.5084301533 10^{-4}	-0.5084301684 10^{-4}	-0.5084301768 10^{-4}	-1.8158
0.1	-0.5940005940 10^{-2}	-0.5940193035 10^{-2}	-0.5940277778 10^{-2}	-2.2078
0.3	-0.7863695937 10^{-1}	-0.7871262099 10^{-1}	-0.7873392857 10^{-1}	-3.5509
0.5	-0.3495145631	-0.3512787293	-0.3515625000	-6.2169
0.8	-0.2706766917 10^{1}	-0.2774459071 10^{1}	-0.27776 10^{1}	-21.5516
1.1	0.1205312673 10^{2}	0.1300416602 10^{2}	0.1298783750 10^{2}	58.2440

Thus, near zero, the partial sum f_4 gives a better estimate of the error than (2/3).

For k=2 we have $r(o) = -127/125 = -1.016\ldots$ and both procedures provide comparable estimates.

For $f(t) = \frac{1}{t}\,\mathrm{Log}(1+t)$ and k=1 we have $r(o) = o$ and

t	(2/3) - [0/1]	t^{-1}Log(1+t)-[0/1]	$f_4(t)$ - [0/1]	r(t)
- 0.9	0.67805	0.74025	0.21529	0.1185
- 0.7	0.18000	0.18150	0.10864	0.2056 10^{-1}
- 0.5	0.52910 10^{-1}	0.52961 10^{-1}	0.43750 10^{-1}	0.5534 10^{-2}
- 0.1	0.97358 10^{-3}	0.97358 10^{-3}	0.97175 10^{-3}	0.2818 10^{-3}
0.1	0.72084 10^{-3}	0.72085 10^{-3}	0.72238 10^{-3}	-0.1660 10^{-3}
0.3	0.49822 10^{-2}	0.49823 10^{-2}	0.53048 10^{-2}	-0.3126 10^{-3}
0.5	0.10929 10^{-1}	0.10930 10^{-1}	0.14583 10^{-1}	-0.3434 10^{-3}
0.9	0.23499 10^{-1}	0.23516 10^{-1}	0.79315 10^{-1}	-0.2994 10^{-3}
3	0.61224 10^{-1}	0.62098 10^{-1}	11.55	-0.7605 10^{-4}
7	0.70342 10^{-1}	0.74841 10^{-1}	408.06	-0.1103 10^{-4}

These results confirm the superiority of Kronrod's procedure. Moreover the estimates obtained with f_4 are bad outside the disc of convergence of the series.

For k=2, r(o) = -1/15680 = -6.37.. 10^{-5} and we get

t	(4/5)-[1/2]	t^{-1}Log(1+t)-[1/2]	$f_7(t)$ - [1/2]	r(t)
-0.9	0.21873	0.21800	- 0.72833 10^{-1}	- 0.2518 10^{-2}
-0.7	0.16935 10^{-1}	0.16904 10^{-1}	- 0.84048 10^{-3}	- 0.1710 10^{-2}
-0.5	0.16794 10^{-2}	0.16790 10^{-2}	0.88499 10^{-3}	- 0.5487 10^{-3}
-0.1	0.72035 10^{-6}	0.72035 10^{-6}	0.71913 10^{-6}	- 0.9224 10^{-4}
0.1	0.43647 10^{-6}	0.43647 10^{-6}	0.43545 10^{-6}	- 0.4503 10^{-4}
0.3	0.22833 10^{-4}	0.22833 10^{-4}	0.17090 10^{-4}	- 0.2361 10^{-4}
0.5	0.11941 10^{-3}	0.11941 10^{-3}	-0.18023 10^{-3}	- 0.1316 10^{-4}
0.9	0.64040 10^{-3}	0.64027 10^{-3}	-0.25843 10^{-1}	- 0.4698 10^{-5}
3	0.75732 10^{-2}	0.75527 10^{-2}	-198.24	- 0.1038 10^{-6}
7	0.18806 10^{-1}	0.18713 10^{-1}	- 88529.	- 0.1048 10^{-8}

Another method for obtaining error estimates is given in [7].

Kronrod's procedure can be, of course, extended to the whole Padé table by means of

$$[n+k/k]_f(t) = \sum_{i=o}^{n} c_i t^i + t^{n+1} [k-1/k]_{f_{n+1}}(t)$$

with $f_{n+1}(t) = c_{n+1}+c_{n+2}t + \ldots$ (and a similar relation for the other half of the Padé table). In other words the linear functional c has to be replaced by $c^{(n+1)}$ defined by

$$c^{(n+1)}(x^i) = c_{n+i+1} \quad i = 0,1,\ldots$$

Shank's transformation [15] (or, equivalently, Wynn's ε-algorithm [17]) is a powerful convergence acceleration method for sequences. It is a generalization of Aitken's Δ^2 process which is related to Padé approximants as follows : if it is applied to the partial sums of a power series then the numbers $\varepsilon_{2k}^{(n)}$ produced by the ε-algorithm are such that

$$\varepsilon_{2k}^{(n)} = [n+k/k]_f(t).$$

Conversely, if the ε-algorithm is applied to a sequence (S_n) converging to S then the numbers $\varepsilon_{2k}^{(n)}$ obtained satisfy

$$\varepsilon_{2k}^{(n)} = [n+k/k]_f(1)$$

where $f(t) = c_o + c_1 t + \ldots$ with $c_o = S_o$ and $c_n = S_n - S_{n-1}$ for $n \geq 1$. Thus the above procedure can be applied to estimate the error $\varepsilon_{2k}^{(n)} - S$.

For example if k=1, $\varepsilon_2^{(n)}$ is Aitken's Δ^2 process and the error estimate is given by

$$e_1 (1 - c_{n+2}/c_{n+1})^{-1} (1+a_o+a_1)^{-1}$$

with $e_1 = (c_{n+1}\, c_{n+3} - c_{n+2}^2)/c_{n+1}$, $e_2 = (c_{n+1} c_{n+4} - c_{n+2} c_{n+3})/c_{n+1}$,

$e_3 = (c_{n+1}\, c_{n+5} - c_{n+2}\, c_{n+4})/c_{n+1}$, $a_1 = -e_2/e_1$ and $a_o = -(e_3+a_1 e_2)/e_1$.

If we consider the sequence $S_o = 1$, $S_{n+1} = \exp(-S_n)$ for n = 0,1,..., which converges to S = 0.567143290409... we obtain the following results

n	error estimate	error
0	$-0.149758\ 10^{-1}$	$-0.150828\ 10^{-1}$
1	$-0.457880\ 10^{-2}$	$-0.456248\ 10^{-2}$
2	$-0.149222\ 10^{-2}$	$-0.149552\ 10^{-2}$
3	$-0.474272\ 10^{-3}$	$-0.473704\ 10^{-3}$
5	$-0.491566\ 10^{-4}$	$-0.491374\ 10^{-4}$
7	$-0.508946\ 10^{-5}$	$-0.508882\ 10^{-5}$
10	$-0.169442\ 10^{-6}$	$-0.169446\ 10^{-6}$
14	$-0.181358\ 10^{-8}$	$-0.181358\ 10^{-8}$

For the sequence $S_n = 0.9^{n+1}/(n+1)$ the results obtained are not as good

n	error estimate	error
0	-0.144791	-0.164189
2	$-0.536296\ 10^{-1}$	$-0.542730\ 10^{-1}$
4	$-0.257294\ 10^{-1}$	$-0.241564\ 10^{-1}$
6	$-0.135000\ 10^{-1}$	$-0.122292\ 10^{-1}$
9	$-0.563656\ 10^{-2}$	$-0.503439\ 10^{-2}$

The general principles studied in the second section can be extended to sequences transformations. Assume we are given a sequence transformation $T : (S_n) \to (T_n)$, then

$$T_n - S_n = (S - S_n) - (S - T_n)$$

where S is the limit of (S_n). We have

$$\frac{T_n - S_n}{S - S_n} = 1 - \frac{T_n - S}{S_n - S}$$

Thus if $T_n - S = o(S_n - S)$, the ratio $(T_n - S_n)/(S - S_n)$ tends to 1 when n goes to infinity which shows that, for n sufficiently large, $T_n - S_n$ is a good estimation of the error $S - S_n$. This is, in particular, true if (S_n) converges superlinearly and if we take $T_n = S_{n+1}$. For example if the fixed point of $x = e^{-x}$ is computed by the secant method with $S_0 = 0$ and $S_1 = 1$, we get

n	$S_{n+1} - S_n$	$S - S_n$
1	-0.3873	-0.4328
2	$-0.4886\ 10^{-1}$	$-0.4556\ 10^{-1}$
3	$0.3332\ 10^{-2}$	$0.3305\ 10^{-2}$
4	$-0.2705\ 10^{-4}$	$-0.2707\ 10^{-4}$
5	$-0.1619\ 10^{-7}$	$-0.1619\ 10^{-7}$
6	$0.7930\ 10^{-13}$	$0.7944\ 10^{-13}$

This method for estimating the error should be compared with the error control proposed in [5] which is based on sequences of intervals asymptotically containing the limit.

REFERENCES.

[1] B. BAILLAUD, H. BOURGET, eds.
Correspondance d'Hermite et de Stieltjes
Gauthier-Villars, Paris, 1905.

[2] C. BREZINSKI
Padé-type approximation and general orthogonal polynomials.
ISNM Vol. 50, Birkhäuser-Verlag, Basel, 1980.

[3] C. BREZINSKI
A general extrapolation algorithm.
Numer. Math., 35 (1980) 175-187.

[4] C. BREZINSKI
Recursive interpolation, extrapolation and projection.
J. Comp. Appl. Math., 9 (1983), 369-376.

[5] C. BREZINSKI
Error control in convergence acceleration processes.
IMA J. Numer. Anal., 3 (1983) 65-80.

[6] C. BREZINSKI
Orthogonal and Stieltjes polynomials with respect to an even functional.
Publication ANO, Univ. de Lille 1, 1986.

[7] C. BREZINSKI
Partial Padé approximation
To appear.

[8] T.S. CHIHARA
An introduction to orthogonal polynomials.
Gordon and Breach, New-York, 1978

[9] A. DRAUX, P. VAN INGELANDT
Polynômes orthogonaux et approximants de Padé. Logiciels.
Editions Technip, Paris, 1987.

[10] A.S. KRONROD
Nodes and weights of quadrature formulas.
Consultants Bureau, New-York, 1965.

[11] G. MONEGATO
On polynomials orthogonal with respect to particular variable-signed weight function.
J. Appl. Math. Phys. (ZAMP), 31 (1980), 449-555.

[12] G. MONEGATO
Stieltjes polynomials and related quadrature rules.
SIAM Rev., 24 (1982), 137-158.

[13] T.N.L. PATTERSON
The optimum addition of points to quadrature formulae.
Math. Comp., 22 (1968) 847-856.

[14] M. PREVOST
Stieltjes and Geronimus type polynomials.
J. Comp. Appl. Math., To appear.

[15] D. SHANKS
Nonlinear transformations of divergent and slowly convergent sequences.
J. Math. Phys., 34 (1955) 1-42.

[16] G. SZÈGÖ
Orthogonal polynomials.
AMS Colloqium publications, Vol. 23, Providence, 1939.

[17] P. WYNN
On a device for computing the $e_m(S_n)$ *transformation.*
MTAC, 10 (1956) 91-96.

ORTHOGONAL POLYNOMIALS IN MONOTONE AND CONVEX INTERPOLATION

Alan Edelman

Department of Mathematics
M. I. T.
Cambridge, M.A. 02139

Charles A. Micchelli

Department of Mathematical Sciences
IBM Thomas J. Watson Research Center
P.O. Box 218
Yorktown Heights, NY 10598

Our purpose here is to describe some results concerning admissible slopes for monotone and convex interpolation by spline functions and rational functions. A detailed presentation will appear alsewhere, [EM].

We begin by recalling the following result of Fritch and Carlson [FC] which is useful for computing monotone C^1 piecewise cubic interpolation.

<u>Proposition 1</u>. (Fritsch, Carlson). Let u be a cubic polynomial on $[a,b]$. Then u is monotone if and only if either

$$u'(a) = u'(b) = 0\,, \qquad \text{if} \quad \Delta = 0$$

or

$$\left(\frac{u'(a)}{\Delta}, \frac{u'(b)}{\Delta}\right) \in \mathcal{M}_3, \quad \text{if } \Delta \neq 0$$

where $\Delta = \dfrac{u(b)-u(a)}{b-a}$ and the monotonicity region $\mathcal{M}_3$ consists of the convex hull of the origin with the ellipse $\frac{1}{4}(x+y-4)^2+\frac{1}{12}(x-y)^2=1$ which is tangent to the coordinate axes.

One for our goals is to explain this result by casting it in a more general framework. Before doing so we review how it is used for the design of monotone interpolation algorithms.

Fritsch and Carlson pointed out that nonmonotonicty occurs if the magnitude of the derivative at each endpoint is too large relative to the slope of the secant line. In the cubic case, if the derivatives are roughly more than three to four times larger than the slope nonmonotonicity will occur. Since the box $[0,3] \times [0,3]$ is contained in $\mathcal{M}_3$ the following sufficient conditions hold for a cubic polynomial

u to be monotone on $[a,b]$:

(1) $\quad 0 \leq su'(a) \leq 3s\Delta$

and

(2) $\quad 0 \leq su'(b) \leq 3s\Delta$

where $s = \text{sign}(\Delta)$.

This observation has been the basis for a number of algorithms proposing the use of piecewise cubics for monotone interpolation. All are some variant on picking derivatives so that (1) and (2) are satisfied.

If we consider the data points $\{(x_i, y_i): 0 \leq i \leq n\}$, with $x_o < x_2 < \ldots < x_n$ and let $\Delta_i = \dfrac{y_{i+1}-y_i}{x_{i+1}-x_i}$. Then (1) translates into requiring that $0 \leq y_i' \leq 3 \min(\Delta_{i-1}, \Delta_i)$ for the derivative at x_i , if the data are locally monotonically increasing. Some of the following choices were mentioned in the literature.

Butland [BU] suggested taking

$$y'_i = \begin{cases} \dfrac{2\Delta_{i-1}\Delta_i}{\Delta_{i-1}+\Delta_i}, & \text{if } \Delta_{i-1}\Delta_i > 0 \\ 0, \text{ otherwise} \end{cases}$$

while [FB] recommended

$$y'_i = H(\Delta_{i-1}, \Delta_i) = \begin{cases} 0, & \text{if } \Delta_{i-1}\Delta_i \leq 0 \\ \text{sign}(\Delta_{i-1}) \dfrac{3\,|\Delta_{i-1}|\,|\Delta_i|}{|\Delta_{i-1}| + 2\,|\Delta_i|}, & \text{if } |\Delta_i| < |\Delta_{i-1}| \\ H(\Delta_i, \Delta_{i-1}), & \text{otherwise} \end{cases}$$

and Brodlie [BR] proposed

$$y'_i = \begin{cases} \dfrac{\Delta_{i-1}\Delta_i}{\alpha\Delta_i + (1-\alpha)\Delta_{i-1}}, & \text{if } \Delta_{i-1}\Delta_i > 0 \\ 0 \quad , & \text{otherwise} \end{cases}$$

where $\quad \alpha = \frac{1}{3}\left(1 + \dfrac{h_i}{h_{i-1} + h_i}\right), \quad h_i = x_{i+1} - x_i .$

In [H], Hyman considers having the derivatives, y'_i available by some other approximation method and then suggest projecting them into the [0,3] x [0,3] box. His formulation does not require the data be monotone but instead his interpolant will automatically be monotone where the data is.

All these algorithms focus on reducing large derivative to slope ratios. We set out to compare various possible interpolants by computing the derivative regions of the interpolants. For a given function class G (such as monotone cubics) the derivative region specifies exactly the admissible slopes at the data points for the existence of a monotone interpolant. We formally describe this procedure below.

Let G be any subset of $C^1(-1,1)$ with $g(1) = 1$ and $g(-1) = -1$ for any $g \in G$. In general, the interpolation problem described above has the form: given $x_0 < x_1 < \dots < x_n < x_{n+1}$ find an $f \in C^1(x_0, x_{n+1})$ such that

(3) $$f(x_i) = y_i, \quad f'(x_i) = y'_i, \qquad i = 0, 1, \dots, n+1$$

and the linear rescaling of f given by

$$f_i(t) = \frac{2f(\frac{(1-t)}{2}x_i + \frac{(1+t)}{2}x_{i+1}) - y_i - y_{i+1}}{y_{i+1} - y_i}, \quad t \in [-1, 1],$$

is in G for all $i=0,1,\dots,n$. In what follows G is used to control the shape of the interpolant between the data points.

It is possible to give a useful condition for the solvability of (3) in terms of the derivative data. To this end, we define

$$D(G) = \{(g'(-1), g'(1)) : g \in G\}.$$

From the definition of f_i, (3) has a solution if and only if for $i = 0, 1, \dots, n$

$$\left(\frac{y'_i}{\Delta_i}, \frac{y'_{i+1}}{\Delta_i}\right) \in D(G), \quad \text{if } \Delta_i \neq 0$$

where as before $\Delta_i = (y_{i+1} - y_i) / (x_{i+1} - x_i)$ (when $\Delta_i = 0$, f_i is undefined but interpolation on $[x_i, x_{i+1}]$ with a constant is appropriate).

The normalization to [-1,1] is a matter of convenience and will not alter the important geometric properties of monotonicity or convexity. We will discuss in turn the following function classes commonly used for shape preserving interpolation: monotone polynomials, C^1 monotone piecewise polynomials, convex polynomials, parametric polynomials curves and rational functions. We pose an open problem in the case of rational functions.

Our first goal is to determine $D(G)$ for G the class of monotonic polynomials mapping $[-1,1]$ onto itself. We denote this set by

$$M_n = \{p : p \in \pi_n,\ p'(t) \geq 0 \text{ for } |t| \leq 1,\ p(\pm 1) = \pm 1\}$$

and so for $n \geq 2$

$$D(M_n) = \{(q(-1), q(1)) : q \in \pi_{n-1}, \frac{1}{2}\int_{-1}^{1} q(t)dt = 1, q(t) \geq 0, |t| \leq 1\},$$

(trivially, $D(M_1) = \{(1,1)\}$). The description of $D(M_n)$ is given below in Theorem 1.

Theorem 1. "For $n \geq 3$, there exists a monotone polynomial u of degree n on $[a,b]$ with specified values $u(a)$, $u(b)$ and derivatives $u'(a)$, $u'(b)$ if and only if

$$u'(a) = u'(b) = 0, \qquad \text{if } \Delta = 0$$

for

$$\left(\frac{u'(a)}{\Delta}, \frac{u'(b)}{\Delta}\right) \in \mathcal{M}_n, \quad \text{if } \Delta \neq 0$$

where $\Delta = \frac{u(b) - u(a)}{b - a}$ and $\mathcal{M}_n$ is the convex hull of $(0,0)$ and an ellipse for n odd or a line segment for n even. If $n = 2m-1$, the equation of the ellipse E_m is

$$\frac{1}{m^2}(x + y - m^2)^2 + \frac{1}{m^2(m^2 - 1)}(x - y)^2 = 1$$

while for $n = 2m$ the line segment L_m is given by

$$x + y = m(m + 1), \qquad x, y \geq 0. \text{"}$$

The proof of Theorem 1 makes use of the elementary

Lemma 1. Let a,b be linearly independent vectors in R^n, $n \geq 3$. The set $E_{a,b} = \{((a.x)^2, (b.x)^2) : \|x\| = 1\}$, $\|.\|$ = Euclidean norm, is the convex hull of $(0,0)$ with the line segment

(4) $$C_{11}x + C_{22}y = 1, \qquad x,y \geq 0,$$

when $a.b = 0$ or the ellipse

(5) $$(C_{11}x + C_{22}y - 1)^2 - 4C_{12}^2 xy = 0,$$

Otherwise where the constants $C = C_{ij}$ are defined by

$$C = A^{-1}, \quad A = \begin{pmatrix} a.a & a.b \\ a.b & b.b \end{pmatrix}$$

For $n = 2$, $E_{a,b}$ is (4) when $a.b = 0$ or (5) otherwise.

This lemma is used in the following special situation.

Let $S_n^{(\alpha,\beta)} = \{p \in \pi_n : \frac{1}{2}\int_{-1}^{1} p^2(x)(1-x)^{\alpha}(1+x)^{\beta}dx = 1\}$ for

for $\alpha,\beta > -1$. Using the notation of [Sz] we introduce $P_n^{(\alpha,\beta)}$ the nth Jacobi polynomial. Then

$$(Hu)(t) = \sum_{k=0}^{n} 2^{1/2}(h_k^{(\alpha,\beta)})^{-1/2} u_k P_k^{(\alpha,\beta)}(t)$$

is an isometry between R^{n+1} with the Euclidean norm and π_n equipped with the inner product

$$(f, g) = \frac{1}{2}\int_{-1}^{1} f(t)g(t)(1-t)^{\alpha}(1+t)^{\beta}dt$$

where

$$h_k^{(\alpha,\beta)} = \int_{-1}^{1} (1-x)^{\alpha}(1+x)^{\beta}(P_k^{(\alpha,\beta)}(x))^2 dx$$

$$= \frac{2^{\alpha+\beta+1}}{2k+\alpha+\beta+1}\,\frac{\Gamma(k+\alpha+1)\,\Gamma(k+\beta+1)}{\Gamma(k+1)\,\Gamma(k+\alpha+\beta+1)}.$$

Thus by introducing the vectors

$$a_k = 2^{1/2}(h_k^{(\alpha,\beta)})^{-1/2} P_k^{(\alpha,\beta)}(-1)$$

$$b_k = 2^{1/2}(h_k^{(\alpha,\beta)})^{-1/2} P_k^{(\alpha,\beta)}(1) \quad k = 0, \dots, n$$

we have $\{(p^2(-1), p^2(1)) : p \in S_n^{(\alpha,\beta)}\} = \{((a \cdot x)^2, (b \cdot x)^2) : \|x\| = 1, x \in \mathbb{R}^{n+1}\}$.

Since

$$\|a\|^2 = \|b\|^2 = \sum_{k=0}^{n} 2(h_k^{(\alpha,\beta)})^{-1}(P_k^{(\alpha,\beta)}(1))^2$$

$$= 2K_n^{(\alpha,\beta)}(1, 1)$$

$$= 2^{\alpha+\beta}\frac{\Gamma(n+\alpha+\beta+2)\,\Gamma(n+\alpha+2)}{\Gamma(\alpha+1)\Gamma(\alpha+2)\Gamma(n+1)\Gamma(n+\beta+1)}$$

and

$$a \cdot b = \sum_{k=0}^{n} 2(h_k^{(\alpha,\beta)})^{-1} P_k^{(\alpha,\beta)}(-1) P_k^{(\alpha,\beta)}(1)$$

$$= 2K_n^{(\alpha,\beta)}(-1, 1)$$

$$= (-1)^n 2^{\alpha+\beta}\frac{\Gamma(n+\alpha+\beta+2)}{\Gamma(\alpha+1)\Gamma(\beta+1)\Gamma(n+1)}$$

so that $C = \begin{pmatrix} \|a\|^2 & a.b \\ a.b & \|a\|^2 \end{pmatrix}^{-1}$ is readily calculated. If $\alpha = \beta = 0$ we get $\|a\|^2 = \|b\|^2 = (n+1)^2$ and $a.b = (-1)^n(n+1)$ which means that $C = \frac{1}{(n+1)((n+1)^2-1)} \begin{pmatrix} n+1 & (-1)^{n-1} \\ (-1)^{n-1} & n+1 \end{pmatrix}$ in this case.

Using these computations we can show that $\{(p^2(-1), p^2(1)) : p \in S_{m-1}^{(0,0)}\}$ gives the convex set $\mathcal{M}_{2m-1}$ described in Theorem 1. So far, we only know that this set is contained in $D(M_{2m-1})$ but from Lukác's lemma [Sz] which states that if $p \in M_{2m-1}$ then $p'(t) = A^2(t)+(1-t^2)B^2(t)$ where $A \in \pi_{m-1}$, $B \in \pi_{m-2}$ we can derive the result that $D(M_{2m-1})$ is in fact equal to $\mathcal{M}_{2m-1}$.

The even case is similar because Lukác's lemma says that for any $p \in M_{2n}$. $p'(t) = (1-t)A^2(t) + (1+t)B^2(t)$ where $A, B \in \pi_{m-1}$. Thus, in this case, it is appropriate to separately identify the line segments $\{(2s(-1), 0) : s \in S_{m-1}^{(1,0)}\}$ and $\{(0, 2s(1)) : s \in S_{m-1}^{(0,1)}\}$ and form their convex hull which is a triangle. Using the previous analysis we therefore obtain $\mathcal{M}_{2m}$.

In the diagram below, we labeled some special points on the ellipse E_m,

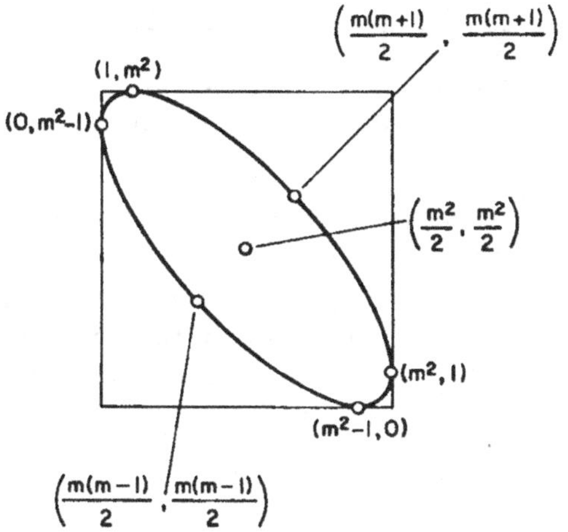

Figure 1

From this diagram it is easy to see that L_m is tangent to E_m and E_{m+1} at $(\frac{m(m+1)}{2}, \frac{m(m+1)}{2})$. These ellipses and line segments appear in an interesting arrangement. The line segment L_m separates the ellipses E_m and E_{m+1} (the latter being above L_m) which are all mutually tangent at $\frac{1}{2} m(m+1)(1,1)$.

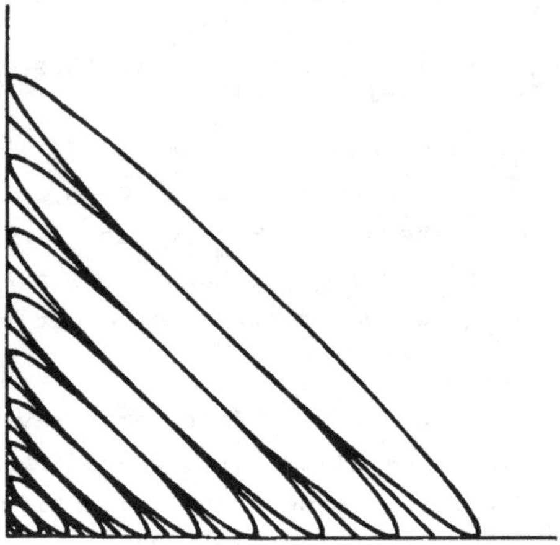

Figure 2

It also may be useful to point out for algorithmic use that

$$\left[0, \frac{(m+1)(m+2)}{2}\right] \times \left[0, \frac{(m+1)(m+2)}{2}\right]$$

is the largest box inscribed in $D(M_{2m+1})$. For $m = 1$, it is the $[0,3] \times [0,3]$ box that was used in all the algorithms we described earlier.

We mention that as a consequence of our discussion there is only one $p \in M_n$ such that $(p'(-1),p'(1)) = (r,s)$ when $(r,s) \in E_m, n = 2m-1$ or $(r,s) \in L_m, n = 2m$. From the above analysis, it is possible to obtain the polynomials corresponding to the boundary curves E_m and L_m . Specially, in case 1 boundary points correspond to those polynomials h of the form $\alpha P_{m-1}^{(1,1)} + \alpha P_{m-2}^{(1,1)}$ that satisfy $\frac{1}{2} \int_{-1}^{1} h^2(t)dt = 1$. We note that since

$$\frac{1}{2}\int_{-1}^{1} (P_{m-1}^{(1,1)}(t))^2 dt = \frac{2m}{m+1} , \quad \int_{-1}^{1} P_{m-1}^{(1,1)}(t)P_{m-2}^{(1,1)}(t)dt = 0$$

the normalization on h becomes

$$\frac{2m}{m+1}\alpha^2 + \frac{2(m-1)}{m}\beta^2 = 1.$$

It is of interest to explicitly display the polynomials corresponding to the special points labeled on Figure 1. The interesection of E_m with the line $y = x$ corresponds to the two polynomials $\frac{m+1}{2m}(P_{m-1}^{(1,1)}(x))^2$ and $\frac{m}{2(m-1)}(P_{m-2}^{(1,1)}(x))^2$ while $p(x) = \frac{m+1}{4(m-1)}(P_{m-2}^{(1,2)}(x))^2(1+x)^2$ gives the point where E_m intersects the y-axis. The corresponding x-axis intersection comes from the polynomials $p(-x)$. To verify the last two claims we note that $(1+x)P_{m-2}^{(1,2)}(x) = P_{m-2}^{(1,1)}(x) + \frac{m-1}{m}P_{m-1}^{(1,1)}(x)$. Finally, $(P_{m-1}^{(1,0)}(x))^2$ and $(P_{m-1}^{(0,1)}(x))^2$ correspond to the topmost and righmost points on the ellipse, respectively. To see the proper form in this case, not that

$$P_{m-1}^{(1,0)}(x) = \frac{m+1}{2m}P_{m-1}^{(1,1)}(x) + \frac{1}{2}P_{m-2}^{(1,1)}(x).$$

Next we proceed to study similar questions for π_{nk}, the class of C^1 piecewise polynomial of degree n with at most k breakpoints in $(-1,1)$. The corresponding set of nondecreasing elements of π_{nk} are denoted by M_{nk}. Now, consider the set $D(M_{nk})$ which so far we have determined for $k = 0$.

As a first elementary observation we have

<u>Lemma 2</u>. For $n,k \geq 2 D(M_{nk}) = R_+^2$.

The main result that we obtain in this context is the exact description of $D(M_{n1})$. To this end, we define

$$\tau_n \begin{cases} m^2, & \text{for } n = 2m-1 \\ m(m+1), & \text{for } n = 2m \end{cases}$$

Note that according to Theorem 1

$$\tau_n = \max\{x:(x,y) \in D(M_n)\} = \max\{y:(x,y) \in D(M_n)\}.$$

Also, we introduce

$$S_n = \{(x,y):0 \leq \min(x,y) < \tau_n\} \cup \{(\tau_n,\tau_n)\}.$$

The general result can now be stated as $D(M_{n1}) = S_n$, $n \geq 3$. We also note the special cases $D(M_{11}) = D(M_{12}) = \{(1,1)\}$ and $D(M_{21}) = \{(x,y):x \geq 0, y \geq 0, x+y \leq 2\}$.

Let us now turn to a description of $D(C_n)$ where C_n is the class of convex polynomials on $[-1,1]$ normalized as follows:

$$C_n = \{p : p \in \pi_n, p''(t) \geq 0 \quad \text{for} \quad |t| \leq 1 \, , \, p(\pm 1) = \pm 1\} \, .$$

An important subset of C_n is CM_n which we define as the set of monotone convex polynomials mapping $[-1,1]$ onto itself. The crux of the matter is a theorem of Chebyshev [Sz] .

Lemma 3. Define

$$a_n = \max\left\{\frac{\int_{-1}^{1} xp(x)dx}{\int_{-1}^{1} p(x)dx} : p \in \pi_n, \, p(t) \not\equiv 0, \; p(t) \geq 0, \; |t| \leq 1\right\}.$$

Then

$$a_n = \begin{cases} \hat{P}^{(0,0)}_{m+1}, & \text{if } n = 2m \\ \hat{P}^{(0,1)}_{m+1}, & \text{if } n = 2m+1 \end{cases}$$

where $\hat{p}$ denotes the largest zero of p .

Obviously, the corresponding minimum value of the ratio above is $-a_n$. Now, let $p_n = 2(1-a_{n-2})^{-1}$ and $q_n = 2(1+a_{n-2})^{-1}$ so that $\frac{1}{p_n} + \frac{1}{q_n} = 1$.

Theorem 2. $D(C_n)$ is a wedge shaped region with vertex $(1,1)$ and intersecting the y-axis at $[q_n, p_n]$. Precisely,

$$D(C_n) = \{(x,y) : \frac{x-y}{x-1} \in [q_n, p_n], \; x < 1, y > 1\} \cup \{(1,1)\}$$

Furthermore, $D(CM_n)$ is the triangular region joining $(0,0), (0,q_n)$, $(0,p_n)$.

The wedge is illustrated in Figure 3 below.

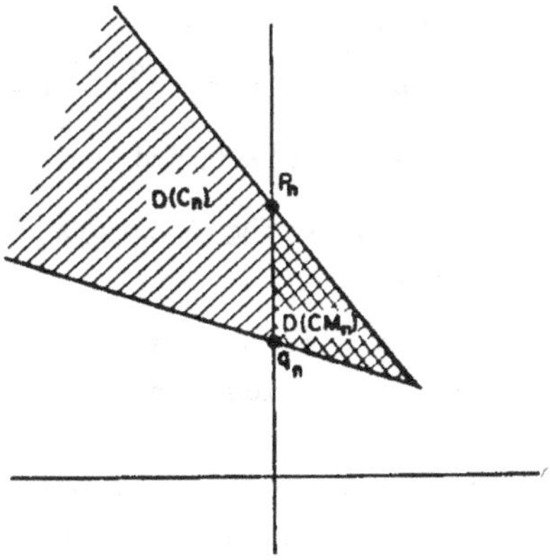

Figure 3

Before investigating rational functions we remark in passing that the complete monotonicity region using monotone parametric cubics is discussed in [EM] . This type of interpolant is commonly used in computer graphics.

Although the case of monotone rational functions presents some new interesting possibilities it also brings several difficulties. The set of functions we now consider is

$$R_{mn} = \left\{ r(t) = \frac{p(t)}{q(t)} : p \in \pi_m, q \in \pi_n, r(\pm 1) = \pm 1, r'(t) \geq 0 \text{ and } q(t) \neq 0 \text{ for } |t| \leq 1 \right\}$$

that is, monotone rationals mapping $[-1,1]$ onto itself.

We have determined $D(R_{mn})$ for every pair (m,n) except the case m odd , $m \geq 5$, $n = 1$. A conjecture as to the solution in this case is given in [EM] . As a start we recall that in [GD] it was shown that $D(R_{mn}) = R_+^2$ for $m,n \geq 2$. The following improves upon this result.

Proposition 2. $D(R_{mn})$ is the entire monotonicity region R_+^2 for $m \geq 1$ and $n \geq 2$.

Now we turn to the difficult case of evaluating $D(R_{m1})$. To this end, it is convenient to define for $|\alpha| > 1$

$$R_{m1}^{\alpha} = \{ r \in R_{m1} : r(x) = \frac{p(x)}{x-a} , p \in \pi_m \} ,$$

i.e. R_{m1}^{α} consists of those members of R_{m1} with a pole at α . Observe that $\alpha = \infty$ makes sense in this context and $R^{\infty}_{m1} = M_m$.

If $p(x) = (x-a)^2 r'(x)$ for some r in R^{α}_{m1} then

(6) $$p \in \pi_m,\ p(x) \geq 0 \text{ for } |x| \leq 1,\ p'(\alpha) = 0, \text{ and } \frac{1}{2}\int_{-1}^{1} \frac{p(x)}{(x-\alpha)^2}dx = 1.$$

Furthermore, any p satisfying these requirements gives rise to an $r \in R^{\alpha}_{m1}$ with $p(x) = (x-\alpha)^2 r'(x)$.

To analyze the set $D(R^{\alpha}_{m1})$ we need a version of Lukâc's for polynomials satisfying (6). This is provided in

<u>Proposition 3</u>. Let $S^{\alpha}_n = \{p \in \pi_n : p'(\alpha) = 0,\ p(t) \geq 0$ for $|t| \leq 1\}$. Every $p \in S^{\alpha}_n$ has the representation $p(t) = p_1(t)+p_2(t)$ where $p_1, p_2 \in S^{\alpha}_n$ and

$$p_1(t) = (a - \varepsilon t)A^2(t), \qquad p_2(t) = (1 - t^2)(b - \varepsilon t)B^2(t), \quad \text{if } n = 2m + 1$$

$$p_1(t) = (1 + t)(c - \varepsilon t)C^2(t), \quad p_2(t) = (1 - t)(d - \varepsilon t)D^2(t), \quad \text{if } n = 2m$$

where $\varepsilon = \text{sign}(\alpha)$, $a,b,c,d > |\alpha|$, and all the roots of the polynomials A,B,C, and D are in the interval $[-1,1]$.

Proposition 3 is a special case of a more general theorem from [KS] concerning so called Chebychev systems.

Proposition 3 gives useful information about $D(R^{\alpha}_{n1})$:

<u>Lemma 4</u>. If n is even, $n > 2$, then $D(R^{\alpha}_{n1})$ is a triangle T_α with vertices $(0,0), (0,\zeta(\alpha))$, and $(\zeta(-\alpha),0)$ where

$$\zeta(\alpha) = \max\{p(1) : p \in \tilde{R}^{\alpha}_{n1},\ p(-1) = 0\}.$$

Here $\tilde{R}^{\alpha}_{n1}$ denotes the set $\{p' : p \in R^{\alpha}_{n1}\}$.

Even though we do not obtain $\zeta(\alpha)$ explicitly Lemma 3 leads us to an exact description of $D(R_{n1})$ for n even. It is of interest to compare this result with that of piecewise polynomials.

<u>Theorem 3</u>. For $n = 2m \geq 4$ we have

$$D(R_{n1}) = \{(x,y) : 0 \leq \min(x,y) < m^2\}.$$

As we mentioned earlier we have not been able to determine $D(R_{n1})$ n odd, $n \geq 5$. Some facts are given about this problem in [EM]. Here we just note the following special cases

(a) $D(R_{11}) = \{(x,y): xy = 1, x,y > 0\}$

(b) $D(R_{21}) = \{(x,y): 0 \leq \min(x,y) < 1 < \max(x,y)\} \cup \{(1,1)\}$

(c) $D(R_{31}) =$ the convex hull of $(0,0)$ with the set $\{(x,y) \in R_+^2 : (x-2)(y-2) = 1\}$.

The first two sets are easily obtained while the third is already quite complicated.

REFERENCES

[BR] K.W. Brodlie, A Review of Methods for Curve and Function Drawing, Mathematical Methods in Computer Graphics and design, K.W. Brodlie, ed. Academic Press, London, 1980, 1-37.

[BU] J. Butland, A Method for Interpolating Reasonable-Shaped Curves through Any Data, Proc. of Computer Graphics 80, Online Publication Ltd., Northwood Hills, Middlesex, U.K., 1980, 409-422.

[EM] Alan Edelman and Charles A. Micchelli, Admissible Slopes for Monotone and Convex Interpolation, IBM Research Report, 1986.

[FB] F.N. Frisch and J. Butland, "A Method for Constructing Local Monotone Piecewise Cubic Interpolants", SIAM J. Sci. Stat. Comput., 5 (1984), 300-304.

[FC] F. N. Fritsch and R.E. Carlson, "Monotone Piecewise Cubic Interpolation", SIAM J. Num. Analysis, 17 (1980), 238-246.

[GD] J.A. Gregory and R. Delbourgo, "Piecewise Rational Quadratic Interpolation to Monotonic Data", IMA Journal of Numerical Analysis, 2 (1982), 123-130.

[H] J.M. Hyman, "Accurate Monotonicity Preserving Cubic Interpolation", SIAM J. Sci. Stat. Comput., 4 (1983), 645-654.

[KS] S. Karlin and W.J. Studden, Tchebycheff Systems: with Appplications in Analysis and Statistics, Interscience Publishers, New York 1966.

[Sc] L.L. Schumaker. "On Shape Preserving Quadratic Spline Interpolation", SIAM J. Num. Analysis, 20 (1983), 854-864.

|Sz| G. Szegö, Orthogonal Polynomials, American Mathematical Society Providence, 1939.

POLYNOMIALS ORTHOGONAL WITH RESPECT TO SINGULAR CONTINUOUS MEASURES

J.S. Geronimo
School of Mathematics
Georgia Institute of Technology
Atlanta, GA 30332

Abstract

Let $T(z)$ be a monic polynomial of degree $d \geq 2$ chosen so that its Julia set J is real. A class of invariant measures and the orthogonal polynomials associated with these measures are constructed and discussed. In particular, the asymptotic properties of the polynomials and the recurrence formula coefficients is presented.

I. Introduction

Let $T(z)$ be a monic polynomial of degree $d \geq 2$ then its Julia set J is given by

$$J = \overline{\{z : T^{on}(z) = z \ , \ |T^{on}(z)'| > 1 \quad n = 1,2,\ldots\}}$$

where $T^{on}(z) = T \circ T^{o(n-1)}(z)$ and $T^{o}(z) = z$ [J], [Fa], [Br]. Thus the Julia set is the closure of the set of repulsive fixed points of T and all its iterates. In the case when $T(z) = z^2 - \lambda$ one knows [Br], [Fa], [J], [BGH1] that when $\lambda = 0$, J is the unit circle, while if $\lambda = 2$, J is the interval $[-2,2]$. For $\lambda \notin [2,\infty]$, J is a set with components not on the real line and in this case J can take beautiful and bizarre shapes [DH], [BGHD]. For $\lambda \in (2,\infty)$, J is a subset of the real line that is Cantorian. It is this case, when J is a Cantor subset of the real line, upon which we will concentrate.

Singular continuous spectra arise in many different physical models such as those connected with fractal structures [DDI] of discrete Schrödinger operators with almost periodic potentials. Operators of the above form have recently appeared in conducting or super-conducting linear chains [TK] and in metal-insulator transitions; the almost Mathiew equation being such an example [AvS]. Very little is known about the nature and behaviour of the wave functions (orthogonal polynomials) belonging to these types of operators. Consequently, polynomials orthogonal with respect to measures supported on Julia sets give us an opportunity to explore various properties of polynomials orthogonal with respect to singular continuous measures.

We begin, in section II, by constructing measures supported on J that are invariant and mixing with respect to T. We then construct the polynomials orthogonal with respect to these measures and show that $p_{nd}(x) = p_n(T(x))$. Next we consider the associated polynomials and show that they are orthogonal with respect to a purely discrete measure whose mass is exterior to the Julia set. In section III, we show that the Jacobi matrices associated with these polynomials satisfy a renormalization group equation. In the special case when $T(z) = \alpha^d C_d(z/\alpha)$, $\alpha > 1$, $C_d(z)$ the monic Chebychev polynomial on $[-2,2]$, it is shown that these matrices are limit periodic for α sufficiently large. Finally, in section IV, we discuss the asym properties of the polynomials.

II. Construction of a class of invariant measures.

Let $T(z) = z^d + K_{d-1}z^{d-1} + \ldots,$ $d \geq 2$ be such that its associated Julia set is real. This will be the case if and only if $|T(z_i)| \geq z_o$, $i = 1,2 \ldots d-1$ and $T(-z_o) \geq z_o$ for d even or $z_d \leq T(z_i) \leq z_o$, $i = 1,2 \ldots d-1$ for d odd. Here z_i, $i = 1,2 \ldots d-1$ are the zeros of $T'(z)$ which must be real, z is the largest fixed point of T and z^d is the smallest fixed point of T when d is odd. Let $\{T_i^{-1}, i=1,2 \ldots d\}$ be a complete assignment of branches of the inverse of T and choose $w_i(x)$, $i = 1,2,\ldots,d$ continuous, strictly positive and such that

$$\sum_{i=1}^{d} w_i(x) = 1 \tag{II.1}$$

for all $x \in J$. We shall further constrain the w_i's so that

$$\sum_{i=1}^{d} \frac{w_i(x)}{z-T_i^{-1}(x)} = \frac{W(z)}{T(z)-x} \tag{II.2}$$

where $W(z) = z^{d-1} + K_{d-2}^{(1)} z^{d-2} + \ldots$. From (II.2) we see that

$$w_i(x) = \frac{W(T_i^{-1}(x))}{T'(T_i^{-1}(x))} \tag{II.3}$$

which implies that W and T' must have the same sign on J in order for $w_i(x) > 0$ for $x \in J$. A case for which the positivity is guaranteed is if $W(z) = \frac{T'(z)}{d}$ which gives $w_i(x) = 1/d$.

From (II.2) we have the elementary but important,

<u>Lemma 1</u>.

$$\sum_{i=1}^{d} w_i(x)(T_1^{-1}(x))^n = \pi_{\left[\frac{n}{d}\right]}(x) \qquad \text{(II.4)}$$

where $\Pi_m(x)$ is a polynomial of degree at most m, and $\left[\frac{n}{d}\right]$ is the largest integer smaller than or equal to $\frac{n}{d}$.

<u>Proof</u>. Expanding (II.2) in powers of $1/z$, then noticing that the lowest power of $1/z$ associated with x^m on the right hand side of (II.2) is $(1/z)^{md}$, yields the result.

With the above we can now construct our invariant measures.

<u>Theorem 1</u>. Given T and W as above there exists a unique Borel probability measure μ having the property that $\forall f \in L(J,\mu)$

$$\int_J f\, d\mu = \sum_{i=1}^{d} \int_J w_i(x) f(T_i^{-1}(x))\, d\mu(x) \quad . \qquad \text{(II.5)}$$

The above theorem can be proved using probabilistic methods [BDEG] or with the help of Lemma 1 using Padé approximants [M].

Choosing $f = \chi_{T_i^{-1}E}(x)$, E a Borel subset of J, we find from the above equation that

$$\mu(T_i^{-1}(E)) = \int_E w_i(x)\, d\mu \quad ,$$

which implies μ is T invariant ; i.e.,

$$\mu(E) = \mu(T^{-1}(E))$$

for all Borel subsets E of J. Furthermore, it can be shown [Br, Thm. 17.1] that μ is mixing, i.e., $\forall f,g \in L^2(J,\mu)$

$$\lim_{n\to\infty} \int_J f(x) g(T^{on}(x))\, d\mu = \int_J f(x)\, d\mu \int_J g(x)\, d\mu \quad .$$

This implies that measures associated with different sets of w_i's but with the same T are mutually singular [Bi, p. 39].

If we set $f(x) = \frac{1}{z-x}$, $x \notin \operatorname{supp} \mu$, then (II.5) gives us the following functional relation for the Stieltjes transform of μ,

$$\int_J \frac{d\mu(x)}{z-x}\; G(z) = W(z) G(T(z)) \qquad \text{(II.6)}$$

Let $p_n(x)$ be the nth degreee monic orthogonal polynomial associated with μ i.e., $p_n(x) = x^n + \gamma_{n-1}x^{n-1} \ldots$ and

$$\int_J p_n(x)\, p_m(x)\, d\mu = 0 \quad , \qquad n \neq m . \tag{II.7}$$

Lemma 1 and (II.5) gives us the following important

<u>Theorem 2</u>. [M] , [BGM]. Let $p_n(x)$ be given as above. Then

$$p_1(x) = x + K_{d-1} - K_{d-2}^{(1)} \tag{II.8}$$

and

$$p_{nd}(x) = p_n(T(x)) \tag{II.9}$$

A sequence $\{p_n(x) : n=0,1,\ldots\}$ of monic orthogonal polynomials on the real line satisfies a three term recurrence formula

$$xp_n(x) = p_{n+1}(x)+b_n p_n(x)+a_n^2 p_{n-1}(x) \quad , \qquad n = 0,1,2,\ldots \tag{II.10}$$

with $p_o(x) = 1$, $p_{-1}(x) = 0$, $a_n > 0$ and $b_n \in R$. The k-th associated polynomials $\{p_n^{(k)}(x) \;, n = 0,1,\ldots\}$ $(k \in N)$ satisfy the following recurrence formula,

$$xp_n^{(k)} = a_{n+k+1}\, p_{n+1}^{(k)}(x)+b_{n+k}p_n^{(k)}(x)+a_{n+k}p_{n-1}^{(k)}(x) \qquad n = 0,1,2,\ldots$$

with $p_o^{(k)} = 1$ and $p_{-1}^{(k)}(x) = 0$.

<u>Theorem 3</u>. $p_{nd-1}^{(1)}(x) = W(x)p_{n-1}^{(1)}(T(x))$, $\qquad n = 0,1,2 \ldots$

<u>Proof</u>. It is well known that

$$p_{n-1}^{(1)}(x) = \int_J \frac{p_n(x)-p_n(y)}{x-y}\, d\mu(x)$$

Consequently

$$p_{nd-1}^{(1)}(x) = \int_J \frac{p_{nd}(x)-p_{nd}(y)}{x-y}\, d\mu(x)$$

and the result now follows from (II.2), (II.5) and (II.9) .

Associated with $\{p_n^{(1)}(x)\}$ is their orthogonality measure $\mu^{(1)}$. The next result shows how when dealing with singular measures even small perturbations can produce spectacular changes.

<u>Theorem 4</u>. [BGH2] , [BGM] Suppose that $W(z)$ does not equal zero for $z \in J$ then for every bounded continuous function f ,

$$\int f(x)\,d\mu^{(1)}(x) = \sum_{i=1}^{d} \int \frac{w_i(x) f(T_i^{-1}(x))}{W[T_i^{-1}(x)]^2}\, d\mu^{(1)}(x) +$$

$$+ \sum_{m=1}^{d-1} f(z_m)\Gamma_m \qquad \text{(II.11)}$$

where z_m, $m=1,2\ldots d-1$ are the zeros of $W(z)$ and

$$\Gamma_m = \frac{R(z_m)}{W'(z_m)} \left[1 + \sum_{i=1}^{\infty} \frac{R(T^{on}(z_m))}{R(z_m)} \prod_{j=1}^{i} \frac{1}{W[T^{oj}(z_m)]} \right] \qquad \text{(II.12)}$$

Here

$$R(z) = \frac{(z+K_{d-1}-K_{d-2}^{(1)})W(z)-(T(z)+K_{d-1}-K_{d-2}^{(1)})}{a_1^2} \qquad \text{(II.12)}$$

Eq (II.11) implies and it can be shown [BGH2] that $\mu^{(1)}$ is purely discrete and has no mass on the Julia set. However, the mass points do accumulate on the Julia set.

III. Jacobi Matrices

$\psi \in l_2$ if and only if $\psi = (\psi(0), \psi(1) \ldots)$ with $\sum_{n=0}^{\infty} (\psi(n))^2 < \infty$. Let $J : l_2 \to l_2$ be the infinite dimensional representation of (II.10); i.e.,

$$J = \begin{pmatrix} b_o & a_1 & & \\ a_1 & b_1 & a_2 & \\ & \cdot & \cdot & \cdot \\ & & \cdot & \cdot & \cdot \\ & & & \cdot & \cdot & \cdot \end{pmatrix}$$

Just a there exists a functional relation for the Stieljes' transform, there also exists a functional relation for the resolvent.

Theorem 5. [B], [BGM]. Let J be the operator associated with (II.10). Then

$$D(z-J)^{-1}D^* = W(z)(T(z)-J)^{-1}, \qquad z \notin J \qquad \text{(III.1)}$$

Proof. The (n,m) entry of $(z-j)^{-1}$ is

$$[(z-J)^{-1}]_{n,m} = \int_J \frac{p_n(x)p_m(x)}{z-x}\, d\mu(x)$$

(S. p. 546). By (II.9) we have

$$[D(z-J)^{-1}D^*]_{n,m} = [(z-J)^{-1}]_{nd,md}$$

$$= \int_J \frac{p_n(T(x))p_m(T(x))}{z-x} d\mu$$

The result now follows using (II.2) and (II.5).

If we now use the Schur decomposition of $D(z-J)^{-1}D^*$, [B] i.e.,

$$D(z-J)^{-1}D^* = [z-DJD^* - DJQ\ (z-\tilde{J})^{-1}QJD^*]^{-1},$$

where $\tilde{J} = QJQ$ with $Q = I-D^*D$ we can obtain with the help of (III.1).

<u>Theorem 6.</u> [B], [BM]. Let $T(z)$ and $W(z)$ be given as above, then

$$b_{nd} = K_{d-1} - K_{d-2}^{(1)} \qquad n = 0, 1, 2,\ldots \tag{III.2}$$

$$\prod_{i=1}^{d} a_{nd+i} = a_{n+1}, \qquad n = 0, 1, 2,\ldots \tag{III.3}$$

$$p_{N-1}^{(nd+1)}(z) = W(z), \qquad n = 0, 1, 2,\ldots \tag{III.4}$$

and

$$a_{nd}^2 p_{N-2}^{(nd-d+1)}(z) + a_{nd+1}^2 p_{N-2}^{(nd+2)}(z)$$

$$= b_n - T(z) + (z-b_o)W(z) \qquad n = 0, 1, 2,\ldots \tag{III.5}$$

In general it is difficult to solve explicitly for the a_n's and b_n's using (III.3-III.5) however in the special case when $T(z) = \alpha^d c_d(z/\alpha)$ and $W(z) = \frac{T'(z)}{d}$ where $c_d(z)$ is the monic Chebychev polynomial of degree $d \geq 2$ on $[-2,2]$ we find that [BGH2]

$$a_{nd+i}^2 = \alpha^2 \quad i=2,3,\ldots,d-1 \qquad n = 0,1,2,\ldots \tag{III.6}$$

$$a_{nd+1}^2 + a_{nd}^2 = 2\alpha^2 \qquad n = 0,1,2,\ldots \tag{III.7}$$

and

$$a^2_{(n+1)d}\, a^2_{nd+1} = \frac{a^2_{n+1}}{\alpha^{2(d-2)}} \qquad n = 0, 1, 2, \ldots \qquad \text{(III.8)}$$

(III.8) is a consequence of (III.6) and (III.3). It is perhaps interesting to note that in these cases for large d the Jacobi matrices have a large number of constant terms. However the spectrum for all finite d and $\alpha > 1$ is a Cantor set.

A bounded sequence $[b^{(m)}(n)]_{n=-\infty}^{\infty}$ is periodic if and only if there exists a positive integer L_m such that

$$b^{(m)}(n + L_m) = b^{(m)}(n) \qquad n = 0, \pm 1, \pm 2, \ldots$$

A bounded sequence $[b(n)]_{n=-\infty}^{\infty}$ is called limit periodic if it is the uniform limit of bounded periodic sequences; that is, there exists periodic sequences $[b^{(m)}(n)]_{n=-\infty}^{\infty}$ such that

$$\lim_{m\to\infty} (\sup_n |b(n) - b^{(m)}(n)|) = 0$$

The coefficients in (II.10) will be said to be in the class L if

$$\lim_{n\to\infty} b_{md^n+s} = b(s) < \infty \qquad s = 0,1,2,\ldots$$

$$\lim_{n\to\infty} a_{md^n+s} = a(s) < \infty \qquad s = 0,1,2,\ldots$$

and

$$a(0) = 0 .$$

<u>Theorem 7.</u> [BBM], [BGH2]. Let $(T(z) = \alpha^d c_d(z/\alpha)$ with $d \geq 2$ and $\alpha > 1$. Then $\{a_n\}_{n=0}^{\infty}$ is in the class L. Moreover $|a_{md^n+s} - a_s| \leq 2\alpha^2\gamma^n$ for all $s \geq 0$, $m \geq 0$, and $n \geq 0$. Here $\gamma = \{(2\alpha^2-2)\alpha^{2(d-2)}\}^{-1}$. In particular $\{a_n\}$ can be extended to $n < 0$ so that it forms a limit periodic sequence whenever $(2\alpha^2-2)\alpha^{2(d-2)} > 1$ (which is always true when $\alpha > \sqrt{3/2}$).

<u>Proof.</u> Both results are immediate for $s = K \bmod(d)$, $K = 2,3,\ldots,d-1$, since all of the corresponding a_n's are equal to α. For $s = 0 \bmod(d)$ we observe from (III.7) and (III.8) that

$$a_{md}^2 = \alpha^{-2(d-2)} \frac{a_m^2}{2\alpha^2 - a_{(m-1)d}^2} \qquad \text{(III.9)}$$

From (III.7) and (III.8) we also see that $a_n^2 \leq 2\alpha^2$ for all n. Since $a_o = 0$ we have from the above equation that for $\alpha > 1$ and $m = 1$, $0 < a_{md}^2 < 1$, and $a_{md}^2 < a_m^2$. Assuming the result is true for $m = 1,2,\ldots K$ we see immediately that for $\alpha > 1$ $a_{(K+1)d}^2 < a_{Kd}^2$. From (III.6) and the fact that $a_{Kd}^2 < 1$ it follows by (III.9) that $a_{(K+1)d}^2 < 1$ for $K \neq 1 \bmod(d)$. For $K = 1 \bmod(d)$ (III.9) becomes

$$a_{(K+1)d}^2 = \alpha^{-2(d-2)} \frac{2\alpha^2 - a_{1d}^2}{2\alpha^2 - a_{1_{d^2}}^2} < 1 \qquad K = 1d + 1 ,$$

since $1 \leq K$. Thus

$$a_{md^n}^2 \leq \frac{\alpha^{-2(d-2)n} a^2(m)}{(2\alpha^2-1)^n} \leq \frac{2\alpha^{-2((d-2)n-1)}}{(2\alpha^2-1)^n}$$

which implies that $a_{md^n}^2 \longrightarrow 0 = a_o^2$. That $a_{md^n+1}^2 \longrightarrow a_1^2$ now follows from (III.7).

To prove the second part of the theorem we note that $|a_i^2 - a_j^2| \leq 2\alpha^2$ for all i and j. Consequently the result is true for $n = 0$ and all $s \geq 0$. We shall now assume the result is true for $n = 0,1,2,\ldots, K$ and for all $s \geq 0$. For $n = K+1$ we shall assume the result holds for all $j < ds$. Consider now the equation

$$a_{md^{K+1}+ds}^2 - a_{ds}^2 = \frac{a_{md^K+s}^2 - a_s^2 + a_s^2 - a_{ds}^2 \beta a_{md^{K+1}+ds-d+1}^2}{\beta a_{md^{K+1}+ds-d+1}^2}$$

where $\beta = \alpha^{2(N-2)}$ and we have used (III.8). Using (III.7) we find

$$a_{md^{K+1}+ds}^2 - a_{ds}^2 = \frac{a_{md^K+s}^2 - a_s^2}{\beta(2\alpha^2 - a_{md^{K+1}+ds-d}^2)}$$

$$+ \frac{a_{ds}^2(a_{ds-d+1}^2 - a_{md^{K+1}+ds-d+1}^2)}{2\alpha^2 - a_{md^{K+1}+ds-d}^2}$$

Therefore

$$|a^2_{md^{K+1}+ds} - a^2_{ds}| \le 2\alpha^2 \frac{(2\alpha^2-2)^{-K}}{\beta^{K+1}(2\alpha^2-1)} \le 2\alpha^2\gamma^{K+1}$$

In the above equation we have used the fact that $a_{ds} < 1$ and $|a_i{}^2 - a_j{}^2| < 2\alpha^2$ for all i and j. The result also hold for all $s = 1 \bmod (d)$, $1 = 2,3,\ldots,d-1$ by (III.6). The case $s = 1 \bmod(d)$ reduces to the case $s = 0 \bmod(d)$ using the fact that $|a^2_{md^n+1} - a^2_1| = |a^2_{md^n} - a^2_o|$ with $a_o = 0$.

We now prove the limit periodicity assertion. We assume $\gamma < 1$ and define a_s for $s < 0$ by

$$a_s = \lim_{K\to\infty} a_{md^K+s}$$

The limit exists for each m because a_{md^K+s} is a Cauchy sequence in K since for $l > K$

$$|a_{md^l+s} - a_{md^K+s}| = |a_{(md^{l-K}-1)d^K+md^K+s} - a_{md^K+s}| < 2\alpha^2\gamma^K$$

We now define sequences $\{a_n^{(d^k)}\}_{n=-\infty}^{\infty}$ for $K = 1,2,3,\ldots$ by $a_n^{(d^K)} = a_{n \bmod d^K}$ form which it follows that $\{a_n\}_{n=-\infty}^{\infty}$ is a limit periodic sequence.

IV. Asymptotic Properties

Let I be the smallest interval containing the spectrum of J and let C be the complex plane. The following is a general result on orthogonal polynomials whose recurrence coefficients are in L. The proof can be found in [BGM lemma III.1].

Theorem 8. Let I be a finite interval, then

$$\lim_{n\to\infty} \frac{p_{md^n+s}(z)}{p_{md^n}(z)} = p_s(z) \qquad s = 0,1,2,\ldots \qquad \text{(IV.1)}$$

uniformly on compact subset of $C \setminus I$ if and only if the coefficients in (II.10) are in L.

We return now to the case of Julia sets and consider more carefully when $W(z) = \frac{T'(z)}{d}$. In this case the spectral measure is the equilibrium measure [Br] associated with J, and in order to distinguish it from other measures we are considering, we will label it μ_e. Since μ_e is the equilibrium measure, $g(z) = \int_J \ln|z-x| \mu_e$ is the Green's function [Br]. Again letting I be the smallest interval containing J we set $B(z) = (e^{g(z)+ih(z)})$, $z \notin I$, where h is a harmonic conjugate of $g(z)$. For $z \notin I$, the Green's star domain $B(z)$, is a conformal map, mapping the exterior of I to the exterior of a "spiky" unit circle ([BGH3], [DH]).

<u>Lemma 2</u>. Let $p_n(z)$ be the monic polynomials associated with the measure μ given by (II.2) and (II.5). Then

$$\lim_{n\to\infty} \frac{p_{md^n}(z)}{B(z)^{md^n}} = 1$$

uniformly on compact subsets of $\hat{C} \setminus I$ where $\hat{C}$ is the extended complex plane.

<u>Proof</u>. It follows from the relation between $B(z)$ and $g(z)$, and (II.2) that

$$B(z)^d = B(T(z)) .$$

Furthermore for $z \notin D_R$, a disk of radius R, where R is chosen sufficiently large, one has the expansion,

$$B(z) = z + \sum_{i=o}^{\infty} c_i z^{-i}$$

Consequently, for $z \notin D_R$ one finds

$$\frac{p_m(z) - B(z)^m}{B(z)^m} = \frac{1}{z} \frac{\sum_{i=o}^{\infty} d_i z^{-i}}{1+ \sum_{i=o}^{\infty} c_i z^{-i}} . \tag{IV.2}$$

Let S be a compact set in $\hat{C} \setminus I$. Since T is expanding on $\hat{C} \setminus I$ there exists an N such that for $n > N$, $T^{on}(z) \notin D_R$ for any $z \in S$. Choosing n sufficiently large, then replacing z by $T^{on}(z)$ in (IV.2), then letting $n \longrightarrow$ yields

$$\lim_{n\to\infty} \left| \frac{p_{md^n}(z)}{B(z)^{md^n}} - 1 \right| = \lim_{n\to\infty} \left| \frac{p_m(T^{on}(z) - B(T^{on}(z))^m}{B(T^{on}(z))} \right| = 0$$

which gives the lemma.

Theorem 8 and lemma 2 gives

Theorem 9. [BGM] If the coefficients in (II.10) are in the class L and p_n are the monic orthogonal polynomials associated with a measure μ given by (II.2) and (II.5), then

$$\lim_{n\to\infty} \left| \frac{p_{md^n+s}(z)}{B(z)^{md^n}} - p_s(z) \right| = 0$$

uniformly on compact subsets of $C \setminus I$.

In order to discuss what happens on the spectrum of the operator J we recast inclusion in the set L as, $J \in L$ if and only if $\forall \phi \in l_2$,

$$\lim_{n\to\infty} \|(JS^{md^n} - S^{md^n} J)\phi\|_{l_2} = 0 \qquad \forall m \geq 0 \tag{IV.3}$$

Here $S: l_2 \longrightarrow l_2$ is given by $Se_m = e_{m+1}$, $m \geq 0$ where $\{e_i\}$ is the natural basis.

An easy approximation argument gives

Lemma 3. If $J \in L$, then for every $f \in C(\sigma(J))$, $\forall m \geq 0$ and $\forall \phi \in l_2$,

$$\lim_{n\to\infty} \|(f(J)S^{md^n} - S^{md^n} f(J))\phi\|_{l_2} = 0$$

This in turn leads to,

Theorem 10. Let $\{\hat{p}_n\}$ be the orthonormal polynomials associated with a measure μ given by (II.2) and (II.5). If the coefficients in (II.10) are in the class L, then

$$\lim_{n\to\infty} \|\hat{p}_{md^n} \hat{p}_s - \hat{p}_{md^n+s}\|_{L^2} = 0, \quad m \geq 0, \ s \geq 0, \tag{IV.4}$$

$$\lim_{m\to\infty} \int_J f\, \hat{p}_{md^n+s}\, \hat{p}_{md^n+l}\, du = \int_J f\, \hat{p}_s \hat{p}_l\, du \tag{IV.5}$$

$\forall f \in L^\infty(\mu)$, $l, s = 0,1,2,\ldots$, $m \geq 0$ and

$$\lim_{n\to\infty} \int_J f\, \hat{p}_{md^n+s}\, \hat{p}_{md^n}\, du = \int_J f\, \hat{p}_s\, du \tag{IV.6}$$

$\forall f \in L^2(\mu)$, $s \geq 0$ and $m \geq 0$

Proof. (IV.4) follows by setting $\phi = e_o$, $f = \hat{p}_s$ and using the isometry between l_2 and $L^2(\mu)$. To show (IV.5) we note that the triangle inequality and Schwartz's inequality imply that

$$\left| \int_J f(\hat{p}_{md^n+1} \hat{p}_{md^n+s} - \hat{p}^2_{md^n} \hat{p}_i \hat{p}_s) du \right|$$

$$\leq ||f||_{L^\infty} \{ ||\hat{p}_{md^n+1} - \hat{p}_{md^n} \hat{p}_1||_{L^2}$$

$$+ ||\hat{p}_{md^n+s} - \hat{p}_{md^n} \hat{p}_s||_{L^2} \} .$$

(IV.5) now follows from the above inequality, (IV.4), and the fact that

$$\lim_{n\to\infty} \int_J f \hat{p}_{md^n} \hat{p}_1 \hat{p}_s \, du = \int_J f \hat{p}_1 \hat{p}_s \, du$$

by the mixing property of μ . Since $||\hat{p}_{md^n}||_{L^\infty} = ||\hat{p}_m||_{L^\infty}$ (IV.5) can be pushed all the way to $f \in L^2(\mu)$ if we take $l = 0$ and this gives (IV.6).

Unfortunately theorem 8 does not imply that

$$\lim_{n\to\infty} \frac{1}{n} \ln|p_n(z)| = g(z) \qquad , z \notin J \tag{IV.7}$$

However it is possible to show that if $W(z) \neq 0$, $z \in J$, then the growth rate of the monic's on J is sufficiently slow so that $\lim_{n\to\infty} \frac{1}{n} \ln||p||_{L^\infty} = 1$ which is also the log of the capacity of J .
Consequently using a theorem of Walsh [W. p. 163] it can be shown [BGM] that if $W(z) \neq 0$, $z \in J$, then (IV.7) holds.

REFERENCES

[AvS] J. Avron and B. Simon , "Singular continuous spectrum for a class of almost periodic Jacobi matrices", Bull. A.M.S. (1982), 6, 81.

[BDEG] M.F. Barnsley, S.G. Demko, J.H. Elton, and J.S. Geronimo, "Invariant measures for Markov processes arising from iterated function systems with place-dependent probabilities".

Submitted to Annals d'Institute Henri Poincare.

[BGH1] M.F. Barnsley, J.S. Geronimo, and A.N. Harrington, "On the invariant sets of a family of quadratic maps", Comm Math. Phys. (1983), 88, 479.

[BGH2] M.F. Barnsley, J.S. Geronimo, and A.N. Harrington, "Almost periodic Jacobi matrices associated with Julia sets", Comm. Math. Phys. (1985), 99, 303.

[BGH3] M.F. Barnsley, J.S. Geronimo, and A.N. Harrington, "Geometrical and electrical properties of some Julia sets", in Classical and Quantum models and Arithmetic Problems, ed. D. Chudnovsky and G. Chudnovsky. Lecture Notes in Pure and Applied Math., Vol. 92, Decker.

[BGHD] M.F. Barnsley, J.S. Geronimo, A.N. Harrington, and L. Drager, "Approximation theory on a snowflake".

[B] J. Bellisard, "Stability and instability in Quantum mechanics", Trend in the Eighties (Ph. Blanchard, ed.) Singapore, 1985.

[BBM] J. Bellisard, D. Bessis, and P. Moussa, "Chaotic States of almost periodic Schrodinger operators", Phys. Rev. Lett. (1982), 49, 701.

[BM] D. Bessis and P. Moussa, "Orthogonality properties of iterated polynomial mappings", Commun. Math. Phys. (1983), 88, 503.

[BGM] D. Bessis, J.S. Geronimo, and P. Moussa, "Function weighted measures and orthogonal polynomials on Julia sets", Accepted Const. Approx.

[Bi] P. Billingsley, Ergodic Theory and Information, Wiley, New York, 1965.

[Br] H. Brolin, "Invariant sets under iteration of rational functions", Arkiv für Matematik (1965), 6, 103.

[BDI] B. Derrida, L. DeSize, and C. Itzykson, "Fractal Structures of zeros in hierarchical lattices", J. Stat. Phys. (1983), 3, 559.

[DH] A. Douady and J. Hubbard. Lecture Notes École Normale Sup.

[Fa] P. Fatou, "Sur les équations fonctionnelles", Bull. Soc. Math. France , (1919), 47, 161, (1920), 48, 33.

[GV] J.S. Geronimo and W. Van Assche, "Orthogonal polynomials on several intervals via a polynomial mapping". Submitted Trans. A.M.S.

[J] G. Julia, "Mémoire sur l'itération des fonctions rationelles", J. de Math. Pures et Appliquées (1918) Ser. 8.1. 47.

[M] P. Moussa, "Itérations des polynômes et proprietés d'orthogonalité"Ann. Inst. H. Poincaré Phys. Theor. (1986), 44, 315.

[PK] T.S. Pitcher and J.R. Kinney, "Some connections between ergodic theory and the iteration of polynomials", Ark. für Math. (1969), 8, 25.

[S] M.H. Stone, Linear Transformations in Hilbert Space, Amer. Math. Soc. Colloq. pub. 15 New York 1928.

[TK] L.A. Turkevitch and R.A. Klemm, "Ginzberg-Landau theory of the upper critical field in filamentary superconductors", Phys. Rev. B. (1979), 19, 2520.

[W] J.L. Walsh, Interpolation and Approximation by Rational Functions in the Complex Plane, Amer. Math. Soc. Vol. IX, New York, 1935.

Group theoretic interpretations of Askey's scheme of hypergeometric orthogonal polynomials

Tom H. Koornwinder*

Centre for Math. and Computer Science, Amsterdam and Math. Institute, University of Leiden

1. Introduction

The concept of *classical orthogonal polynomials* changed quite drastically during the last decade (cf. Andrews and Askey [1]). The present consensus about what should be included in this class can be neatly summarized in the *Askey scheme of hypergeometric orthogonal polynomials*, cf. Askey and Wilson [4, Appendix], Labelle [30] and §5, Table 3 in the present paper. A remarkable aspect of this scheme are the arrows indicating limit transitions between the various families of orthogonal polynomials, starting at the top with the Wilson and Racah polynomials and ending at the bottom with the Hermite polynomials.

Almost all members of the Askey scheme have group theoreretic interpretations. The present paper gives a partial survey of these connections with group theory, but, at the same time, it tries to illustrate a philosophy and a research program. The philosophy consists of the following:

1. Almost any identity for special functions in, say, the Bateman project [14], [16] has a deeper meaning in (i) the harmonic analysis of orthogonal systems of special functions and (ii) some group theoretic context.

2. The Askey scheme is a subgraph of a much bigger scheme of families of unitary integral transforms with hypergeometric kernels. From this point of view the selection criterium that these kernels should be of polynomial nature is quite arbitrary.

3. The integral transforms occurring in the (extended) Askey scheme can be grouped together in triples (in many different ways) such that the three transforms in a triple, when performed successively, yield identity. The limit transitions in the scheme often extend to limit transitions for these triples.

The research program naturally follows from this philosophy. Its main task is to make the three above items concrete and to fill them in as completely as possible. Partly this can be done by a search through the literature, partly new research is needed. However, there are also related problems of more fundamental nature, for instance the following one:

The spectral theory of second order difference operators on $\mathbf{Z}_+$ and of second order differential operators on an interval (possibly with singularities at the end points) is quite well understood nowadays. The first theory is closely related to the general theory of orthogonal polynomials and the second theory to the integral transform theory with respect to the eigenfunctions of a second order differential operator, cf. for instance Trimèche [35]. Classical orthogonal polynomials $p_n(x)$ are not only eigenfunctions of a difference operator in n but also of a second order diferential or difference operator in x. "Classical" eigenfunctions $\phi_\lambda(t)$ of a second order differential operator in t are also eigenfunctions of a second order differential or difference operator in λ. What would be helpful, but is still missing, is a general spectral theory of second order difference operators, not just on $\mathbf{Z}_+$, but also on $\mathbf{C}$, with the difference operator being analytic, possibly with singularities. Recent work by Ruijsenaars [32], related to completely integrable systems in relativistic quantum mechanics, also shows the need for such a theory.

The guiding formula for this paper is an identity yielding Wilson polynomials as Jacobi function transforms of Jacobi polynomials, cf. Koornwinder [27, §9], [29] and formula (4.20) in the present paper. We gradually approach this identity in sections 2, 3 and 4, starting with formulas which contain a differential operator $p_k(i\,d/dt)$, where p_k is an orthogonal polynomial (based on joint work with E. Badertscher [6]). At the end of §4 we have already seen some illustration of the

* Mail address: CWI, P.O.Box 4079, 1009 AB Amsterdam, The Netherlands

philosophy developed above and we are now ready to discuss in §5 the Askey scheme and its possible extensions. Until here, not much group theory did enter, except some in §3. In section 6 we discuss some fundamental ways to construct orthogonal systems from group representations. The rest of the paper gives examples of these constructions which yield orthogonal systems of hypergeometric functions fitting into the (extended) Askey scheme: at the end of §6 for $SU(2)$ and the Euclidean motion group $I_0(\mathbb{R}^2)$, in §7 for the discrete series representations of $SL(2,\mathbb{R})$ (following Basu and Wolf [9]), in §8 the Hahn and Racah polynomials and, finally, in §9 the Wilson polynomials, where the above-mentioned identity finds its proper setting.

The group theoretic interpretations discussed in this paper only use $SU(2)$, $SL(2,\mathbb{R})$, $I_0(\mathbb{R}^2)$, the rotation groups and the generalized Lorentz groups. By lack of space and time many other groups relevant for the Askey scheme are not discussed here, for instance the Heisenberg group and the symmetric group. We also avoided discussion of q-Wilson polynomials, mainly because known group theoretic interpretations for orthogonal polynomials of q-hypergeometric type are restricted to Chevalley groups and homogeneous trees.

The analysis in this paper remains at the formal level. The convergence of series and integrals, the precise sense in which one family of the Askey scheme approximates another family and the interpretation of inner products of distributions are not considerd here. However, the author is preparing a paper, where these questions will be dealt with.

2. Continuous Hahn polynomials: Playing around with some remarkable formulas

Consider *Bessel functions*

$$J_k(t) := \frac{(\frac{1}{2}t)^k}{k!}\ {}_0F_1(k+1\,;\,-\tfrac{1}{4}t^2) \tag{2.1}$$

of order $k = 0,1,2,\cdots$ with integral representation

$$J_k(t) = \pi^{-1} i^{-k} \int_0^{\pi} e^{it\cos\psi} \cos(k\psi)\, d\psi \tag{2.2}$$

(cf. [14, Vol.2, 7.3(2)]). Substitute the expression for the *Chebyshev polynomial*

$$T_k(\cos\psi) := \cos(k\psi) \tag{2.3}$$

in (2.2) and make the change of variable $x := \cos\psi$ in the integral. Then we obtain

$$J_k(t) = \pi^{-1} i^{-k} \int_{-1}^{1} e^{itx}\, T_k(x)(1-x^2)^{-\frac{1}{2}}\, dx. \tag{2.4}$$

In particular, for $k = 0$:

$$J_0(t) = \pi^{-1} \int_{-1}^{1} e^{itx} (1-x^2)^{-\frac{1}{2}}\, dx. \tag{2.5}$$

Let us now build from the polynomial T_k the differential operator with constant coefficients $T_k(-i\,d\,/\,dt)$. If we let this operate on (2.5) then we obtain

$$T_k(-i\,d\,/\,dt) J_0(t) = \pi^{-1} \int_{-1}^{1} e^{itx}\, T_k(x)(1-x^2)^{-\frac{1}{2}}\, dx.$$

Hence, in view of (2.4), we get the curious formula

$$T_k(-i\,d\,/\,dt) J_0(t) = i^k J_k(t). \tag{2.6}$$

Observe that, by (2.5), J_0 is the Fourier transform of the orthogonality measure for which the polynomials T_k are orthogonal.

We try to play the same game with the *associated Legendre functions*

$$P_{i\lambda-\frac12}^k(\mathrm{ch}\, t) := \frac{2^{-k}\,\Gamma(i\lambda+k+\frac12)}{k!\,\Gamma(i\lambda-k+\frac12)}\,(\mathrm{sh}\, t)^k$$
$$\times\, {}_2F_1(i\lambda+k+\tfrac12,\, -i\lambda+k+\tfrac12\,;\, k+1\,;\, -(\mathrm{sh}(\tfrac12 t))^2), \tag{2.7}$$

where $t\in\mathbb{R}$, $\lambda\in\mathbb{C}$, $k\in\mathbb{Z}_+$, with integral representation

$$P_{i\lambda-\frac12}^k(\mathrm{ch}\, t) = \frac{(i\lambda+\frac12)_k}{\pi k!}\int_0^\pi (\mathrm{ch}\, t + \mathrm{sh}\, t\cos\psi)^{i\lambda-\frac12}\cos(k\psi)\,d\psi \tag{2.8}$$

(cf. [14, 3.6(1), 3.7(14)]). Make again the substitution (2.3) in (2.8) and make the change of variable $\mathrm{th}\, y := -\cos\psi$ in the integral. Then (2.8) becomes

$$P_{i\lambda-\frac12}^k(\mathrm{ch}\, t) = \frac{(-1)^k\,(i\lambda+\frac12)_k}{\pi k!}\int_{-\infty}^{\infty} (\mathrm{ch}(t-y))^{i\lambda-\frac12}\, T_k(\mathrm{th}\, y)(\mathrm{ch}\, y)^{-i\lambda-\frac12}\,dy, \tag{2.9}$$

which we recognize as a convolution product on $\mathbb{R}$. In particular, for the *Legendre function* $P_{i\lambda-\frac12} := P_{i\lambda-\frac12}^0$ we obtain

$$P_{i\lambda-\frac12}(\mathrm{ch}\, t) = \pi^{-1}\int_{-\infty}^{\infty} (\mathrm{ch}(t-y))^{i\lambda-\frac12}\,(\mathrm{ch}\, y)^{-i\lambda-\frac12}\,dy. \tag{2.10}$$

Hence

$$(i\,d/dt)^k\,P_{i\lambda-\frac12}(\mathrm{ch}\, t) = \pi^{-1}\int_{-\infty}^{\infty} (\mathrm{ch}(t-y))^{i\lambda-\frac12}\,(i\,d/dy)^k\,(\mathrm{ch}\, y)^{-i\lambda-\frac12}\,dy.$$

One easily proves by induction with respect to k that

$$(\mathrm{ch}\, y)^{i\lambda+\frac12}\,(i\,d/dy)^k\,(\mathrm{ch} y)^{-i\lambda-\frac12} = (-i)^k\,(i\lambda+\tfrac12)_k\,(\mathrm{th}\, y)^k + \text{lower degree in th } y.$$

It follows that, for each $\lambda\in\mathbb{C}$ with $-i\lambda-\frac12\notin\mathbb{Z}_+$ there are unique polynomials p_k of degree k such that

$$p_k(i\,d/dy)(\mathrm{ch}\, y)^{-i\lambda-\frac12} = \text{const. } T_k(\mathrm{th}\, y)(\mathrm{ch}\, y)^{-i\lambda-\frac12}. \tag{2.11}$$

For these polynomials we conclude:

$$p_k(i\,d/dt)\,P_{i\lambda-\frac12}(\mathrm{ch}\, t) = \text{const. } P_{i\lambda-\frac12}^k(\mathrm{ch}\, t). \tag{2.12}$$

We will show that the polynomials p_k are orthogonal on $\mathbb{R}$ with respect to a weight function w which can be explicitly determined. Observe from (2.7) that the function $t\mapsto P_{i\lambda-\frac12}^k(\mathrm{ch} t)$ has a zero of order k at 0. Let $l<k$. Then

$$\begin{aligned} 0 &= \text{const. } (i\,d/dt)^l\,P_{i\lambda-\frac12}^k(\mathrm{ch}\, t)\Big|_{t=0} \\ &= (i\,d/dt)^l\,p_k(i\,d/dt)\,P_{i\lambda-\frac12}(\mathrm{ch}\, t)\Big|_{t=0} \\ &= (2\pi)^{-1}\int_{-\infty}^{\infty}\mu^l\,p_k(\mu)\,w(\mu)\,d\mu, \end{aligned}$$

where

$$\begin{aligned} w(\mu) &= \int_{-\infty}^{\infty} P_{i\lambda-\frac12}(\mathrm{ch}\, t)\,e^{i\mu t}\,dt \\ &= (\pi)^{-1}\left[\int_{-\infty}^{\infty}(\mathrm{ch}\, t)^{i\lambda-\frac12}\,e^{i\mu t}\,dt\right]\left[\int_{-\infty}^{\infty}(\mathrm{ch}\, t)^{-i\lambda-\frac12}\,e^{i\mu t}\,dt\right] \end{aligned}$$

by (2.10). Hence, by [14, 1.5(26)] we obtain

$$w(\mu) = \text{const. } \Gamma(\tfrac12(i\mu+i\lambda+\tfrac12))\,\Gamma(\tfrac12(-i\mu+i\lambda+\tfrac12))$$

$$\times\,\Gamma(\tfrac12(i\mu-i\lambda+\tfrac12))\,\Gamma(\tfrac12(-i\mu-i\lambda+\tfrac12)). \tag{2.13}$$

Here, for convergence of the integrals, we require that $|\mathrm{Im}\lambda|<\frac12$. It follows that the polynomials p_k are orthogonal on $\mathbb{R}$ with respect to the weight function w. In particular, if $\lambda\in\mathbb{R}\cup(-\frac12 i,\frac12 i)$ then this weight function is positive. The polynomials orthogonal with respect to the weight function given by (2.13) are particular cases of the *continuous symmetric Hahn polynomials*

$$p_k(x;a,b) := i^k\,{}_3F_2\left[\begin{matrix}-k,\ k+2a+2b-1,\ a-ix\\ a+b,\ 2a\end{matrix}\,\middle|\,1\right], \tag{2.14}$$

which are orthogonal on $\mathbb{R}$ with respect to the weight function $x\mapsto|\Gamma(a+ix)\Gamma(b+ix)|^2$ $(a,b>0$ or $b=\bar a$, $\mathrm{Re}\,a>0)$, cf. Askey and Wilson [3]. From this reference we find that our polynomials p_k can be written as

$$\begin{aligned} p_k(\mu) &= \text{const.}\ p_k(\tfrac12\mu;\tfrac14+\tfrac12 i\lambda,\tfrac14-\tfrac12 i\lambda)\\ &= \text{const.}\ i^k\,{}_3F_2\left[\begin{matrix}-k,\ k,\ \tfrac14+\tfrac12 i\lambda-\tfrac12 i\mu\\ \tfrac12,\ \tfrac12+i\lambda\end{matrix}\,\middle|\,1\right]. \end{aligned} \tag{2.15}$$

So, similarly as with (2.6), the function $t\mapsto P_{i\lambda-\frac12}(\mathrm{ch}\,t)$ in (2.12) is the Fourier transform of the orthogonality measure for which the polynomials p_k in (2.12) are orthogonal.

We now have three formulas of the same structure:

$$J_k(t) = \text{const.}\ T_k(i\,d/dt)J_0(t), \tag{2.16}$$

$$P^k_{i\lambda-\frac12}(\mathrm{ch}\,t) = \text{const.}\ p_k(\tfrac12 i\,d/dt;\tfrac14+\tfrac12 i\lambda,\tfrac14-\tfrac12 i\lambda)\,P_{i\lambda-\frac12}(\mathrm{ch}\,t), \tag{2.17}$$

$$P^k_n(\cos\theta) = \text{const.}\ Q_k(\tfrac12 n+\tfrac12 i\,d/d\theta;-\tfrac12,-\tfrac12,n)\,P_n(\cos\theta). \tag{2.18}$$

Here (2.17) is obtained from (2.12) and (2.15), while (2.18) is obtained from (2.17) by taking $i\lambda-\frac12$ as an integer $\geqslant k$ and by taking t purely imaginary. Then P_n in (2.18) becomes the *Legendre polynomial* while

$$Q_k(\tfrac12 n+\tfrac12 m;-\tfrac12,-\tfrac12,n) := {}_3F_2\left[\begin{matrix}-k,\ k,\ -\tfrac12 n-\tfrac12 m\\ \tfrac12,\ -n\end{matrix}\,\middle|\,1\right] \tag{2.19}$$

is a special *Hahn polynomial*, orthogonal on $\{-n,-n+2,\ldots,n\}$ with respect to the weights

$$w_m = \frac{(\frac12)_{\frac12 n+\frac12 m}(\frac12)_{\frac12 n-\frac12 m}}{(\frac12 n+\frac12 m)!\,(\frac12 n-\frac12 m)!}, \tag{2.20}$$

cf. Karlin and McGregor [22] except for the standardized notation, which is given in Askey and Wilson [4]. Note that the weights w_m are the Fourier coefficients of $P_n(\cos\theta)$:

$$P_n(\cos\theta) = \sum_{m=-n,-n+2,\ldots,n} w_m\,e^{im\theta}, \tag{2.21}$$

cf. Szegö [34, (4.9.19)].

Formula (2.16) is a limit case of both (2.17) and (2.18): Replace t (or θ) by $\lambda^{-1}t$ (or $n^{-1}\theta$) and let $\lambda\to\infty$ (or $n\to\infty$).

3. Orthogonal polynomials of argument *id/dt*: A group theoretic interpretation

Corresponding to the formulas (2.16), (2.17), (2.18), respectively, we consider the Lie groups $G=I_0(\mathbb{R}^2)$, $SO_0(1,2)$ and $SO(3)$. Here $I_0(\mathbb{R}^2)$ is the group of orientation preserving Euclidean motions of $\mathbb{R}^2$, $SO_0(1,2)$ is the connected component of the group of linear transformations of $\mathbb{R}^3$ which leave invariant the quadratic form $-x^2+y^2+z^2$ and $SO(3)$ the group of rotations around the origin in $\mathbb{R}^3$. These groups G have natural transtive actions on certain spaces Ω, i.e., respectively, on $\mathbb{R}^2$, on a sheet of the hyperboloid $-x^2+y^2+z^2=1$ and on the unit sphere S^2 in $\mathbb{R}^3$. In all cases the subgroup K of G leaving some point of Ω fixed is isomorphic to $SO(2)$, the group of rotations of the circle. So Ω can be identified with the homogeneous space G/K.

By a *unitary representation* π of G on a Hilbert space $\mathcal{H}(=\mathcal{H}(\pi))$ we mean a mapping π of G into the group of unitary transformations of $\mathcal{H}$ which is a *group homomorphism* (i.e. $\pi(g_1g_2)=\pi(g_1)\pi(g_2)$ for all $g_1,g_2\in G$) and which is *strongly continuous* (i.e. for all $v\in\mathcal{H}$ the mapping $g \mapsto \pi(g)v$: $G \to \mathcal{H}$ is continuous).

The representation π of G is called *irreducible* if the only closed G-invariant subspaces of $\mathcal{H}(\pi)$ are $\mathcal{H}(\pi)$ and $\{0\}$. The set of all irreducible unitary representations of G (more precisely, of all equivalence classes of irreducible unitary representations of G) is denoted by $\hat{G}$.

We can identify the group K with the group $SO(2)$ of matrices

$$u_\theta := \begin{pmatrix} \cos\theta & \sin\theta \\ -\sin\theta & \cos\theta \end{pmatrix}, \quad \theta \in \mathbf{R}.$$

It can be shown that the three pairs (G,K) as considered above share the following property: For each $\pi \in \hat{G}$ there is a subset I of $\mathbf{Z}$ and an orthonormal basis $\{e_k\}_{k\in I}$ of $\mathcal{H}(\pi)$ such that

$$\pi(u_\theta)e_k = e^{ik\theta}e_k, \quad k \in I,\ \theta \in \mathbf{R}.$$

Otherwise stated: For each $\pi \in \hat{G}$ the restriction of π to K is multiplicity free.

Definition. Let $\pi \in \hat{G}$ and denote the inner product on $\mathcal{H}(\pi)$ by $(.,.)$. Suppose that e_0 occurs in $\mathcal{H}$. Then

$$\phi(g) := (\pi(g)e_0, e_0), \quad g \in G, \tag{3.1}$$

denotes the *spherical function* ϕ for the pair (G,K) and the representation π. If, for some $k \in \mathbf{Z}$, e_k also occurs in $\mathcal{H}$ then

$$\phi^k(g) := (\pi(g)e_0, e_k), \quad g \in G, \tag{3.2}$$

denotes an *associated spherical function* for (G,K) and π.

Note that ϕ and ϕ^k are right invariant with respect to K, so they can be considered as functions on G/K. Moreover, the spherical function ϕ is left invariant with respect to K.

For each group G as considered above there is a one-parameter subgroup $A = \{a_t\}_{t\in\mathbf{R}}$ of G such that we have the *Cartan decomposition* $G = KAK$, i.e., each $g \in G$ can be written as $g = k_1 a_t k_2$ for some $k_1, k_2 \in K$ and $t \in \mathbf{R}$. Now it can be shown that, after possibly rescaling of $\{a_t\}$, the associated spherical functions ϕ^k, when restricted to A, only change by a constant factor if k is replaced by $-k$ and that, up to a constant factor, $\phi^k(a_t)$ with $k \geqslant 0$ equals

$J_k(\lambda t)$ for some $\lambda \in \mathbf{R}$ or

$P^k_{i\lambda-\frac{1}{2}}(\mathrm{ch}\, t)$ for some $\lambda \in \mathbf{R} \cup [-\tfrac{1}{2}i, \tfrac{1}{2}i]$ or

$P^k_n(\cos t)$ for some $n \in \mathbf{Z}_+$,

according to whether $G = I_0(\mathbf{R}^2)$, $SO_0(1,2)$ or $SO(3)$. These are essentially the left hand sides of (2.16), (2.17) and (2.18). So, roughly stated, we have

$$\phi^k(a_t) = \text{const.}\ p_k(i\,d/dt)\phi(a_t), \tag{3.3}$$

where p_k is an orthogonal polynomial with respect to a measure which is the Fourier transform of $t \mapsto \phi(a_t)$.

In Badertscher and Koornwinder [7] it is shown that, with e_0, e_k and p_k as above, we have

$$e_k = \text{const.}\ p_k(i\,d/dt)\pi(a_t)e_0\Big|_{t=0} \tag{3.4}$$

(the function $t \mapsto \pi(a_t)e_0$: $\mathbf{R} \to \mathcal{H}$ is C^∞). Hence, in view of (3.1) and (3.2), we get

$$\phi^k(g) = \text{const.}\ p_k(i\,d/dt)\phi(a_{-t}g)\Big|_{t=0}, \quad g \in G,$$

which implies (3.3). Formula (3.50) provides us with an explicit right-invariant differential operator on G which sends the spherical function ϕ to the associated spherical function ϕ^k.

In the case that G is an arbitrary connected semisimple Lie group with finite center and K a maximal compact subgroup, Helgason [20, section 4] has a definition of associated spherical function (there called generalized spherical function) which generalizes our definition (3.2) and it is proved in [20, Theorem 4.1] that, under certain conditions for the corresponding representation, there exists a right-invariant differential operator on G which sends the spherical function to the associated spherical function. However, Helgason does not specify this differential operator in any sense.

4. Integral transforms mapping systems of orthogonal polynomials onto each other

It follows from (2.11) and (2.15) that

$$T_k(\operatorname{th} y)(\operatorname{ch} y)^{-i\lambda-\frac12} = \text{const.}\, p_k(\tfrac12 i\, d/dy\,; \tfrac14+\tfrac12 i\lambda, \tfrac14-\tfrac12 i\lambda)\, (\operatorname{ch} y)^{-i\lambda-\frac12}. \tag{4.1}$$

Take Fourier transforms on both sides and use [14, 1.5(26)]. It follows that

$$\int_{-\infty}^{\infty} T_k(\operatorname{th} y)(\operatorname{ch} y)^{-i\lambda-\frac12} e^{i\mu y}\, dy = \text{const.}\, \Gamma(\tfrac12(i\mu+i\lambda+\tfrac12))$$
$$\times\, \Gamma(\tfrac12(-i\mu+i\lambda+\tfrac12))\, p_k(\tfrac12\mu; \tfrac14+\tfrac12 i\lambda, \tfrac14-\tfrac12 i\lambda). \tag{4.2}$$

It is clear from the orthogonality relations for the Chebyshev and continuous symmetric Hahn polynomials that the functions

$$y \mapsto T_k(\operatorname{th} y)(\operatorname{ch} y)^{-i\lambda-\frac12}, \quad k=0,1,2,\cdots$$

form an orthogonal basis of $L^2(\mathbb{R},dy)$ and the functions

$$\mu \mapsto \Gamma(\tfrac12(i\mu+i\lambda+\tfrac12))\,\Gamma(\tfrac12(-i\mu+i\lambda+\tfrac12))\, p_k(\tfrac12\mu; \tfrac14+\tfrac12 i\lambda, \tfrac14-\tfrac12 i\lambda), \quad k=0,1,2,\cdots$$

an orthogonal basis of $L^2(\mathbb{R},d\mu)$ and that the Fourier transform maps the first basis onto the second basis.

More generally there is an identity of type (4.2) for *Gegenbauer polynomials*

$$C_k^\alpha(x) := \frac{(2\alpha)_k}{k!}\, {}_2F_1(-k, k+2\alpha\,; \alpha+\tfrac12\,; \tfrac12-\tfrac12 x), \tag{4.3}$$

which are orthogonal on $(-1,1)$ with respect to the weight function $x \mapsto (1-x^2)^{\alpha-\frac12}$. We have (cf. [8])

$$\int_{-\infty}^{\infty} [C_k^\alpha(\operatorname{th} y)(\operatorname{ch} y)^{-i\lambda-\alpha-\frac12}]\, e^{i\mu y}\, dy = \text{const.}\ \Gamma(\tfrac12(i\mu+i\lambda+\alpha+\tfrac12))$$
$$\times\, \Gamma(\tfrac12(-i\mu+i\lambda+\alpha+\tfrac12))\; p_k(\tfrac12\mu; \tfrac14+\tfrac12\alpha+\tfrac12 i\lambda, \tfrac14+\tfrac12\alpha-\tfrac12 i\lambda). \tag{4.4}$$

Here the functions of y occurring in square brackets at the left hand side form an orthogonal basis of $L^2(\mathbb{R},dy)$, while the functions of μ at the right hand side form an orthogonal basis of $L^2(\mathbb{R},d\mu)$, and the two bases are mapped onto each other by the Fourier transform.

A much more elementary result of this type is the well-known formula

$$(2\pi)^{-\frac12} \int_{-\infty}^{\infty} H_k(y)\, e^{-\frac12 y^2}\, e^{i\mu y}\, dy = i^{-k}\, H_k(\mu)\, e^{-\frac12\mu^2}, \tag{4.5}$$

where H_k is the *Hermite polynomial*, orthogonal on $\mathbb{R}$ with respect to the weight function $y \mapsto e^{-y^2}$. Actually, (4.5) can be obtained as a limit case of (4.4). Just replace y by $\alpha^{-\frac12}y$ and μ by $\alpha^{\frac12}\mu$ in (4.4) and let $\alpha\to\infty$. Schematically we display the result in Table 1 below.

There is a further generalization of (4.4) which takes the Jacobi function transform of Jacobi polynomials rather than the Fourier transform of Gegenbauer polynomials. *Jacobi polynomials* are given by

$$P_n^{(\alpha,\beta)}(x) := \frac{(\alpha+1)_n}{n!}\, {}_2F_1(-n, n+\alpha+\beta+1\,; \alpha+1\,; \tfrac12(1-x)), \quad n=0,1,2,\cdots. \tag{4.6}$$

For $\alpha,\beta>-1$ these are orthogonal polynomials on $(-1,1)$ with respect to the weight function

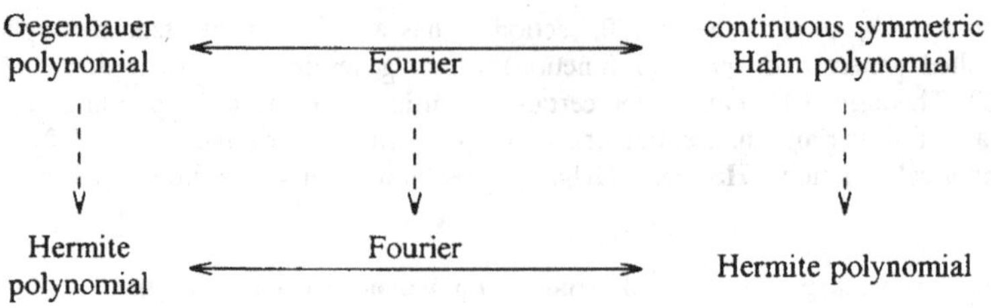

Table 1

$x \mapsto (1-x)^\alpha (1+x)^\beta$.

Jacobi functions are defined by

$$\phi_\lambda^{(\alpha,\beta)}(t) := {}_2F_1(\tfrac{1}{2}(\alpha+\beta+1+i\lambda), \tfrac{1}{2}(\alpha+\beta+1-i\lambda); \alpha+1; -\mathrm{sh}^2 t). \tag{4.7}$$

We will always assume that $\alpha > -1$ and $\beta \in \mathbb{R} \cup i\mathbb{R}$. Let

$$\Delta_{\alpha,\beta}(t) := (2\,\mathrm{sh}\,t)^{2\alpha+1}(2\,\mathrm{ch}\,t)^{2\beta+1}. \tag{4.8}$$

Then $\phi_\lambda^{(\alpha,\beta)}$ is an even C^∞-function on $\mathbb{R}$ which satisfies

$$\phi_\lambda^{(\alpha,\beta)}(0) = 1 \tag{4.9}$$

and

$$\left[\frac{d^2}{dt^2} + \frac{\Delta'_{\alpha,\beta}(t)}{\Delta_{\alpha,\beta}(t)}\frac{d}{dt} + \lambda^2 + (\alpha+\beta+1)^2\right]\phi_\lambda^{(\alpha,\beta)}(t) = 0, \tag{4.10}$$

and $\phi_\lambda^{(\alpha,\beta)}$ is uniquely defined by these properties. Let

$$c_{\alpha,\beta}(\lambda) := \frac{2^{\alpha+\beta+1-i\lambda}\,\Gamma(\alpha+1)\,\Gamma(i\lambda)}{\Gamma(\tfrac{1}{2}(i\lambda+\alpha+\beta+1))\,\Gamma(\tfrac{1}{2}(i\lambda+\alpha-\beta+1))}. \tag{4.11}$$

Then, for $\alpha > -1$, $\beta \in [-\alpha-1, \alpha+1] \cup i\mathbb{R}$ we have the integral transform pair

$$\left.\begin{aligned} g(\lambda) &= \int_0^\infty f(t)\,\phi_\lambda^{(\alpha,\beta)}(t)\,\Delta_{\alpha,\beta}(t)\,dt \\ f(t) &= (2\pi)^{-1}\int_0^\infty g(\lambda)\,\phi_\lambda^{(\alpha,\beta)}(t)\,|c_{\alpha,\beta}(\lambda)|^{-2}\,d\lambda \end{aligned}\right\} \tag{4.12}$$

This establishes a 1-1 correspondence between the space of even C^∞-functions on $\mathbb{R}$ with compact support and its image under the Fourier-cosine transform, as characterized by the classical Paley-Wiener theorem. The mapping $f \mapsto g$ (which we call the *Jacobi function transform*) extends to an isometry of Hilbert spaces between $L^2(\mathbb{R}_+, \Delta_{\alpha,\beta}(t)\,dt)$ and $L^2(\mathbb{R}_+, (2\pi)^{-1}|c_{\alpha,\beta}(\lambda)|^{-2}\,d\lambda)$. In the case that $\alpha > -1$ and $|\beta| > \alpha+1$, (4.12) remains valid except that we have to add to the right hand side of the second formula a term

$$\textstyle\sum_{\lambda \in D_{\alpha,\beta}} g(\lambda)\,d_{\alpha,\beta}(\lambda),$$

where the finite subset $D_{\alpha,\beta}$ of $i\mathbb{R}_+$ and the weights $d_{\alpha,\beta}(\lambda)$ can be explicitly given. See [24], [27] and [19, Appendix 1] for further details about Jacobi functions.

Noteworthy special cases of Jacobi functions are

$$\phi_\lambda^{(-\frac{1}{2},-\frac{1}{2})}(t) = \cos(\lambda t), \tag{4.13}$$

$$\phi_\lambda^{(\frac{1}{2},-\frac{1}{2})}(t) = \frac{\sin(\lambda t)}{\lambda\,\mathrm{sh}\,t}, \tag{4.14}$$

$$\phi_\lambda^{(0,-\frac{1}{2})}(t) = P_{i\lambda-\frac{1}{2}}(\mathrm{ch}\,t). \tag{4.15}$$

Thus, for $\beta=-\frac{1}{2}$, (4.12) reduces to the Fourier-cosine transform pair if $\alpha=-\frac{1}{2}$, to the Fourier-sine transform pair if $\alpha=\frac{1}{2}$ and to the Mehler-Fock transform pair if $\alpha=0$.

As a generalization of (4.4) we now want to evaluate

$$\int_0^\infty [(\operatorname{th} t)^{\alpha+\frac{1}{2}} (\operatorname{ch} t)^{-\delta-i\mu-1} P_n^{(\alpha,\delta)}(1-2\operatorname{th}^2 t)]$$

$$\times \left[\frac{\Gamma(\frac{1}{2}(\alpha+\beta+1+i\lambda))\,\Gamma(\frac{1}{2}(\alpha-\beta+1+i\lambda))}{\Gamma(i\lambda)} (\operatorname{sh} t)^{\alpha+\frac{1}{2}} (\operatorname{ch} t)^{\beta+\frac{1}{2}} \phi_\lambda^{(\alpha,\beta)}(t) \right] dt, \tag{4.16}$$

where $\beta,\lambda,\mu\in\mathbb{R}$ and $\alpha,\delta>-1$. Indeed, (4.16) reduces to the left hand side of (4.4), up to a constant factor, if $\beta=-\frac{1}{2}$, $\alpha=\pm\frac{1}{2}$ and δ is replaced by $\alpha-\frac{1}{2}$. (Use (4.13),(4.14) and the quadratic transformations for Jacobi polynomials.)

Let moreover $|\beta|>\alpha+1$ in (4.16). In view of the orthogonality relations for Jacobi polynomials and the isometry property of the Jacobi function transform, the functions of λ defined by (4.16) for $n=0,1,2,\cdots$ will form an orthogonal basis of $L^2(\mathbb{R}_+)$. One can evaluate (4.16) (cf. [29]) as

$$\text{const.}\ \frac{\Gamma(\frac{1}{2}(\alpha+\beta+1+i\lambda))\,\Gamma(\frac{1}{2}(\alpha-\beta+1+i\lambda))\,\Gamma(\frac{1}{2}(\delta+i\mu+1+i\lambda))\,\Gamma(\frac{1}{2}(\delta-i\mu+1+i\lambda))}{\Gamma(i\lambda)}$$

$$\times W_n(\tfrac{1}{4}\lambda^2\,;\tfrac{1}{2}(\delta+i\mu+1),\tfrac{1}{2}(\delta-i\mu+1),\tfrac{1}{2}(\alpha+\beta+1),\tfrac{1}{2}(\alpha-\beta+1)), \tag{4.17}$$

where

$$W_n(x^2\,;a,b,c,d) := (a+b)_n\,(a+c)_n\,(a+d)_n$$

$$\times {}_4F_3\left[\begin{matrix} -n\,,n+a+b+c+d-1\,,a+ix\,,a-ix \\ a+b\,,a+c\,,a+d \end{matrix}\,\middle|\,1\right] \tag{4.18}$$

is a *Wilson polynomial*, a polynomial of degree n in x^2 (cf. Wilson [37] except for the standardized notation which is taken from Askey and Wilson [4, Appendix]). If the parameters a,b,c,d all have positive real parts and if they are all real or one or both pairs of them consist of complex conjugates then the functions $x\mapsto W_n(x^2)$ form a complete orthogonal system of functions on $\mathbb{R}_+$ with respect to the weight function

$$x\mapsto \left| \frac{\Gamma(a+ix)\,\Gamma(b+ix)\,\Gamma(c+ix)\,\Gamma(d+ix)}{\Gamma(2ix)} \right|^2. \tag{4.19}$$

So we have the identity

$$(4.16) = (4.17), \tag{4.20}$$

where the constant in (4.17) is independent of λ, but does depend on $n,\alpha,\beta,\delta,\mu$. Observe that the orthogonality of the Jacobi polynomials and the isometry property of the Jacobi function transform imply the orthogonality relations for the Wilson polynomials. On the other hand the isometry property of the Jacobi function transform is implied by the orthogonality properties for the Jacobi and Wilson polynomials.

If $|\beta|>\alpha+1$ then one of the parameters of W_n in (4.17) becomes negative. Then the discrete terms in the inversion formula for the Jacobi function transform will correspond to discrete components of the orthogonality measure for the Wilson polynomials. These discrete mass points do indeed occur if one of the four parameters in (4.18) is negative, cf. Wilson [37, (3.3)].

Just as (4.5) was a limit case of (4.4), we can take limits in the identity (4.20) by replacing t by $\delta^{-\frac{1}{2}}t$ and λ by $\delta^{\frac{1}{2}}\lambda$ and by letting $\delta\to\infty$. Then both the Jacobi polynomials and the Wilson polynomials tend to *Laguerre polynomials*, which are defined by

$$L_n^\alpha(x) := \frac{(\alpha+1)_n}{n!}\,{}_1F_1(-n\,,\alpha+1\,;x) \tag{4.21}$$

and which are orthogonal polynomials on $\mathbb{R}_+$ with respect to the weight function $x\mapsto x^\alpha e^{-x}$. Furthermore, Jacobi functions tend to *Bessel functions*

$$J_\alpha(x) := \frac{(\tfrac12 x)^\alpha}{\Gamma(\alpha+1)}\, {}_0F_1(\alpha+1\,;\,-\tfrac14 x^2) \tag{4.22}$$

and the Jacobi function transform pair to the *Hankel transform pair*

$$\left.\begin{aligned} g(\lambda) &= \int_0^\infty f(t) J_\alpha(\lambda t)(\lambda t)^{\frac12}\, dt \\ f(t) &= \int_0^\infty g(\lambda) J_\alpha(\lambda t)(\lambda t)^{\frac12}\, d\lambda \end{aligned}\right\} \tag{4.23}$$

The limit identity (cf. [29]) is

$$\int_0^\infty t^{\alpha+\frac12} e^{-\frac12 t^2} L_n^\alpha(t^2) J_\alpha(\lambda t)(\lambda t)^{\frac12}\, dt = (-1)^n \lambda^{\alpha+\frac12} e^{-\frac12\lambda^2} L_n^\alpha(\lambda^2). \tag{4.24}$$

This identity is quite classical, cf. [15, Vol.2, 8.9(3)].

We can summarize our results in Table 2 below.

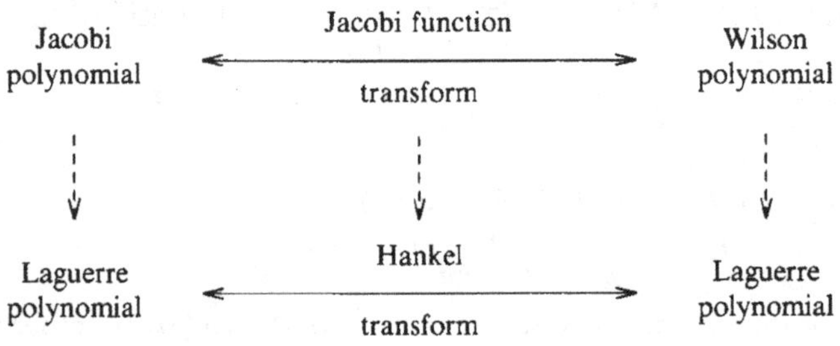

Table 2

Remark. There is another limit case of (4.20), in which Jacobi polynomials, Jacobi functions and Wilson polynomials tend to Laguerre polynomials, Whittaker functions of the second kind and continuous dual Hahn polynomials, respectively (cf. [29, §5]).

5. The Askey scheme of hypergeometric orthogonal polynomials

There are many indications that the Wilson polynomials, introduced in (4.18), are the most general orthogonal polynomials which can be written in terms of a ${}_pF_q$-hypergeometric function. In Table 3 below we reproduce Askey's scheme of hypergeometric orthogonal polynomials, as (almost) given in Askey and Wilson [4, Appendix]. The arrows denote limit transitions. Of course, shortcuts can be made in these limit transitions by at once going down more than one level.

Let us give the explicit expressions and weight functions of the orthogonal polynomials in the Askey scheme.

(a) *Wilson polynomials.* See (4.18), (4.19).

(b) *Racah polynomials.*

$$R_n(\lambda(x);\alpha,\beta,\gamma,\delta) := {}_4F_3\left[\begin{matrix}-n\,,\,n+\alpha+\beta+1\,,\,-x\,,\,x+\gamma+\delta+1\\ \alpha+1\,,\,\beta+\delta+1\,,\,\gamma+1\end{matrix}\,\middle|\,1\right], \tag{5.1}$$

$$\lambda(x) := x(x+\gamma+\delta+1),\ \beta+\delta+1 = -N,\ n = 0,1,\ldots,N.$$

The functions $x \mapsto R_n(\lambda(x))$ are orthogonal on $\{0,1,\ldots,N\}$ with respect to weights which can be explicitly given.

(c) *Continuous dual Hahn polynomials.*

$$S_n(x^2;a,b,c) := (a+b)_n(a+c)_n\, {}_3F_2\left[\begin{matrix}-n\,,\,a+ix\,,\,a-ix\\ a+b\,,\,a+c\end{matrix}\,\middle|\,1\right], \tag{5.2}$$

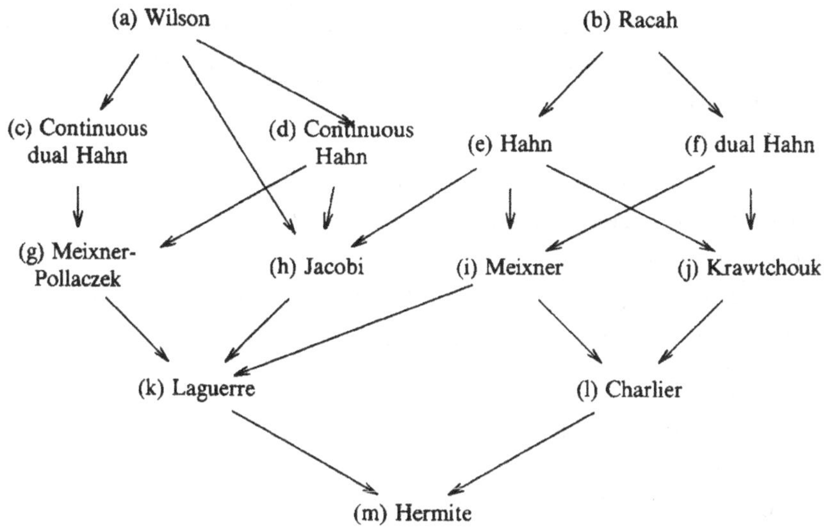

Table 3

where a,b,c have positive real parts; if one of these parameters is not real then one of the other parameters is its complex conjugate. The functions $x \mapsto S_n(x^2)$ are orthogonal on $(0,\infty)$ with respect to the weight function

$$x \mapsto \left| \frac{\Gamma(a+ix)\,\Gamma(b+ix)\,\Gamma(c+ix)}{\Gamma(2ix)} \right|^2. \tag{5.3}$$

(d) *Continuous Hahn polynomials.*

$$p_n(x\,;a,b,\bar{a},\bar{b}) := i^n \frac{(a+\bar{a})_n\,(a+\bar{b})_n}{n!}\; {}_3F_2\left[\begin{matrix}-n\,,\, n+a+\bar{a}+b+\bar{b}-1\,,\, a+ix\\ a+\bar{a}\,,\, a+\bar{b}\end{matrix}\middle|\, 1\right], \tag{5.4}$$

where a,b have positive real part. The polynomials p_n are orthogonal on $\mathbb{R}$ with respect to the weight function $x \mapsto |\Gamma(a+ix)\Gamma(b+ix)|^2$. In Askey and Wilson [4, Appendix] only the symmetric case ($a,b>0$ or $a=b$) of these polynomials occurs. Under these restrictions on a,b, the polynomials (5.4) reduce to (2.14), up to a constant factor. The general continuous Hahn polynomials were discovered by Atakishiyev and Suslov [5]. See Askey [2] for notation, but read $a+ix$ instead of $a-ix$ in his formula (3).

(e) *Hahn polynomials.*

$$Q_n(x\,;\alpha,\beta,N) := {}_3F_2\left[\begin{matrix}-n\,,\, n+\alpha+\beta+1\,,\, -x\\ \alpha+1\,,\, -N\end{matrix}\middle|\, 1\right], \quad n=0,1,\ldots,N, \tag{5.5}$$

orthogonal on $\{0,1,\ldots,N\}$ with respect to the weights

$$x \mapsto \binom{\alpha+x}{x}\binom{N-x+\beta}{N-x}. \tag{5.6}$$

(f) *Dual Hahn polynomials.*

$$R_n(x(x+\alpha+\beta+1);\alpha,\beta,N) := {}_3F_2\left[\begin{matrix}-x\,,\, x+\alpha+\beta+1\,,\, -n\\ \alpha+1\,,\, -N\end{matrix}\middle|\, 1\right], \quad n=0,1,\ldots,N, \tag{5.7}$$

i.e., as (5.5) but with n and x interchanged. The functions $x \mapsto R_n(x(x+\alpha+\beta+1))$ are orthogonal on $\{0,1,\ldots,N\}$ with respect to weights which can be explicitly given.

(g) *Meixner-Pollaczek polynomials.*

$$P_n^{(a)}(x;\phi) := e^{in\phi}\,{}_2F_1(-n,a+ix;2a;1-e^{-2i\phi}),\quad a>0,\ 0<\phi<\pi, \tag{5.8}$$

orthogonal on $\mathbb{R}$ with respect to the weight function $x \mapsto e^{(2\phi-\pi)x}\,|\Gamma(a+ix)|^2$.

(h) *Jacobi polynomials.* See (4.6).

(i) *Meixner polynomials.*

$$M_n(x;\beta,c) := {}_2F_1(-n,-x;\beta;1-c^{-1}),\quad 0<c<1,\ \beta>0, \tag{5.9}$$

orthogonal on $\mathbb{Z}_+$ with respect to the weights $x \mapsto (\beta)_x\, c^x / x!$.

(j) *Krawtchouk polynomials.*

$$K_n(x;p,N) := {}_2F_1(-n,-x;-N;p^{-1}),\quad 0<p<1,\ n=0,1,\ldots,N, \tag{5.10}$$

orthogonal on $\{0,1,\ldots,N\}$ with respect to the weights

$$x \mapsto \begin{bmatrix} N \\ x \end{bmatrix} p^x (1-p)^{N-x}. \tag{5.11}$$

(k) *Laguerre polynomials.* See (4.21).

(l) *Charlier polynomials.*

$$C_n(x;a) := {}_2F_0(-n,-x;-a^{-1}),\quad a>0, \tag{5.12}$$

orthogonal on $\mathbb{Z}_+$ with respect to the weights $x \mapsto a^x / x!$.

(m) *Hermite polynomials.*

$$H_n(x) := (2x)^n\,{}_2F_0(-\tfrac{1}{2}n, \tfrac{1}{2}(1-n); -x^{-2}), \tag{5.13}$$

orthogonal on $\mathbb{R}$ with respect to the weight function $x \mapsto e^{-x^2}$.

Note that the number of parameters on which the families of orthogonal polynomials depend, declines from 4 to 0 as one goes down in Table 3. The families in the left half of the table (including the Hermite polynomials) have continuous weight functions, while the ones in the right half have discrete orthogonality measure. Jacobi, Laguerre and Hermite polynomials are the hard-core classical orthogonal polynomials, but most of the other families also have a long history. Only Wilson, Racah, continuous Hahn and continuous dual Hahn polynomials are recent inventions. See Andrews and Askey [1] for more information about the concept "classical orthogonal polynomial".

The limit behaviour as shown in Table 3 is oversimplified, since, in fact, there are three types of Wilson polynomials $W_n(x^2;a,b,c,d)$ (cf. (4.18)), namely the type (a,b,c,d), where a,b,c,d are real, the type $(a,\bar{a},c,d)$, where c,d are real, and the type $(a,\bar{a},c,\bar{c})$, where we have two pairs of complex conjugates. (Of course there are overlaps.) There are also two types of continuous dual Hahn polynomials $S_n(x^2;a,b,c)$ (cf. (5.2)): the type (a,b,c) with a,b,c real and the type $(a,\bar{a},c)$. With this refinement the left half of Table 3 can be rewritten as Table 4 below.

Each system of orthogonal polynomials $p_n(x)$ in the Askey scheme can be considered as an orthogonal basis $\{p_n\}$ of some Hilbert space. After trivial transformations these Hilbert spaces take the form $L^2(\mathbb{R})$, $L^2(\mathbb{Z}_+)$ or $L^2(\{1,\ldots,N\})$. Expansion of an arbitrary element of the Hilbert space in terms of the basis can be considered as an isometry of $L^2(\mathbb{R})$ or $L^2(\mathbb{Z}_+)$ onto $L^2(\mathbb{Z}_+)$ or of $L^2(\{1,\ldots,N\})$ onto itself. The limit transitions in Tables 3 and 4 can now be considered as taking weak limits of the isometries. Sometimes the two Hilbert spaces do not change in this limit transition, but sometimes they do, cf. in Table 3:

(e)→(i), (f)→(i), (j)→(l) and (e)→(h) (from $L^2(\{1,\ldots,N\})$ to $L^2(\mathbb{Z}_+)$ once or twice);

(e)→(h) (from $L^2(\{1,\ldots,N\})$ to $L^2(\mathbb{R})$);

(l)→(m) and (i)→(k) (from $L^2(\mathbb{Z}_+)$ to $L^2(\mathbb{R})$).

Schematically the possibilities for types of isometries and limit transitions are given in Table 5 below.

This table strongly suggests that isometries of $L^2(\mathbb{R})$ onto itself might also be obtained as limit cases of isometries occurring in the Askey scheme. The following example shows that this is indeed

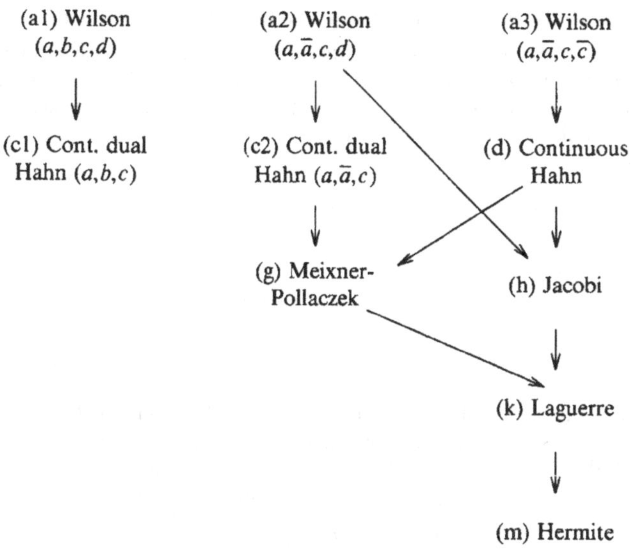

Table 4

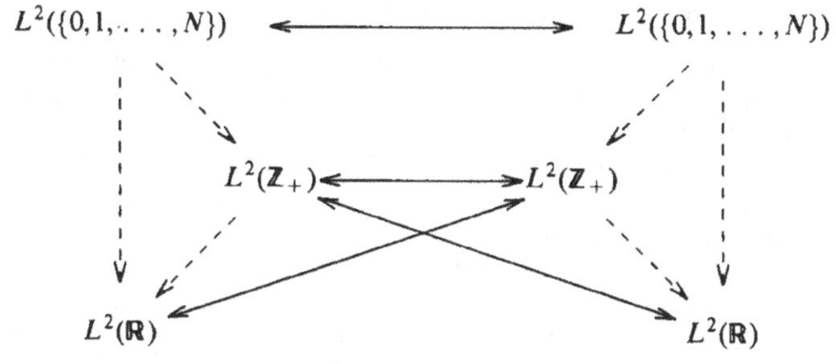

Table 5

possible, at least formally. We have

$$\lim_{n\to\infty} {}_3F_2\left[\begin{matrix} -n\,,\,\tfrac{1}{2}(\alpha+\beta+1)+i\lambda\,,\,\tfrac{1}{2}(\alpha+\beta+1)-i\lambda \\ \tfrac{1}{2}(\alpha+\beta+1)+n\,\mathrm{sh}^{-2}t\,,\,\alpha+1 \end{matrix}\;\middle|\; 1\right]$$
$$= {}_2F_1(\tfrac{1}{2}(\alpha+\beta+1)+i\lambda\,,\,\tfrac{1}{2}(\alpha+\beta+1)-i\lambda\,;\,\alpha+1\,;\,-\mathrm{sh}^2 t),$$

where the limit is taken formally by power series expansion and termwise limits. Hence, in view of (5.2) and (4.7) we obtain

$$\lim_{n\to\infty} \frac{S_n(\tfrac{1}{4}\lambda^2\,;\,\tfrac{1}{2}(\alpha+\beta+1)\,,\,n\,\mathrm{sh}^{-2}t\,,\,\tfrac{1}{2}(\alpha-\beta+1))}{(\tfrac{1}{2}(\alpha+\beta+1)+n\,\mathrm{sh}^{-2}t)_n\,(\alpha+1)_n} = \phi_\lambda^{(\alpha,\beta)}(t), \tag{5.14}$$

i.e., Jacobi functions are limits of continuous dual Hahn polynomials. In §7 we will find many more examples of unitary integral transforms with hypergeometric kernel as limits of hypergeometric orthogonal polynomial transforms and we will be able to give an extension of the Askey scheme such that these integral transforms are also incorporated. Moreover, in §6 we will find isometries of $L^2(\mathbb{Z})$ and $L^2(\mathbb{Z}_+)$ which are of hypergeometric but not of orthogonal polynomial type.

6. Orthogonal special functions from group theory: two fundamental constructions

Let G be a Lie group and π a unitary representation of G on a Hilbert space $\mathcal{H}(\pi)$ (cf. §3). Let H and L be closed subgroups of G and suppose that the restrictions of π to H and L are multiplicity free. Then we can write π restricted to H as a direct sum or integral of irreducible representations γ of H uniquely, and similarly for δ, cf. Table 6 below.

$$\begin{array}{ccc} & G,\pi & \\ / & & \backslash \\ H,\gamma & & L,\delta \end{array}$$

Table 6

In the very special situation that all these representations γ of H and δ of L are one-dimensional, we have two orthonormal bases for $\mathcal{H}(\pi)$ (in the ordinary sense in case of a direct sum and in the generalized sense, with distribution vectors, in case of a direct integral). We write these bases as $\{v_{\pi,\gamma}\}$ with γ in some subset of $\hat{H}$ and $\{w_{\pi,\delta}\}$ with δ in some subset of $\hat{L}$, such that

$$\pi(h)v_{\pi,\gamma} = \gamma(h)v_{\pi,\gamma}, \quad \pi(l)w_{\pi,\delta} = \delta(l)w_{\pi,\delta}. \tag{6.1}$$

There are now two ways to construct orthogonal special functions adapted to the group G and its subgroups H and L.

Method I.

$$\phi^{\gamma,\delta}(g,\pi) := (\pi(g)v_{\pi,\gamma}, w_{\pi,\delta}), \quad g\in G,\ \pi\in\hat{G},\ \gamma\in\hat{H},\ \delta\in\hat{L}. \tag{6.2}$$

Here γ,δ are considered as parameters which are held fixed, while we get orthogonality and dual orthogonality with respect to the variables g and π, respectively. In case of a compact group G, orthogonality of the functions $\phi^{\gamma,\delta}(\,.\,,\pi)$ is just the Schur orthogonality for matrix elements of irreducible representations and completeness of the system follows from the Peter-Weyl theorem. In case of a noncompact group G, the integral transform with $\phi^{\gamma,\delta}(g,\pi)$ as kernel is a specialization of the Fourier transform on the group and the inversion formula is a particular case of the Plancherel formula on the group. Part of the game in all these cases is the restriction of $\phi^{\gamma,\delta}(\,.\,,\pi)$ to a suitable subset of G which (for the particular γ,δ) already completely determines the function. If this subset can be identified with a subset of Euclidean space and the essential part of the Haar measure for G can be determined on this subset, then we have made the identification of the orthogonal system (6.2) with a system which can be described without group theoretic terminology.

If $v_{\pi,\gamma}$ and $w_{\pi,\delta}$ are generalized elements of $\mathcal{H}(\pi)$ then the inner product in (6.2) can be given a meaning as a distribution by testing it against C^∞-functions f with compact support on G. This testing yields $(\pi(f)v_{\pi,\gamma}, w_{\pi,\delta})$, then $\pi(f)v_{\pi,\gamma}$ is a C^∞-vector and its inner product with the distribution vector $w_{\pi,\delta}$ is well-defined.

Method II

$$\psi^{g,\pi}(\gamma,\delta) := (\pi(g)v_{\pi,\gamma}, w_{\pi,\delta}), \quad g\in G,\ \pi\in\hat{G},\ \gamma\in\hat{H},\ \delta\in\hat{L}. \tag{6.3}$$

Now g and π are held fixed as parameters, while γ and δ are the variables with respect to which orthogonality and dual orthogonality are obtained. Note that $\{\pi(g)v_{\pi,\gamma}\}$ and $\{w_{\pi,\delta}\}$ are both (generalized) orthonormal bases of $\mathcal{H}(\pi)$, so $\psi^{g,\pi}(\gamma,\delta)$ is the matrix element or integral kernel of the unitary transformation which maps the one basis onto the other.

Our requirement that the representations γ and δ are one-dimensional is too restrictive, although it is surprising how many special functions already come out with this assumption, mainly for $G=SL(2,\mathbb{R})$, $SU(2)$ or $I_0(\mathbb{R}^2)$. More generally we assume that we have two chains of closed subgroups and corresponding irreducible representations, cf. Table 7 below.

Here we assume that the representation γ_i of H_i, when restricted to H_{i+1}, has a multiplicity free decomposition in terms of irreducible representations γ_{i+1} of H_{i+1}, and the irreducible

$$G, \pi$$

$$H_1, \gamma_1 \qquad L_1, \delta_1$$
$$| \qquad |$$
$$H_2, \gamma_2 \qquad L_2, \delta_2$$

$$\vdots \qquad \vdots$$

$$H_p, \gamma_p \qquad L_q, \delta_q$$

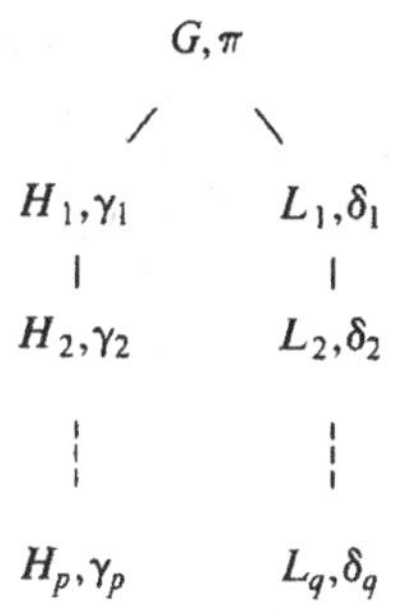

Table 7

representations γ_p of H_p which occur are one-dimensional, and similarly for the other chain. Thus we obtain (generalized) orthonormal bases $\{v_{\pi, \gamma_1, \ldots, \gamma_p}\}$ and $\{w_{\pi, \delta_1, \ldots, \delta_q}\}$ for $\mathfrak{K}(\pi)$ and we can write, on the one hand,

$$\phi^{\gamma_1, \ldots, \gamma_p; \delta_1, \ldots, \delta_q}(g, \pi) := (\pi(g) v_{\pi, \gamma_1, \ldots, \gamma_p}, w_{\pi, \delta_1, \ldots, \delta_q}), \qquad (6.4)$$

where $\gamma_1, \ldots, \gamma_p, \delta_1, \ldots, \delta_q$ are considered as parameters, and, on the other hand,

$$\psi^{g, \pi}(\gamma_1, \ldots, \gamma_p; \delta_1, \ldots, \delta_q) := (\pi(g) v_{\pi, \gamma_1, \ldots, \gamma_p}, w_{\pi, \delta_1, \ldots, \delta_q}), \qquad (6.5)$$

where g, π are considered as parameters.

Let us discuss a number of special cases of the constructions I and II.

Ia. Let $K := H = L$ be compact and let (G, K) be a *Gelfand pair*, i.e., in each $\pi \in \hat{G}$ the trivial representation 1 of K occurs at most once. Take $\gamma = \delta = 1$ in (6.2). Then $\phi^{1,1}(\cdot, \pi)$ becomes a *spherical function* for (G, K), cf. (3.1).

Ib. Let $K := H_1 = L_1$ be compact and take $p = 1$, $\gamma_1 = 1$ in (6.4). Then $\phi^{1; \delta_1, \ldots, \delta_q}$ becomes an *associated spherical function* for (G, K), cf. (3.2).

Ic. Let H and L be non-conjugate subgroups of G, let H be compact and take $\gamma = 1 = \delta$ in (6.2). Then $\phi^{1,1}(\cdot, \pi)$ is a so-called *intertwining function* for the triple (G, H, L).

Id. Let $H_i = L_i$ $(i = 1, \ldots, p; p = q)$. Then $\phi^{\gamma_1, \ldots, \gamma_p; \delta_1, \ldots, \delta_q}(\cdot, \pi)$ as given by (6.4) denotes a (generalized) *matrix element* of π with respect to the basis $\{v_{\pi, \gamma_1, \ldots, \gamma_p}\}$, with the dependence on g emphasized.

IIa. Assumptions as in case Id. The function $\psi^{g, \pi}$ in (6.5) is again a (generalized) *matrix element* of π, but here with the dependence on $\gamma_1, \ldots, \gamma_p$ and $\delta_1, \ldots, \delta_p$ emphasized.

IIb. Let H and L be non-conjugate subgroups. Then $\psi^{g, \pi}$ as given by (6.5) is called a *mixed basis matrix element* (possibly generalized) of π. Note that, without loss of generality, we may assume $g = e$ in (6.5). Then the functions $\psi^{g, \pi}$ in (6.5) are called the *overlap functions* for the bases $\{v_{\pi, \gamma_1, \ldots, \gamma_p}\}$ and $\{w_{\pi, \delta_1, \ldots, \delta_q}\}$ of $\mathfrak{K}(\pi)$.

The spherical function case Ia is already so rich that whole books can be written only about this case. See for instance Helgason [21], Faraut [18] and Koornwinder [27]. Let us here just make the observation that spherical functions always satisfy a *product formula*

$$\phi(g_1, \pi) \phi(g_2, \pi) = \int_K \phi(g_1 k g_2) \, dk \qquad (6.6)$$

with corresponding commutative positive convolution structure for the K-biinvariant functions on G and a *linearization formula*

$$\phi(g, \pi_1) \phi(g, \pi_2) = \int_{\hat{G}} \phi(g, \pi_3) c(\pi_1, \pi_2, \pi_3) \, d\mu(\pi_3) \qquad (6.7)$$

with $c(\pi_1,\pi_2,\pi_3)\geqslant 0$ and corresponding positive dual convolution structure. If we can identify the spherical functions with special functions then we have, of course, also two positive convolution structures associated with the expansions in terms of these special functions.

In case II we may have the situation of a third subgroup chain. In the simple situation we started with in this section, for instance, we may have a diagram as in Table 8 below and three corresponding (generalized) orthonormal bases $\{u_{\pi,\beta}\}$, $\{v_{\pi,\gamma}\}$ and $\{w_{\pi,\delta}\}$.

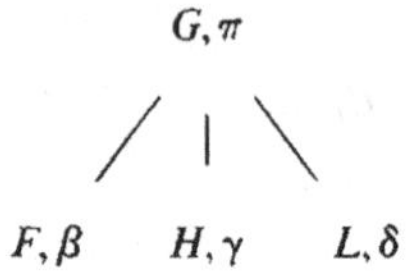

Table 8

Then the three functions of type (6.3) (with $g=e$) related to these bases satisfy an identity of the form

$$(u_{\pi,\beta},w_{\pi,\delta}) = \int_{\hat{H}} (u_{\pi,\beta},v_{\pi,\gamma})(v_{\pi,\gamma},w_{\pi,\delta})\,d\mu(\gamma). \tag{6.8}$$

In section 7 we will meet concrete examples of this.

Mixed case. There is a third way to obtain orthogonal systems from the inner products $(\pi(g)v_{\pi,\gamma},w_{\pi,\delta})$. Fix π and γ as parameters and consider

$$\chi^{\pi,\gamma}(g,\delta) := (\pi(g)v_{\pi,\gamma},w_{\pi,\delta}). \tag{6.9}$$

We might as well fix π and some $u\in\mathcal{H}(\pi)$ and then consider

$$\chi_1^{\pi,u}(g,\gamma) := (\pi(g)u,v_{\pi,\gamma}), \quad \chi_2^{\pi,u}(g,\delta) := (\pi(g)u,w_{\pi,\delta}). \tag{6.10}$$

Then a Schur type orthogonality yields

$$\int_G \chi_1^{\pi,u}(g,\gamma)\overline{\chi_2^{\pi,u}(g,\delta)}\,dg = \text{const.}\,(w_{\pi,\delta},v_{\pi,\gamma}). \tag{6.11}$$

Again we have produced an identity involving three orthogonal systems: two of mixed type and one of type II. It will turn out later that (4.16), (4.17) can be brought into this form.

As an example of (6.2) versus (6.3) consider $G:=SU(2)$, the group of 2×2 unitary matrices with determinant 1. (See Vilenkin [36, Ch. 3] as a reference for this example.) Let $K:=H=L=U(1)$, the subgroup of diagonal matrices in $SU(2)$. For each $l\in\frac{1}{2}\mathbb{Z}_+$ there is, up to equivalence, a unique irreducible unitary representation T^l of G with $(2l+1)$-dimensional representation space $\mathcal{H}(T^l)$. We can choose an orthonormal K-basis $\{v_{-l},v_{-l+1},\ldots,v_l\}$ of $\mathcal{H}(T^l)$ such that

$$T^l(\operatorname{diag}(e^{-\frac{1}{2}i\theta},e^{\frac{1}{2}i\theta}))v_n = e^{in\theta}v_n, \quad n=-l,-l+1,\ldots,l. \tag{6.12}$$

Then $(T^l(g)v_n,v_m)$ as a function of $g\in G$ is already completely determined if we know it for g restricted to the matrices

$$u_\gamma := \begin{pmatrix} \cos\frac{1}{2}\gamma & -\sin\frac{1}{2}\gamma \\ \sin\frac{1}{2}\gamma & \cos\frac{1}{2}\gamma \end{pmatrix}.$$

So write

$$\phi^{n,m}(u_\gamma,T^l) := (T^l(u_\gamma)v_n,v_m) =: \psi^{\gamma,l}(n,m). \tag{6.13}$$

Then the functions $\phi^{n,m}$ (fitting into case Id) can be evaluated in terms of Jacobi polynomials (cf. (4.6)) by

$$\phi^{n,m}(u_\gamma,T^l) = \left[\frac{(l-m)!\,(l+m)!}{(l-n)!\,(l+n)!}\right]^{\frac{1}{2}}$$

$$\times(\sin \tfrac{1}{2}\gamma)^{m-n}(\cos \tfrac{1}{2}\gamma)^{m+n}P_{l-m}^{(m-n,m+n)}(\cos\gamma),\quad m+n,m-n\geq 0. \tag{6.14}$$

Orthogonality of the Jacobi polynomials is here an immediate consequence of Schur's orthogonality relations for matrix elements of the inequivalent irreducible representations T^l.

On the other hand, as was shown in [26, §2], the functions $\psi^{\gamma,l}$ (fitting into case IIa) can be evaluated in terms of Krawtchouk polynomials (cf. (5.10)) by

$$\psi^{\gamma,l}(n,m) = \begin{bmatrix}2l\\ l+m\end{bmatrix}^{\frac{1}{2}}\begin{bmatrix}2l\\ l+n\end{bmatrix}^{\frac{1}{2}}$$
$$\times(-1)^{l+n}(\sin\tfrac{1}{2}\psi)^{n+m+2l}(\cos\tfrac{1}{2}\psi)^{-n-m}K_{m+l}(n+l;\sin^2\tfrac{1}{2}\psi,2l). \tag{6.15}$$

The orthogonality of the Krawtchouk polynomials is now an immediate consequence of the fact that $T^l(u_\gamma)$is a unitary operator and the basis $\{v_n\}$ is orthonormal.

It is possible to deform the group $SU(2)$ such that it tends to the Euclidean motion group $I_0(\mathbb{R}^2)$, while the subgroup $U(1)$ of $SU(2)$ becomes the subgroup $SO(2)$ of $I_0(\mathbb{R}^2)$ and the matrices u_γ tend to the translations t_x. Then the irreducible representations T^l of $SU(2)$ also tend to an irreducible unitary representation π_λ of $I_0(\mathbb{R}^2)$ (cf. the representation π of $I_0(\mathbb{R}^2)$ which we met in section 3). The orthonormal basis $\{v_{-l},v_{-l+1},\ldots,v_l\}$ of $\mathcal{H}(T^l)$ tends to an orthonormal basis $\{v_n\}_{n\in\mathbb{Z}}$ of $\mathcal{H}(\pi_\lambda)$ which behaves nicely with respect to the subgroup $SO(2)$ of $I_0(\mathbb{R}^2)$. We obtain, by taking limits in (6.14) or (6.15) that

$$\lim_{l\to\infty}(T^{l\lambda}(u_{x/l})v_n,v_m) = (\pi_\lambda(t_x)v_n,v_m) = (-1)^{m-n}J_{n+m}(\lambda x), \tag{6.16}$$

where J_{n+m} is a Bessel function (cf. (2.1)). When we emphasize the dependence on λ,x in $J_{n+m}(\lambda x)$ then we get the generalized orthogonal system which yields the integral kernel for the Hankel transform (cf. (4.23)). However, when we emphasize the dependence on n and m then we get the extremely simple orthogonality relations

$$\sum_{n=-\infty}^{\infty} J_{n+k}(x)J_{n+l}(x) = \delta_{k,l},\quad k,l\in\mathbb{Z},\ x\in\mathbb{R}, \tag{6.17}$$

which are not widely known, although they are hidden in the formula books. Indeed, start with

$$e^{ix\sin\phi} = \sum_{n=-\infty}^{\infty} e^{in\phi}J_n(x) \tag{6.18}$$

(the Fourier series expansion implied by (2.2)). Hence

$$e^{ix\sin\phi}e^{-ik\phi} = \sum_{n=-\infty}^{\infty} e^{in\phi}J_{n+k}(x).$$

So, by Parseval's formula for Fourier series,

$$\sum_{n=-\infty}^{\infty} J_{n+k}(x)J_{n+l}(x) = (2\pi)^{-1}\int_0^{2\pi} e^{ix\sin\phi}e^{-ik\phi}\,\overline{e^{ix\sin\phi}e^{-il\phi}}\,d\phi = \delta_{k,l}.$$

By (6.17) we have, for each $x\in\mathbb{R}$, an orthonormal system $\{n\mapsto J_{n+k}(x)\}_{k\in\mathbb{Z}}$ in $L^2(\mathbb{Z})$, which consists of ${}_0F_1$-hypergeometric functions and is a limit case of the Krawtchouk polynomials, but is definitely not a system of orthogonal polynomials. Still it naturally fits into an extended concept of the Askey scheme.

One may wonder what else might be included in such an extended scheme. Here is another example, not (yet) coming from group theory. From the Jacobi polynomials we build two orthogonal bases of $L^2((0,1))$. A simple computation yields the transition matrix:

$$\int_0^1 x^{\frac{1}{2}\alpha}(1-x)^{\frac{1}{2}\beta}P_n^{(\alpha,\beta)}(1-2x)\,x^{\frac{1}{2}\alpha}(1-x)^{\frac{1}{2}\delta}P_m^{(\alpha,\delta)}(1-2x)\,dx$$

$$= \frac{(-1)^n(\beta+1)_n\,\Gamma(\alpha+m+1)\Gamma(\tfrac{1}{2}(\beta+\delta)+1)(-m+\tfrac{1}{2}(\beta-\delta)+1)_m}{n!\,m!\,\Gamma(\alpha+\tfrac{1}{2}(\beta+\delta)+2+m)}$$

$$\times {}_4F_3\left[\begin{matrix}-n, n+\alpha+\beta+\delta+1, \tfrac{1}{2}(\beta+\delta)+1, \tfrac{1}{2}(\beta-\delta)+1\\ \beta+1, m+\alpha+\tfrac{1}{2}(\beta+\delta)+2, -m+\tfrac{1}{2}(\beta-\delta)+1\end{matrix}\;\middle|\; 1\right] \tag{6.19}$$

This yields an orthogonal system on $L^2(\mathbb{Z}_+)$ of hypergeometric but non-polynomial type. One can derive a limit case of (2.8) with Laguerre polynomials at the left hand side and a ${}_3F_2$ of argument 1 at the right hand side.

7. Orthogonal special functions related to the discrete series of $SL(2,\mathbb{R})$

The paper [9] by Basu and Wolf is a rich source of special functions related to representations of $SL(2,\mathbb{R})$ (the group of real 2×2 matrices of determinant 1). In particular, the special functions related to the discrete series of $SL(2,\mathbb{R})$, discussed in [9, section 3], are very neat, much related to the Askey scheme and a good illustration of the framework we developed in §6. I feel justified to reproduce here some of the formulas in [9], since the orthogonality properties, limit relations and identifications with known special functions are not much emphasized there.

The *discrete series representations* D_k^+ $(k=\tfrac{1}{2},1,3/2,\cdots)$ of $SL(2,\mathbb{R})$ are realized in [9, (2.5)] as unitary representations on $L^2(\mathbb{R}_+)$:

$$\left(D_k^+\begin{bmatrix}a & b\\ c & d\end{bmatrix}f\right)(r) := \int_0^\infty f(s)\, e^{-i\pi k}\, b^{-1}(rs)^{\frac{1}{2}}$$

$$\times \exp(\tfrac{1}{2}ib^{-1}(dr^2+as^2))\, J_{2k-1}(b^{-1}rs)\, ds, \quad f\in L^2(\mathbb{R}_+),\ b\neq 0, \tag{7.1}$$

and, by taking the limit as $b\to 0$:

$$\left(D_k^+\begin{bmatrix}a & 0\\ c & d\end{bmatrix}f\right)(r) := (\operatorname{sgn} a)^{2k}\, |a|^{-\frac{1}{2}}\, e^{\frac{1}{2}ia^{-1}cr^2} f(|a|^{-1}r), \quad f\in L^2(\mathbb{R}_+). \tag{7.2}$$

Further specializations of (7.1) and (7.2) yield:

$$\left(D_k^+\begin{bmatrix}1 & 0\\ c & 1\end{bmatrix}f\right)(r) = e^{\frac{1}{2}icr^2} f(r), \quad f\in L^2(\mathbb{R}_+), \tag{7.3}$$

$$\left(D_k^+\begin{bmatrix}a & 0\\ 0 & a^{-1}\end{bmatrix}f\right)(r) = (\operatorname{sgn} a)^{2k}\, |a|^{-\frac{1}{2}} f(|a|^{-1}r), \quad f\in L^2(\mathbb{R}_+), \tag{7.4}$$

$$\left(D_k^+\begin{bmatrix}\cos\tfrac{1}{2}\gamma & -\sin\tfrac{1}{2}\gamma\\ \sin\tfrac{1}{2}\gamma & \cos\tfrac{1}{2}\gamma\end{bmatrix}f\right)(r) = \int_0^\infty f(s)\, e^{i\pi k}(\sin\tfrac{1}{2}\gamma)^{-1}(rs)^{\frac{1}{2}}$$

$$\times e^{-\frac{1}{2}i(\operatorname{cotg}\frac{1}{2}\gamma)(r^2+s^2)}\, J_{2k-1}((\sin\tfrac{1}{2}\gamma)^{-1}rs)\, ds, \quad f\in L^2(\mathbb{R}_+),\ \gamma\notin 2\pi\mathbb{Z}, \tag{7.5}$$

$$\left(D_k^+\begin{bmatrix}0 & -1\\ 1 & 0\end{bmatrix}f\right)(r) = \int_0^\infty f(s)\, e^{i\pi k} J_{2k-1}(rs)(rs)^{\frac{1}{2}}\, ds, \quad f\in L^2(\mathbb{R}_+). \tag{7.6}$$

The transformation (7.1) is built up from the Hankel transform (7.6) (cf. (4.23)) and the elementary transforms (7.3) and (7.4) according to the *Bruhat decomposition* of $SL(2,\mathbb{R})$:

$$\begin{bmatrix}a & b\\ c & d\end{bmatrix} = \begin{bmatrix}-b & 0\\ 0 & -b^{-1}\end{bmatrix}\begin{bmatrix}1 & 0\\ bd & 1\end{bmatrix}\begin{bmatrix}0 & -1\\ 1 & 0\end{bmatrix}\begin{bmatrix}1 & 0\\ b^{-1}a & 1\end{bmatrix}, \quad ad-bc=1,\ b\neq 0. \tag{7.7}$$

Write $G:=SL(2,\mathbb{R})$. There are three conjugacy classes of 1-parameter subgroups in G: the parabolic, hyperbolic and elliptic ones. We choose representatives of these conjugacy classes as follows:

$$\textit{parabolic}:\ N := \left\{n_c := \begin{bmatrix}1 & 0\\ c & 1\end{bmatrix} \;\middle|\; c\in\mathbb{R}\right\};$$

$$\textit{hyperbolic}:\ H := \left\{h_\beta := \begin{bmatrix}e^{-\frac{1}{2}\beta} & 0\\ 0 & e^{\frac{1}{2}\beta}\end{bmatrix} \;\middle|\; \beta\in\mathbb{R}\right\};$$

$$\textit{elliptic}:\ K := \{u_\gamma := \begin{bmatrix} \cos \frac12\gamma & -\sin \frac12\gamma \\ \sin \frac12\gamma & \cos \frac12\gamma \end{bmatrix} \mid \gamma\in\mathbb{R}\}.$$

Following [9] we choose for $L^2(\mathbb{R}_+)=\mathcal{H}(D_k^+)$ a generalized orthonormal N-basis $\{v_{k,\rho}^N \mid \rho\in\mathbb{R}_+\}$ and H-basis $\{v_{k,\mu}^H \mid \mu\in\mathbb{R}\}$ and an orthonormal K-basis $\{v_{k,m}^K \mid m=k,k+1,\cdots\}$ such that

$$D_k^+(n_c)v_{k,\rho}^N = e^{ic\rho^2}\, v_{k,\rho}^N,\quad \rho\in\mathbb{R}_+,\ n_c\in N, \tag{7.8}$$

$$D_k^+(h_\beta)v_{k,\mu}^H = e^{i\beta\mu}\, v_{k,\mu}^H,\quad \mu\in\mathbb{R},\ h_\beta\in H, \tag{7.9}$$

$$D_k^+(u_\gamma)v_{k,m}^K = e^{i\gamma m}\, v_{k,m}^K,\quad m=k,k+1,\cdots,u_\gamma\in K. \tag{7.10}$$

Thus we are in the situation of Table 8 and the equations (7.8),(7.9),(7.10) are examples of (6.1). See Table 9 below.

	G,D_k^+		
subgroups:	N	H	K
labels of representations:	ρ	μ	m
corresponding basis vectors:	$v_{k,\rho}^N$	$v_{k,\mu}^H$	$v_{k,m}^K$

Table 9

For the N-basis we have

$$v_{k,\rho}^N := \delta_\rho,\quad \rho\in\mathbb{R}_+, \tag{7.14}$$

where δ_ρ denotes the delta distribution with support $\{\rho\}$. Then (7.8) follows from (7.3).

For the H-basis we have

$$v_{k,\mu}^H(r) := \pi^{-\frac12}\, r^{-\frac12+2i\mu},\quad r>0,\ \mu\in\mathbb{R}. \tag{7.15}$$

Then (7.9) follows from (7.2).

For the K-basis we have an expression in terms of Laguerre polynomials:

$$v_{k,m}^K(r) := \left[\frac{2(m-k)!}{(m+k+1)!}\right]^{\frac12} r^{2k-\frac12}\, e^{-\frac12 r^2}\, L_{m-k}^{2k-1}(r^2),\quad m=k,k+1,\cdots. \tag{7.16}$$

Then (7.10) can be obtained from the following generalization of (4.24) (cf. [16, 8.9(5)]):

$$\int_0^\infty t^{\alpha+\frac12}\, e^{-\frac12 t^2}\, L_n^\alpha(t^2)\, e^{\frac12 i(\operatorname{cotg}\gamma)(t^2+\lambda^2)}\, J_\alpha\!\left(\frac{\lambda t}{\sin\frac12\gamma}\right)\frac{(\lambda t)^{\frac12}}{\sin\frac12\gamma}\,dt$$
$$= (-1)^n\, e^{i(\pi-\gamma)(n+\frac12\alpha+\frac12)}\, \lambda^{\alpha+\frac12}\, e^{-\frac12\lambda^2}\, L_n^\alpha(\lambda^2). \tag{7.17}$$

Consider now the matrix elements $(D_k^+(g)v_{k,\lambda'}^L, v_{k,\lambda}^L)$ and mixed basis matrix elements $(D_k^+(g)v_{k,\lambda'}^{L'}, v_{k,\lambda}^L)$, where L,L' are two of the subgroups N,H,K and where k and g are held fixed as parameters, while the orthogonality is considered in the labels λ,λ' of the subgroup representations (cf. cases IIa and IIb of §6). In Table 10 below we summarize which orthogonal special functions arise in this way.

Let us quote these results more explicitly from the paper by Basu and Wolf. In each of the following formulas g denotes the matrix element $\begin{pmatrix} a & b \\ c & d\end{pmatrix}$ of G.

***N*-*N* matrix elements**: *Bessel functions*:

$$(D_k^+(g)v_{k,\rho'}^N, v_{k,\rho}^N) = e^{-i\pi k}\, b^{-1}(\rho\rho')^{\frac12}\, e^{\frac12 i b^{-1}(d\rho^2+a\rho'^2)}\, J_{2k-1}(b^{-1}\rho\rho'),\quad b\neq 0. \tag{7.18}$$

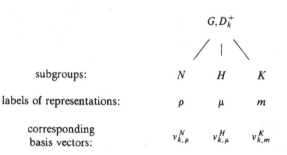

	$D_k^+(g)v_{k,\rho'}^N$	$D_k^+(g)v_{k,\mu'}^H$	$D_k^+(g)v_{k,m'}^K$
$v_{k,\rho}^N$	Bessel functions	Laguerre functions (Mellin if $g=e$)	Laguerre polynomials
$v_{k,\mu}^H$		Meixner-Pollaczek functions	Meixner-Pollaczek polynomials
$v_{k,m}^K$			Meixner polynomials

Table 10

This follows immediately from (7.14) and (7.1). The corresponding integral transform is essentially the Hankel transform.

H-H **matrix elements**: *Meixner-Pollaczek functions*:

$$(D_k^+(g)v_{k,\mu'}^H, v_{k,\mu}^H) = \text{elementary factors} \times {}_2F_1(k-i\mu, k+i\mu'; 2k; (ad)^{-1}). \tag{7.19}$$

We call the right hand side, in its dependence on μ or μ', a Meixner-Pollaczek function because it bears the same relationship to Meixner-Pollaczek polynomials (cf. (5.8)) as Jacobi functions do to Jacobi polynomials (cf. (4.7),(4.6')). The corresponding unitary integral transform does not seem to be widely known in literature.

K-K **matrix elements**: *Meixner polynomials*:

$$(D_k^+(g)v_{k,m'}^K, v_{k,m}^K) = \text{elementary factors} \times M_{m-k}(m'-k, 2k; \frac{a^2+b^2+c^2+d^2-2}{a^2+b^2+c^2+d^2+2}). \tag{7.20}$$

N-H **mixed basis matrix elements**: *Laguerre functions*:

$$(D_k^+(g)v_{k,\mu}^H, v_{k,\rho}^N) = \text{elementary factors} \times {}_1F_1(k+i\mu; 2k; -(2ab)^{-1}i\rho^2), \quad a,b\neq 0. \tag{7.21}$$

Consider, for each $\mu\in\mathbf{R}$, the right hand side as a function of ρ. It bears the same relationship to Laguerre polynomials as Jacobi functions do to Jacobi polynomials. The limit case of (7.21) for $g\to e$ follows from (7.14) and (7.15):

$$(v_{k,\mu}^H, v_{k,\rho}^N) = \pi^{-1/2}\rho^{-1/2+1/2 i\mu}. \tag{7.22}$$

This is the integral kernel of the *Mellin transform*, which is, by a trivial change of variables, just the Fourier transform.

N-K **mixed basis matrix elements**: *Laguerre polynomials*.
In view of the *Iwasawa decomposition* $G=NHK$, these matrix elements, in their dependence on g, are already determined when they are given for $g\in H$. As H acts quite trivially on $v_{k,\rho}^N$ (in view of (7.4),(7.14)), it is sufficient to consider the matrix element at $g=e$. By (7.16) this equals

$$(v_{k,m}^K, v_{k,\rho}^N) = v_{k,m}^K(\rho) = \text{elementary factors} \times L_{m-k}^{2k-1}(\rho^2). \tag{7.23}$$

H-K **mixed basis matrix elements**: *Meixner-Pollaczek polynomials*:

$$(D_k^+(g)v_{k,m}^K, v_{k,\mu}^H) = \text{elementary factors} \times P_{m-k}^{(k)}(\mu; \tfrac{1}{2}\arccos(\frac{(ac+bd)^2-1}{(ac+bd)^2+1})). \tag{7.24}$$

Some remarks can be made at this stage. Of the six cases just discussed there are three orthogonal polynomial cases (*K-K*, *K-H* and *K-N*), in correspondence with the fact that K is the only compact group among K,H,N. The matrix elements satisfy second order differential equations in the ρ-variable but second order difference equations in the μ- and m-variables, where the differences in the case of μ are taken in imaginary direction. In case of a second order differential equation (first row of Table 10), the spectral theory of differential equations with regular singularities can be used, but, in particular in the *H-H* case, there is missing an analogous spectral theory of second order difference equations.

Both the conjugacy class of elliptic and of hyperbolic subgroups have N in their closure:

$$\begin{pmatrix}\lambda^{1/2} & 0\\ 0 & \lambda^{-1/2}\end{pmatrix}\begin{pmatrix}\cos(c\lambda) & -\sin(c\lambda)\\ \sin(c\lambda) & \cos(c\lambda)\end{pmatrix}\begin{pmatrix}\lambda^{1/2} & 0\\ 0 & \lambda^{-1/2}\end{pmatrix}^{-1} = \begin{pmatrix}\cos(c\lambda) & -\lambda\sin(c\lambda)\\ \lambda^{-1}\sin(c\lambda) & \cos(c\lambda)\end{pmatrix}$$

and

$$\begin{bmatrix} \tfrac12\sqrt{2}\,\lambda^{1/2} & -\tfrac12\sqrt{2}\,\lambda^{1/2} \\ \tfrac12\sqrt{2}\,\lambda^{-1/2} & \tfrac12\sqrt{2}\,\lambda^{-1/2} \end{bmatrix} \begin{bmatrix} e^{\lambda c} & 0 \\ 0 & e^{-\lambda c} \end{bmatrix} \begin{bmatrix} \tfrac12\sqrt{2}\,\lambda^{1/2} & -\tfrac12\sqrt{2}\,\lambda^{1/2} \\ \tfrac12\sqrt{2}\,\lambda^{-1/2} & \tfrac12\sqrt{2}\,\lambda^{-1/2} \end{bmatrix}^{-1} = \begin{bmatrix} \mathrm{ch}(c\lambda) & \lambda\,\mathrm{sh}(c\lambda) \\ \lambda^{-1}\mathrm{sh}(c\lambda) & \mathrm{ch}(c\lambda) \end{bmatrix}$$

tend both to $\begin{bmatrix} 1 & 0 \\ c & 1 \end{bmatrix}$ as $\lambda\downarrow 0$. There are corresponding limit transitions for the matrix elements of D_k^+ with respect to the various subgroups. We display this in Table 11 below.

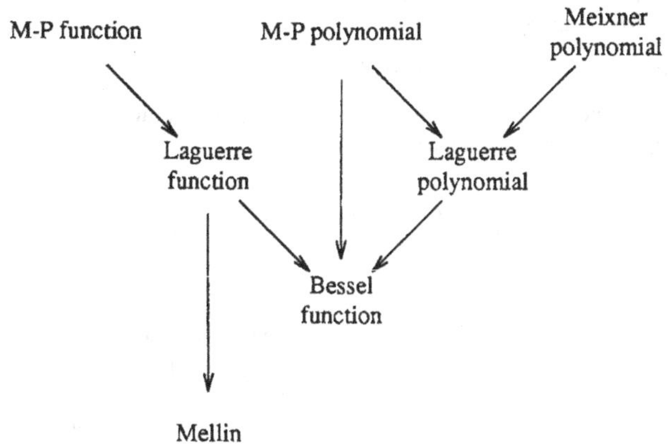

Table 11

Of course, the various matrix elements encountered here offer ample illustration of formula (6.8). In particular, we obtain integral transforms sending Laguerre or Meixner-Pollaczek polynomials to polynomials of similar kind. Moreover, the various cases are connected by limit transitions, cf. Table 12 below.

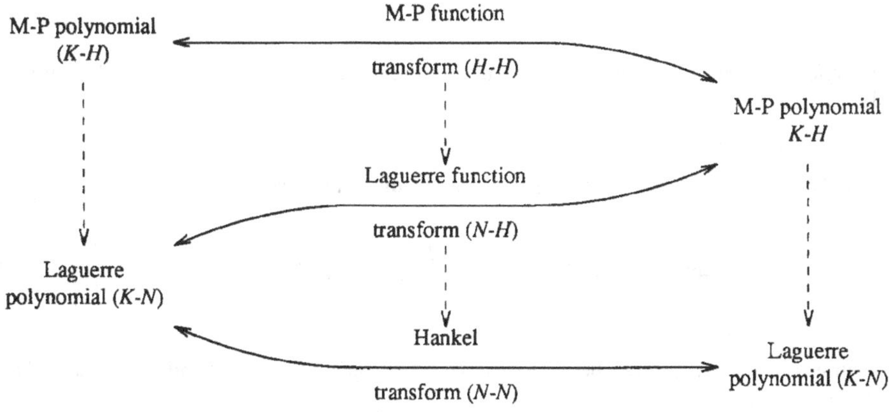

Table 12

By (7.22) the Laguerre function transform simplifies into the Mellin transform for $g=e$. That particular case in Table 12 then corresponds to the well-known formula

$$\int_0^\infty \frac{n!\,e^{-in\phi}}{(2a)_n}\, e^{-\frac12 x(1+i\,\mathrm{cotg}\,\phi)}\, x^a\, L_n^{2a-1}(x)\, x^{-1-i\lambda}\, dx$$

$$= e^{(ia - \lambda)(\phi - \frac{1}{2}\pi)} (2\sin\phi)^{a - i\lambda} \Gamma(a - i\lambda) P_n^{(a)}(\lambda;\phi). \tag{7.25}$$

8. Group theoretic interpretations of Hahn and Racah polynomials

Hahn polynomials have a group theoretic interpretation as Clebsch-Gordan coefficients for $SU(2)$. We can understand these coefficients as overlap coefficients for bases with respect to two different subgroup reductions of $SU(2) \times SU(2)$. Let T^l be the $(2l + 1)$-dimensional irreducible unitary representation of $SU(2)$ (cf. end of §6). Denote the (1-dimensional) irreducible representations of the subgroup $U(1)$ of $SU(2)$ by δ_n:

$$\delta_n(\text{diag}(e^{-\frac{1}{2}i\theta}, e^{\frac{1}{2}i\theta})) := e^{in\theta}, \quad n \in \frac{1}{2}\mathbf{Z}. \tag{8.1}$$

For $SU(2) \times SU(2)$, together with its irreducible unitary representations $T^{l_1} \otimes T^{l_2}$, Table 13 below presents two subgroup reductions similar to Table 7.

$$SU(2) \times SU(2), \ T^{l_1} \otimes T^{l_2}$$

$$U(1) \times U(1), \ \delta_{j_1} \otimes \delta_{j_2} \qquad \qquad \text{diag}(SU(2) \times SU(2)), \ T^l$$

$$\text{diag}(U(1) \times U(1)), \ \delta_j$$

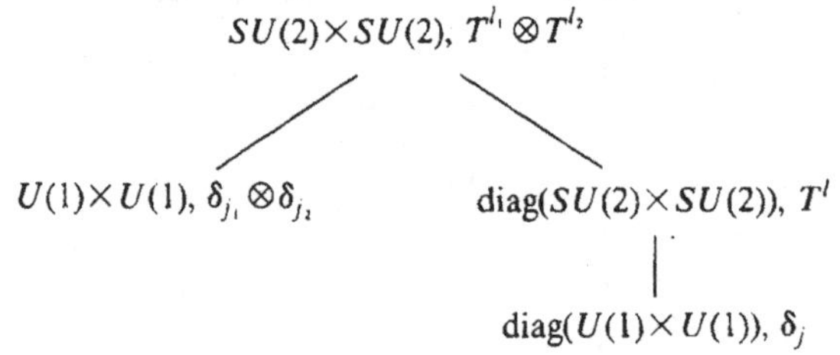

Table 13

There are corresponding orthonormal bases $\{v_{l_1,l_2;j_1,j_2}\}$ and $\{w_{l_1,l_2;l,j}\}$ of $\mathcal{K}(T^{l_1} \otimes T^{l_2})$. Consider the overlap coefficients (cf. case IIb in §6)

$$\psi^{l_1,l_2}(j_1,j_2;l,j) := (w_{l_1,l_2;l,j}, v_{l_1,l_2;j_1,j_2}).$$

It turns out that these overlap coefficients vanish if $j \neq j_1 + j_2$, so we can fix j as a parameter, put $j_2 := j - j_1$ and only consider

$$\psi^{l_1,l_2,j}(j_1,l) := (w_{l_1,l_2;l,j}, v_{l_1,l_2;j_1,j-j_1}).$$

By construction, the left hand side yields the matrix coefficient of a unitary matrix. It is usually denoted by

$$C_{j_1,j_2,j}^{l_1,l_2,l} := (w_{l_1,l_2;l,j}, v_{l_1,l_2;j_1,j_2}), \quad j = j_1 + j_2, \tag{8.2}$$

a so-called *Clebsch-Gordan coefficient* or (with slightly different notation) a *3j-symbol*. It can be evaluated as

$$C_{j_1,j_2,j}^{l_1,l_2,l} = \text{elementary factors} \times {}_3F_2\left[\begin{matrix} -l_1 + l_2 - l, \ -l_1 + l_2 + l + 1, \ -l_1 + j_1 \\ -l_1 + l_2 + j + 1, \ -2l_1 \end{matrix}\middle| 1\right], \tag{8.3}$$

$$l_1 - l_2 \leqslant j \leqslant l_2 - l_1 \leqslant l \leqslant l_1 + l_2; \ -l_1 \leqslant j_1 \leqslant l_1; \ j_1 + j_2 = j.$$

By symmetries the general case can be reduced to (8.3). It follows from (8.3) and (5.5) that we have obtained Hahn polynomials:

$$C_{\frac{1}{2}N - x, \frac{1}{2}(\alpha - \beta - N) + x, \frac{1}{2}(\alpha - \beta)}^{\frac{1}{2}N, \frac{1}{2}(N + \alpha + \beta), n + \frac{1}{2}(\alpha + \beta)} = \text{elementary factors} \times Q_n(x; \alpha, \beta, N). \tag{8.4}$$

The row and column orthogonality of the Clebsch-Gordan matrix yield the orthogonality relations for Hahn and dual Hahn polynomials. See, for instance, Vilenkin [36, Ch.3] and Koornwinder [25] for a more detailed treatment of Clebsch-Gordan coefficients.

Racah polynomials (cf. (5.1)) were first obtained when it was recognized that the Racah coefficients for $SU(2)$ can be expressed in terms of orthogonal polynomials (see Wilson [37]). Racah coefficients are the overlap coefficients for certain subgroup reductions of $SU(2)\times SU(2)\times SU(2)$. Write

$$G_1\times G_2\times G_3 := SU(2)\times SU(2)\times SU(2),$$
$$G_{i,j} := \mathrm{diag}(G_i\times G_j),\quad i\neq j,$$
$$G_0 := \mathrm{diag}(G_1\times G_2\times G_3),$$
$$K_0 := \mathrm{diag}(U(1)\times U(1)\times U(1)).$$

Use again the irreducible unitary representations T^l of $SU(2)$ and δ_j of $U(1)$. Then we have three subgroup reductions for the irreducible unitary representations $T^{l_1}\otimes T^{l_2}\otimes T^{l_3}$ of $G_1\times G_2\times G_3$ as given in Table 14 below.

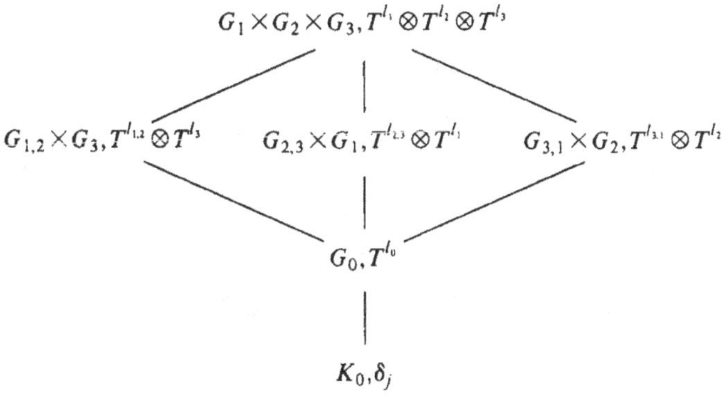

Table 14

There are three corresponding orthonormal bases $u_{l_1,l_2,l_3;l_{1,2},l_0,j}$, $v_{l_1,l_2,l_3;l_{2,3},l_0,j}$, $w_{l_1,l_2,l_3;l_{3,1},l_0,j}$. Now consider l_1,l_2,l_3,l_0 as parameters and define the *Racah coefficients* or *6j-symbols* as the overlap coefficients

$$R^{l_1,l_2,l_3,l_0}_{l_{1,2},l_{1,3}} := (u_{l_1,l_2,l_3;l_{1,2},l_0,j}, v_{l_1,l_2,l_3;l_{2,3},l_0,j}). \tag{8.5}$$

These coefficients can be expressed in terms of ${}_4F_3$-hypergeometric functions of argument 1, and can next be written in terms of Racah polynomials (5.1). The row and column orthogonality of the Racah matrix yield the orthogonality relations for the Racah polynomials. See Biedenharn and Louck [10, Ch.3,§18] for the general theory of Racah coefficients and Wilson [37, §5] for a proof that these matrix coefficients can be expressed in terms of Racah polynomials.

There is a fourth subgroup reduction from $G_1\times G_2\times G_3$, $T^{l_1}\otimes T^{l_2}\otimes T^{l_3}$ to K_0,δ_j via the intermediate stage $K_1\times K_2\times K_3$, $\delta_{j_1},\delta_{j_2},\delta_{j_3}$ (K_i being the subgroup $U(1)$ in G_i). This determines an orthonormal basis $\{e_{l_1,j_1}\otimes e_{l_2,j_2}\otimes e_{l_3,j_3}\}$ $(j_1+j_2+j_3=j)$ for the representation space and we have

$$R^{l_1,l_2,l_3,l_0}_{l_{1,2},l_{1,3}} = \sum_{j_1,j_2,j_3} (u_{l_1,l_2,l_3;l_{1,2},l_0,j}, e_{l_1,j_1}\otimes e_{l_2,j_2}\otimes e_{l_3,j_3}) \times (e_{l_1,j_1}\otimes e_{l_2,j_2}\otimes e_{l_3,j_3}, v_{l_1,l_2,l_3;l_{2,3},l_0,j}) \tag{8.6}$$

(in the spirit of (6.8)). The two overlap coefficients occurring as factors in the sum at the right hand side of (8.6) can be written as products of two Clebsch-Gordan coefficients and they must yield orthogonal systems. These orthogonal systems can be written as functions depending on two discrete variables which are products of elementary factors and two Hahn polynomials. It would be interesting to see if these functions coincide with the orthogonal polynomials in two variables built from Hahn polynomials which are obtained by Dunkl [12, §4] in connection with harmonic analysis on the symmetric group. Anyhow, (8.6) shows that the two orthogonal systems of functions in two variables can be expanded in terms of each other by means of Racah coefficients. We will shortly meet other examples of this phenomenon.

It is natural to expect similar group theoretic interpretations for continuous Hahn and dual Hahn polynomials and for Wilson polynomials. One might speculate about decompositions of two- and threefold tensor products of representations of $SL(2,\mathbb{R})$. As far as I know, it is still an open problem to find such interpretations.

However, there is another group theoretic interpretation of Racah polynomials for which we know how to imitate it in the case of Wilson polynomials. This interpretation was earlier given in [28, §4] and Nikiforov, Suslov and Uvarov [31, Ch.5, §3]. Let harm(n,p) be the space of *spherical harmonics* of degree n on the unit sphere S^{p-1} in $\mathbb{R}^p$. Let $O(p)$ be the group of real $p\times p$ orthogonal matrices and let π_n^p be the natural irreducible unitary representation of $O(p)$ on harm(n,p). Consider in this way the representation π_n^{p+q+r} of $O(p+q+r)$. We have three subgroup reductions as given in Table 15 below.

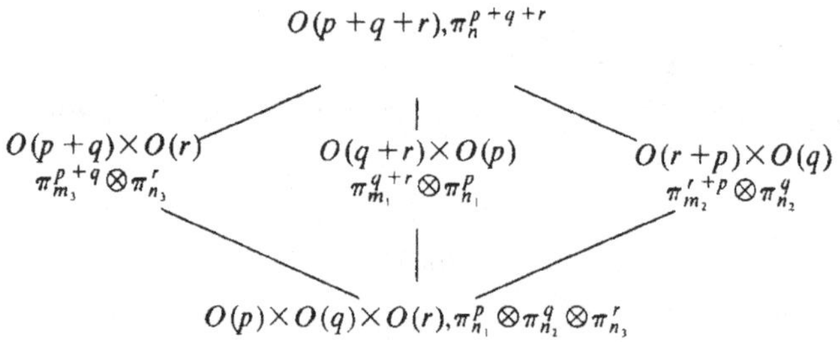

Table 15

Here

$$n-m_1,\, n-m_2,\, n-m_3,\, m_1-n_2-n_3,\, m_2-n_3-n_1,\, m_3-n_1-n_2,\, n-n_1-n_2-n_3 \in 2\mathbb{Z}_+. \tag{8.7}$$

Fix n,n_1,n_2,n_3 in Table 15. Corresponding to the three subgroup reductions there are three orthonormal bases for $\mathcal{H}(\pi_n^{p+q+r})$ and the overlap coefficients turn out to be Racah coefficients again. This can be shown by realizing π_n^{p+q+r} as a representation on harm$(n,p+q+r)$ and then giving explicit expressions for the three bases in terms of products of Jacobi polynomials.

In fact, by iteration of [23, Theorem 4.2], we obtain that harm$(n,p+q+r)$ is spanned by the functions

$$\mathbb{R}^p\times\mathbb{R}^q\times\mathbb{R}^r\supset S^{p+q+r-1}\ni(x,y,z)\mapsto f_{n,m_3}^{n_1,n_2,n_3}(x,y,z)\,Y_{n_1}^p(x)\,Y_{n_2}^q(y)\,Y_{n_3}^r(z), \tag{8.8}$$

where n_1,n_2,n_3 satisfy the inequalities (8.7), $Y_{n_1}^p$, $Y_{n_2}^q$, $Y_{n_3}^r$ are harmonic homogeneous polynomials of degree n_1,n_2,n_3, respectively, and

$$f_{n,m_3}^{n_1,n_2,n_3}(x,y,z) := (|x|^2+|y|^2)^{\frac12(m_3-n_1-n_2)}\,P_{\frac12(m_3-n_1-n_2)}^{(\frac12 q-1+n_2,\frac12 p-1+n_1)}\Big(1-2\frac{|y|^2}{|x|^2+|y|^2}\Big)$$
$$\times P_{\frac12(n-m_3-n_3)}^{(\frac12 r-1+n_3,\frac12(p+q)-1+m_3)}(1-2|z|^2),\quad x\in\mathbb{R}^p,\ y\in\mathbb{R}^q,\ z\in\mathbb{R}^r,\ |x|^2+|y|^2+|z|^2=1, \tag{8.9}$$

with $P_n^{(\alpha,\beta)}$ being a Jacobi polynomial. By cyclic permutation in (8.8),(8.9) we get, in correspondence with Table 15, two other families $\{g_{n,m_1}^{n_1,n_2,n_3}\,Y_{n_1}^p\,Y_{n_2}^q\,Y_{n_3}^r\}$ and $\{h_{n,m_2}^{n_1,n_2,n_3}\,Y_{n_1}^p\,Y_{n_2}^q\,Y_{n_3}^r\}$ of functions which span harm$(n,p+q+r)$.

Fix n_1,n_2,n_3, $Y_{n_1}^p, Y_{n_2}^q, Y_{n_3}^r$ and write

$$u:=|z|^2,\ v:=|y|^2,\ \nu:=\tfrac12(n-n_1-n_2-n_3),\ \mu:=\tfrac12(m_3-n_1-n_2),$$
$$\mu':=\tfrac12(m_2-n_3-n_1),\ \alpha:=\tfrac12 r-1+n_3,\ \beta:=\tfrac12 q-1+n_2,\ \gamma:=\tfrac12 p-1+n_1.$$

Then $f_{n,m_3}^{n_1,n_2,n_3}(x,y,z)$ can be written as

$$P_{\nu,\mu}^{\alpha,\beta,\gamma}(u,v) := P_{\nu-\mu}^{(\alpha,\beta+\gamma+2\mu+1)}(1-2u)(1-u)^{\mu}\,P_{\mu}^{(\beta,\gamma)}(1-2\frac{v}{1-u}) \tag{8.10}$$

and $h_{n,m_2}^{n_1,n_2,n_3}(x,y,z)$ as

$$Q_{\nu,\mu'}^{\alpha,\beta,\gamma}(u,v) := P_{\nu-\mu'}^{(\beta,\gamma+\alpha+2\mu'+1)}(1-2v)(1-v)^{\mu'}\,P_{\mu'}^{(\gamma,\alpha)}(1-2\frac{1-u-v}{1-v}). \tag{8.11}$$

Computation of the overlap coefficients now amounts to expanding $Q_{\nu,\mu'}^{\alpha,\beta,\gamma}$ in terms of the functions $P_{\nu,\mu}^{\alpha,\beta,\gamma}$.

Observe that, for arbitrary $\alpha,\beta,\gamma>-1$, both $\{P_{n,m}^{\alpha,\beta,\gamma}\}_{m=0,1,\ldots,n}$ and $\{Q_{n,m}^{\alpha,\beta,\gamma}\}_{m=0,1,\ldots,n}$ form an orthogonal basis of the space of all orthogonal polynomials of degree n on the triangular region $\{(u,v)\in\mathbb{R}^2 \mid u,v,\ 1-u-v>0\}$ with respect to the measure $u^{\alpha}v^{\beta}(1-u-v)^{\gamma}du\,dv$. The two families are obtained by choosing two different ways of orthogonalizing monomials. (In fact, there are three such families, reflecting the symmetry of order three of the triangle.)

Thus, for arbitrary $\alpha,\beta,\gamma>-1$ there is an expansion

$$Q_{\nu,\mu'}^{\alpha,\beta,\gamma}(u,v) = \sum_{\mu=0}^{\nu} c_{\nu,\mu,\mu'}^{\alpha,\beta,\gamma}\,P_{\nu,\mu}^{\alpha,\beta,\gamma}(u,v). \tag{8.12}$$

The expansion coefficients are then obtained by elementary computations after putting $u:=0$ in (8.12). The result is

$$c_{\nu,\mu,\mu'}^{\alpha,\beta,\gamma} := \text{elementary factors} \times R_{\mu}(\mu'(\mu'+\alpha+\gamma+1);\gamma,\beta,-\nu-1,\alpha+\gamma+\nu+1), \tag{8.13}$$

where R_{μ} is a Racah polynomial. This result was first obtained by Dunkl [13, Theorem 1.7] as a limit case of a similar formula for Hahn polynomials in two variables.

Following the limit transitions in the Askey scheme (Table 3) we can take limits in (8.12) and (8.13). Thus we obtain that the two orthogonal bases

$$(u,v)\mapsto L_{n-k}^{\alpha}(u)\,L_k^{\beta}(v)$$

and

$$(u,v)\mapsto L_{n-m}^{\alpha+\beta+2m+1}(u+v)(u+v)^m\,P_m^{(\alpha,\beta)}(1-2\frac{u}{u+v})$$

for orthogonal polynomials of degree n on $\mathbb{R}_+\times\mathbb{R}_+$ with respect to the measure $u^{\alpha}v^{\beta}e^{-u-v}du\,dv$ can be expanded in terms of each other with Hahn polynomials as coefficients. This was obtained by Suslov [33] in connection with the Schrödinger equation for the Coulomb problem.

There are similar expansions for two orthogonal bases of the orthogonal polynomials on the unit disk with respect to the measure $(1-x^2-y^2)^{\alpha}\,dx\,dy$ (using symmetric Hahn polynomials) and for two orthogonal bases of the orthogonal polynomials on $\mathbb{R}^2$ with respect to the measure $\exp(-x^2-y^2)\,dx\,dy$ (using symmetric Krawtchouk polynomials).

9. Group theoretic interpretation of Wilson polynomials

The group theoretic interpretation of Racah polynomials related to Table 15 in §8 admits a (formal) analytic continuation such that it becomes a group theoretic interpretation of Wilson polynomials. I will now summarize the results of [28, §5], where this interpretation was given.

Let $\mathrm{Hyp}(p,q) := \{(x,y)\in\mathbb{R}^p\times\mathbb{R}^q \mid -|x|^2+|y|^2=1\}$, an hyperboloid, and denote by $\mathrm{harm}(\lambda;p,q)$ the class of *hyperboloid harmonics* of degree $i\lambda-\frac{1}{2}(p+q)+1$, i.e. the class of functions on $\mathrm{Hyp}(p,q)$ which are restrictions of C^{∞}-functions on $\{(x,y)\in\mathbb{R}^p\times\mathbb{R}^q \mid -|x|^2+|y|^2>0\}$ that are even, homogeneous of degree $i\lambda-\frac{1}{2}(p+q)+1$ and annihilated by the differential operator

$$\frac{\partial^2}{\partial x_1^2}+\cdots+\frac{\partial^2}{\partial x_p^2}-\frac{\partial^2}{\partial y_1^2}-\cdots-\frac{\partial^2}{\partial y_q^2}.$$

Let $O(p,q)$ be the group of linear transformations of $\mathbb{R}^{p+q}$ which leave the form $-|x|^2+|y|^2$ on $\mathbb{R}^p\times\mathbb{R}^q$ invariant. Then $O(p,q)$ naturally acts on $\mathrm{harm}(\lambda;p,q)$. If $\lambda>0$ then we can associate an irreducible unitary representation $\tau_{\lambda}^{p,q}$ of $O(p,q)$ with this action (cf. Faraut [17]).

Consider in this way the representation $\tau_\lambda^{p+q,r}$ of $O(p+q,r)$ on $\mathrm{harm}(\lambda;p+q,r)$. By formal analytic continuation of Table 15 one is led to three remarkable subgroup reductions of this representation. See Table 16 below.

$$O(p+q,r),\tau_\lambda^{p+q,r}$$

$$O(p+q)\times O(r) \qquad O(q,r)\times O(p) \qquad O(p,r)\times O(q)$$
$$\pi_{2m}^{p+q}\otimes\pi_0^r \qquad \tau_\mu^{q,r}\otimes\pi_0^p \qquad \tau_\gamma^{p,r}\otimes\pi_0^q$$

$$O(p)\times O(q)\times O(r),\pi_0^p\otimes\pi_0^q\otimes\pi_0^r$$

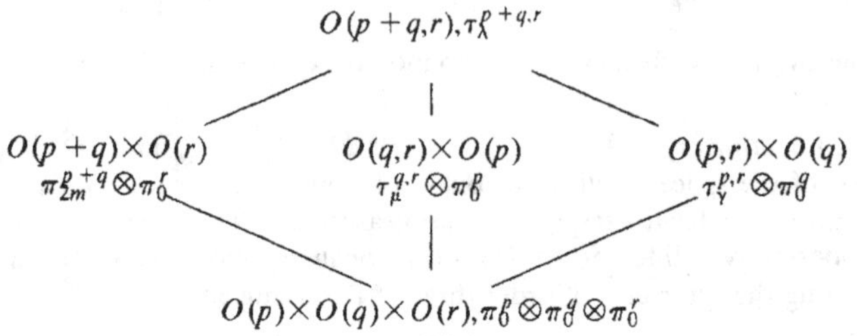

Table 16

For simplicity we have restricted ourselves to the case that the representation of the subgroup $O(p)\times O(q)\times O(r)$ at the bottom of Table 16 is trivial. (The representations π in Table 16 are representations on spaces of spherical harmonics as in §8.)

Now fix $\lambda>0$ and p,q,r in Table 16. The three subgroup chains determine three different (generalized) orthonormal bases for the $O(p)\times O(q)\times O(r)$-invariant elements in $\mathrm{harm}(\lambda;p+q,r)$. These basis elements have explicit expressions which are analytic continuations of (8.8),(8.9) (with the Y-factors trivial). In fact, the basis functions $\phi_{\lambda,m}$ $(m\in\mathbb{Z}_+)$ with respect to the first subgroup chain and $\psi_{\lambda,\mu}$ $(\mu\in\mathbb{R}_+)$ with respect to the second subgroup chain are given,up to a constant factor, by

$$\phi_{\lambda,m}(u,v) := \phi_\lambda^{(\frac12(p+q)+2m-1,\frac12 r-1)}(\mathrm{arcsh}\,(u+v)^{\frac12})(u+v)^m\, P_m^{(\frac12 p-1,\frac12 q-1)}(\frac{-u+v}{u+v}), \tag{9.1}$$

$$\psi_{\lambda,\mu}(u,v) := \phi_\lambda^{(\frac12 p-1,i\mu)}(\mathrm{arcsh}\,u^{\frac12})(u+1)^{\frac12 i\mu-\frac14(q+r)+\frac12}$$
$$\times \phi_\mu(\frac12 q-1,\frac12 r-1)(\mathrm{arcsh}\,\frac{v^{\frac12}}{(u+1)^{\frac12}}), \tag{9.2}$$

where $(x,y,z)\in\mathrm{Hyp}(p+q,r)\subset\mathbb{R}^p\times\mathbb{R}^q\times\mathbb{R}^r$ and $u:=|x|^2$, $v:=|y|^2$. The functions $P_n^{(\alpha,\beta)}$ and $\phi_\lambda^{(\alpha,\beta)}$ are Jacobi polynomials (4.6) and Jacobi functions (4.7), respectively.

The systems $\{\phi_{\lambda,m}\}_{\lambda\in\mathbb{R}_+,m\in\mathbb{Z}_+}$ and $\{\psi_{\lambda,\mu}\}_{\lambda,\mu\in\mathbb{R}_+}$ can be viewed as generalized orthogonal bases for L^2 on $(\mathbb{R}_+)^2$ with respect to the measure

$$u^{\frac12 p-1}\,v^{\frac12 q-1}\,(1+u+v)^{\frac12 r-1}\,du\,dv.$$

They are built as a kind of semidirect products from two orthogonal systems in one variable, just as the orthogonal polynomials (8.10),(8.11) on the triangle are built from Jacobi polynomials. This reduction to the one-variable case also makes it easy to invert the integral transform with (9.1),(9.2) as kernel. In (9.2) one of the contributing one-variable transforms is the Jacobi function transform (4.12) with β imaginary. This case of the Jacobi function transform seemed to be unobserved before it occurred here.

The overlap coefficients between the systems in (9.1) and (9.2) can be given by Wilson polynomials:

$$\psi_{\lambda,\mu} = \sum_{m=0}^{\infty} \text{elementary factors}$$
$$\times W_m(\tfrac14\mu^2\,;\frac{p+2i\lambda}{4},\frac{p-2i\lambda}{4},\frac{q+r-2}{4},\frac{q-r+2}{4})\phi_{\lambda,m}, \tag{9.3}$$

with weak convergence, in analogy with (8.12) and (8.13). Essentially, the proof is also similar to that of (8.13), but much more complicated. Instead of restricting the arguments in (9.3) to one of the boundary lines of $(\mathbb{R}_+)^2$, one has to restrict arguments to a boundary line of $(\mathbb{R}_+)^2$ at infinity, i.e., one has to study the asymptotics of (9.3) for (u,v) large. In this way the identity (4.16)=(4.17) can be interpreted as an asymptotic boundary case of (9.3) and the proof of (9.3) can be reduced to this identity.

Remarks.

1. The case $p=q=1$ of (9.3) and its group theoretic interpretation were earlier obtained by Boyer and Ardalan [11].

2. One remarkable difference of (9.3) with (8.12) is that (8.12) restricted to $u=0$ does not have an interpretation as a unitary transform between two orthogonal bases for L^2-spaces.

3. There is a limit case of (9.100, (9.2), (9.3) in which $\psi_{\lambda,\mu}$ tends to a product of two Laguerre functions (cf. (7.21)) and $\phi_{\lambda,m}$ to a product of a Laguerre function and a Jacobi polynomial. The Wilson polynomial in (9.3) tends to a continuous Hahn polynomial. See Suslov [33].

4. There is a challenging problem left by Table 16: What are the overlap coefficients relating the second and the third subgroup chain in Table 16 to each other? They might be naturally called Wilson functions, but what are they like? Maybe of ${}_4F_3$-type, but probably of an even higher hypergeometric type. One may expect that everything in this paper can be obtained as a limit case of these yet virtual Wilson functions.

References

1. G.E. Andrews and R. Askey, "Classical orthogonal polynomials," pp. 36-62 in *Polynômes Orthogonaux et Applications*, ed. C. Brezinski, A. Draux, A.P. Magnus, P. Maroni, A. Ronveaux, Lecture Notes in Math. 1171, Springer (1985).

2. R. Askey, "Continuous Hahn polynomials," *J. Phys. A: Math. Gen.* **18**, pp. L1017-L1019 (1985).

3. R. Askey and J. Wilson, "A set of hypergeometric orthogonal polynomials," *SIAM J. Math. Anal.* **13**, pp. 651-655 (1982).

4. R. Askey and J. Wilson, *Some basic hypergeometric orthogonal polynomials that generalize Jacobi polynomials,* Memoirs Amer. Math. Soc. 319 (1985).

5. N.M. Atakishiyev and S.K. Suslov, "The Hahn and Meixner polynomials of an imaginary argument and some of their applications," *J.Phys. A: Math. Gen.* **18**, pp. 1583-1596 (1985).

6,7,8. E. Badertscher and T.H. Koornwinder, "Continuous Hahn polynomials of argument $i\,d/dt$ and analysis on Riemannian symmetric spaces of constant curvature," CWI preprint, to appear.

9. D. Basu and K.B. Wolf, "The unitary irreducible representations of SL(2,R) in all subgroup reductions," *J. Math. Phys.* **23**, pp. 189-205 (1982).

10. L.C. Biedenharn and J.D. Louck, *Angular momentum in quantum physics,* Encyclopedia of Math., Vol. 8, Addison-Wesley (1981).

11. C.P. Boyer and F. Ardalan, "On the decomposition $SO(p,1)\supset SO(p-1,1)$ for most degenerate representations," *J. Math. Phys.* **12**, pp. 2070-2075 (1971).

12. C.F. Dunkl, "A difference equation and Hahn polynomials in two variables," *Pac. J. Math.* **92**, pp. 57-71 (1981).

13. C.F. Dunkl, "Orthogonal polynomials with symmetry of order three," *Can. J. Math.* **36**, pp. 685-717 (1984).

14. A. Erdélyi, W. Magnus, F. Oberhettinger, and F.G. Tricomi, *Higher transcendental functions, Vols. I,II,III,* McGraw-Hill (1953,1953,1955).

15,16. A. Erdélyi, W. Magnus, F. Oberhettinger, and F.G. Tricomi, *Tables of integral transforms, Vols. I,II,* McGraw-Hill (1954).

17. J. Faraut, "Distributions sphériques sur les espaces hyperboliques," *J. Math. Pures Appl.* **58**, pp. 369-444 (1979).

18. J. Faraut, "Analyse harmonique sur les paires de Guelfand et les espaces hyperboliques," pp. 315-446 in *Analyse harmonique*, CIMPA, Nice (1982).

19. M. Flensted-Jensen, "Spherical functions on a simply connected semisimple Lie group. II. The Paley-Wiener theorem for the rank one case," *Math. Ann.* **228**, pp. 65-92 (1977).

20. S. Helgason, "A duality for symmetric spaces with applications to group representations, II. Differential equations and eigenspace representations," *Advances in Math.* **22**, pp. 187-219 (1976).

21. S. Helgason, *Groups and geometric analysis*, Academic Press (1984).

22. S. Karlin and J.L. McGregor, "The Hahn polynomials, formulas and an application," *Scripta Math.* **26**, pp. 33-46 (1963).

23. T.H. Koornwinder, "The addition formula for Jacobi polynomials and spherical harmonics," *SIAM J. Appl. Math.* **25**, pp. 236-246 (1973).

24. T.H. Koornwinder, "A new proof of a Paley-Wiener type theorem for the Jacobi transform," *Ark. Mat.* **13**, pp. 145-159 (1975).

25. T.H. Koornwinder, "Clebsch-Gordan coefficients for $SU(2)$ and Hahn polynomials," *Nieuw Archief voor Wiskunde (3)* **29**, pp. 140-155 (1981).

26. T.H. Koornwinder, "Krawtchouk polynomials, a unification of two different group theoretic interpretations," *SIAM J. Math. Anal.* **13**, pp. 1011-1023 (1982).

27. T.H. Koornwinder, "Jacobi functions and analysis on noncompact semisimple Lie groups," pp. 1-85 in *Special functions: Group theoretical aspects and applications*, (R.A. Askey, T.H. Koornwinder, W. Schempp, eds.), Reidel (1984).

28. T.H. Koornwinder, "A group theoretic interpretation of Wilson polynomials," Report PM-R8504, Centre for Math. and Computer Science, Amsterdam (1985).

29. T.H. Koornwinder, "Special orthogonal polynomial systems mapped onto each other by the Fourier-Jacobi transform," pp. 174-183 in *Polynômes Orthogonaux et Applications*, (C. Brezinski, A. Draux, A.P. Magnus, P. Maroni, A. Ronveaux, eds.), Lecture Notes in Math. 1171, Springer (1985).

30. J. Labelle, "Tableau d'Askey," pp. xxxvi-xxxvii in *Polynômes Orthogonaux et Applications*, ed. C. Brezinski, A. Draux, A.P. Magnus, P. Maroni, A. Ronveaux, Lecture Notes in Math. 1171, Springer (1985).

31. A.F. Nikiforov, S.K. Suslov, and V.B. Uvarov, *Classical orthogonal poynomials of a discrete variable (Russian)*, Nauka, Moscow (1985).

32. S.N.M. Ruijsenaars, "Complete integrability of relativistic Calogero-Moser systems and elliptic function identities," preprint (1986).

33. S.K. Suslov, "The Hahn polynomials in the Coulomb problem (Russian)," *Yadernaya Fiz.* **40, no.1**, pp. 126-132 (1984).

34. G. Szegö, *Orthogonal polynomials*, American Math. Society Colloquium Publications, Vol. 23, Fourth ed. (1975).

35. K. Trimèche, "Transformation intégrale de Weyl et théorème de Paley-Wiener associés à un opérateur différentiel singulier sur $(0,\infty)$," *J. Math. Pures Appl.* **(9) 60**, pp. 51-98 (1981).

36. N.Ja. Vilenkin, *Special functions and the theory of group representations*, Amer. Math. Soc. Transl. of Math. Monographs, Vol. 22 (1968).

37. J.A. Wilson, "Some hypergeometric orthogonal polynomials," *SIAM J. Math. Anal.* **11**, pp. 690-701 (1980).

A REVIEW OF ORTHOGONAL POLYNOMIALS SATISFYING BOUNDARY VALUE PROBLEMS

Allan M. Krall

ABSTRACT. There are four sets of orthogonal polynomials satisfying differential equations of second order: The Jacobi, Laguerre, Hermite and Bessel polynomials. The first three are classical, with well known properties, including weights, orthogonality, moments. The fourth is less well known. A real weight has not been found.

All, however, are orthogonal with respect to a distributional weight

$$w = \sum_{n=0}^{\infty} (-1)^n \mu_n \delta^{(n)}(x)/n!$$

where μ_n is the nth moment associated with the polynomials and $\delta^{(n)}(x)$ is the nth distributional derivative of the Dirac delta function. The first two are even distributionally orthogonal when their parameters A, B and α are less than -1.

The first three satisfy a singular Sturm-Liouville problem, complete with singular boundary conditions involving Wronskians.

The first, second and fourth sometimes generate indefinite inner product spaces.

There are three new sets of orthogonal polynomials which satisfy singular Sturm-Liouville problems of fourth order. There are two known satisfying problems of sixth order, one known satisfying a problem of eighth order. They also have some interesting properties, including a distributional weight, the generation of indefinite inner product spaces, satisfying differential equations of second order in which the coefficients vary with n.

I. INTRODUCTION. This paper is divided into two sections. The first is devoted to a discussion of orthogonal polynomials satisfying a differential equation of second order. Three of the four polynomial sets to be discussed have been known for over a century, and the fourth has been encountered various times within the past fifty years. As a result we shall focus our attention on recent, new results concerning them, including a new generalized kind of orthogonality and a new description of the boundary value problems the first three satisfy.

The second section is devoted to a discussion of orthogonal polynomials satisfying differential equations of fourth, sixth and eighth order. These have been studied seriously only in the past five years, and are not well known. In addition to the usual properties associated with orthogonal polynomials, they satisfy some very unusual boundary value problems.

Since the proofs of results way be found elsewhere, they will be omitted here in the interests of clarity and brevity. We hope this is not an imposition.

ORTHOGONAL POLYNOMIALS SATISFYING DIFFERENTIAL EQUATIONS OF SECOND ORDER.

II. FUNDAMENTALS. Orthogonal polynomials can conveniently be introduced through the use of moments. Let $\{\mu_n\}_{n=0}^{\infty}$ be an infinite collection of real numbers satisfying

$$(1) \qquad \Delta_n = \begin{vmatrix} \mu_0 & \cdots & \mu_n \\ \vdots & & \vdots \\ \mu_n & \cdots & \mu_{2n} \end{vmatrix} \neq 0 \quad , \quad n=0,1,\ldots .$$

Then the "Tchebycheff" polynomials $\{p_n\}_{n=0}^{\infty}$ are defined by setting $p_0 = 1$,

$$(2) \qquad p_n(x) = \begin{vmatrix} \mu_0 & \cdots & \mu_n \\ \vdots & & \vdots \\ \mu_{n-1} & \cdots & \mu_{2n-1} \\ 1 & \cdots & x^n \end{vmatrix} / \Delta_{n-1}, \quad n=1,2,\ldots .$$

These polynomials are orthogonal with respect to the distributional weight function [15]

$$(3) \qquad w = \sum_{n=0}^{\infty} (-1)^n \mu_n \delta^{(n)}(x)/n!$$

That is,

$$(4) \qquad \begin{aligned} \langle w, p_n p_m \rangle &= 0 \quad , \quad n \neq m \\ &= \Delta_n/\Delta_{n-1} \quad , \quad n = m. \end{aligned}$$

They satisfy a three term recurrence relation

$$(5) \qquad p_{n+1} = (x+B_n)p_n - C_n p_{n-1},$$

where, if

$$(6) \qquad p_n = x^n - S_n x^{n-1} + \ldots,$$

$B_n = S_n - S_{n+1}$ and $C_n = \Delta_n \Delta_{n-2}/\Delta_{n-1}^2$.

If such a collection $\{p_n\}_{n=0}^{\infty}$ satisfies a differential equation of the form

(7) $$P\,y'' + Q\,y' + Ry = \lambda_n Wy$$

where P,Q,R and W are independent of n, it is easy to show that the differential equation can be rewritten as

(8) $$(\ell_{22}x^2+\ell_{21}x+\ell_{20})y'' + (\ell_{11}x+\ell_{10})y' = \lambda_n y,$$

where

(9) $$\lambda_n = \ell_{11}n + \ell_{22}n(n-1).$$

II.1. THEOREM. *If the collection* $\{p_n\}_{n=0}^{\infty}$ *satisfies* (8), (9), *then* w, *given by* (3) *satisfies*

(10) $$-((\ell_{22}x^2+\ell_{21}x+\ell_{20})w)' + (\ell_{11}x+\ell_{10})w = 0,$$

Further, we find

(11) $$wLy = w[(\ell_{22}x^2+\ell_{21}x+\ell_{20})y'' + (\ell_{21}x+\ell_{10})y']$$

is formally symmetric over polynomials, and the moments $\{\mu_n\}_{n=0}^{\infty}$ *satisfy*

(12) $$\ell_{11}\mu_n + \ell_{10}\mu_{n-1} + (n-1)[\ell_{22}\mu_n+\ell_{21}\mu_{n-1}+\ell_{20}\mu_{n-2}] = 0, \qquad n = 1,2,...$$

Conversely, if the moments $\{\mu_n\}_{n=0}^{\infty}$ *satisfy* (12) *then* (11) *is symmetric, and* p_n *satisfies* (8), (9), $n = 0,1,\dots$. [9].

There are four situations which can arise. The coefficient $\ell_{22}x^2+\ell_{21}x+\ell_{20}$ can have two different, one, no zeros, or a double zero can occur. We shall discuss each in turn [3], [15].

III. THE JACOBI POLYNOMIALS. If the second derivative coefficient has two different zeros, then by an appropriate linear transformation (8) can be put in the form

(13) $$(1-x^2)y'' + (B-A-[2+A+B]x)y' + n(n+A+B+1)y = 0,$$

the Jacobi equation. The moments satisfy

(14) $$(A+B+n+1)\mu_n + (A-B)\mu_{n-1} - (n-1)\mu_{n-2} = 0,$$

and are

$$\mu_n = \sum_{j=0}^{n} \binom{n}{j} (-1)^j 2^j (A+1)_j/(A+B+2)_j$$
$$= \sum_{j=0}^{n} \binom{n}{j} (-1)^{n-j} 2^j (B+1)_j/(A+B+2)_j ,$$

$n = 0,1,\ldots,$ where $(a)_j = a(a+1)\ldots(a+j-1)$.

The weight can be represented by (3) or the weight differential equation (10) can be solved. In this instance (10) is

$$(1-x^2)w' - ([B-A]-[A+B]x)w = 0. \tag{16}$$

When $A > -1$ and $B > -1$, (16) has the distributional solution, vanishing at both $\pm\infty$,

$$w = (1-x)^A(1+x)^B[H(x+1)-H(x-1)], \tag{17}$$

where $H(x)$ is the Heaviside function

$$H(x) = \begin{cases} 1, & x \geq 0 \\ 0, & x < 0. \end{cases} \tag{18}$$

When either $A < -1$ or $B < -1$ or both, a Cauchy regularization of (17) is required. If $-N-1 < A < -N$, $-M-1 < B < -M$ and $A+B+1$ is also not a negative integer, then

$$\begin{aligned} \langle w,\phi\rangle = \left[\frac{\Gamma(A+B+2)}{\Gamma(A+1)\Gamma(B+1)2^{A+B+1}}\right] \Bigg[\int_0^1 (1-x)^A\Big\{(1+x)^B\phi(x) \; - \\ - \sum_{j=0}^{N-1} \frac{[(1+x)^B\phi(x)]^{(j)}}{j!}\Bigg|_{x=1} (-1)^j(1-x)^j\Big\}dx \\ + \int_{-1}^{0} (1+x)^B\Big\{(1-x)^A\phi(x) \; - \\ - \sum_{k=0}^{M-1} \frac{[(1-x)^A\phi(x)]^{(k)}}{k!}\Bigg|_{x=-1} (1+x)^k\Big\}dx \\ + \sum_{k=0}^{N-1} \frac{[(1-x)^B\phi(x)]^{(k)}}{j!}\Bigg|_{x=1} \frac{(-1)^j}{(A+1+j)} \\ + \sum_{k=0}^{M-1} \frac{[(1-x)^A\phi(x)]^{(k)}}{k!}\Bigg|_{x=-1} \frac{1}{(B+1+k)}\Bigg] . \end{aligned} \tag{19}$$

Formula (19) generates an indefinite inner product space through the inner product

$$(f,g) = \langle w, f\bar{g}\rangle . \tag{20}$$

In particular

(21) $$\langle w, P_n^{A,B} P_m^{A,B} \rangle = 0 \quad , \quad n \neq m,$$
$$= \frac{\Gamma(A+B+2)\Gamma(A+n+1)\Gamma(B+n+1)}{\Gamma(A+1)\Gamma(B+1)\Gamma(A+B+n+1)\Gamma(A+B+2n+1)n!} \;, \quad n = m.$$

When $A = B = -\frac{1}{2}$, the Tchebycheff polynomials of the first kind are generated. When $A = B = \frac{1}{2}$, the Tchebycheff polynomials of the second kind are found. When $A = B = 0$, the Legendre polynomials are discovered. In the Legendre case the weight

(22) $$w = \sum_{n=0}^{\infty} \frac{2}{2(n+1)!}\, \delta^{(2n)}(x)$$

has the same Fourier transform, $\sin t/\pi t$, as the classical weight

(23) $$w = [H(x+1)-H(x-1)].$$

They are in some sense the same.

IV. THE LAGUERRE POLYNOMIALS. If the second derivative coefficient has a single zero, it can be put at 0. The resulting equation is

(24) $$xy'' + (\alpha+1-x)y' + ny = 0,$$

the Laguerre differential equation. The moments satisfy

(25) $$\mu_n - (n+\alpha)\mu_{n-1} = 0,$$

and are

(26) $$\mu_n = \Gamma(\alpha+n+1)/\Gamma(\alpha+1) = (\alpha+1)_n \quad , \quad n = 0,1,\ldots\,.$$

The weight can be represented by

(27) $$w = \sum_{n=0}^{\infty} \frac{(-1)^n\Gamma(\alpha+n+1)}{\Gamma(\alpha+1)\; n!}\, \delta^{(n)}(x)$$
$$= \sum_{n=0}^{\infty} \frac{(-1)^n(\alpha+1)_n}{n!}\, \delta^{(n)}(x),$$

or the weight differential equation

(28) $$xw' + (x-\alpha)w = 0$$

can be solved. If $\alpha > -1$, then

(29) $$w = x^{\alpha}e^{-x}H(x)/\Gamma(\alpha+1).$$

If $\alpha < -1$, then a Cauchy regularization is needed. When $-j-1 < \alpha < -j$,

(30) $$\langle w,\phi\rangle = \frac{(-1)^{\alpha}}{\Gamma(\alpha+j+1)}\int_0^{\infty} x^{\alpha+j}[e^{-x}\phi(x)]^{(j)}dx.$$

As in the Jacobi case, (30) generates an indefinite inner product space through the inner product

(31) $$(f,g) = \langle w,f\bar{g}\rangle.$$

since

(32) $$\begin{aligned}\langle w,L_n^{\alpha}L_m^{\alpha}\rangle &= 0 , & n \neq m,\\ &= n!(\alpha+1)_n, & n = m,\end{aligned}$$

the signs of $\langle w,{L_n^{\alpha}}^2\rangle$ alternate up to $n = j$ beginning with (-), ending with $(-)^j$. From $n = j+1,...$ they all have the same sign as $(-)^j$. Arbitrary elements in the indefinite space can be expanded as a series of Laguerre polynomials, just as in the classical situations $\alpha > -1$.

It is interesting to note that when $\alpha = 0$, the weight

(33) $$w = \sum_{n=0}^{\infty} (-1)^n\delta^{(n)}(x)$$

serves as a weight for the ordinary Laguerre polynomials.

V. THE HERMITE POLYNOMIALS. If the coefficient of the second derivative has no zeros, the differential equation can be written as

(34) $$y'' - 2xy' + 2ny = 0,$$

the Hermite differential equation. The moments satisfy

(35) $$2\mu_n - (n-1)\mu_{n-2} = 0,$$

and are

(36) $$\begin{aligned}\mu_{2n} &= \sqrt{\pi}\ (2n)!/4^n n!,\\ \mu_{2n+1} &= 0,\end{aligned}$$

$n = 0,1,...$.

The weight function can be expressed by

$$w = \sum_{n=0}^{\infty} \sqrt{\pi}\, \delta^{(2n)}(x)/4^n n! \tag{37}$$

Or the weight differential equation

$$w' + 2xw = 0 \tag{38}$$

can be solved to find

$$w = e^{-x^2} \quad , \quad -\infty < x < \infty . \tag{39}$$

VI. THE BESSEL POLYNOMIALS. If the coefficient of the second derivative has a double zero, the differential equation can be expressed as

$$x^2y'' + (ax+b)y' - n(n+a-1)y = 0. \tag{40}$$

If $b = 0$, this is the Euler-Cauchy equation. The polynomial solutions $\{x^n\}_{n=0}^{\infty}$ cannot be orthogonal. If $b \neq 0$, the moments satisfy

$$(n+a-1)\mu_n + b\,\mu_{n-1} = 0 \tag{41}$$

and are

$$\mu_n = (-b)^{n=1}/(a)_n \quad , \quad n=0,1,\ldots .$$

The weight may be expressed by

$$w = -\sum_{n=0}^{\infty} \frac{b^{n+1}}{(a)_n\, n!}\, \delta^{(n)}(x), \tag{42}$$

which works in a combinatoric sense.

If the weight equation

$$x^2w' + [(2-a)x-b]w = 0 \tag{43}$$

is solved, the solutions (42) and

$$w = x^{a-2}e^{-b/x} \tag{44}$$

are all that have been found. (44) serves as a complex weight, but not as a real weight. So far no real weight has been found.

In addition [7], the distribution

(45) $$\langle w,\phi\rangle = \lim_{\varepsilon\to 0} -\frac{1}{\pi}\int_{\alpha}^{\beta}\phi(x)\mathrm{Im}[(-b/z)\,{}_1F_1(1;a;-b/z)]dx$$

is another variant of (42) when $\alpha < 0 < \beta$, $z = x+i\varepsilon$. The relation between (42)-(45) and (44), if any, has not been established, however, and the problem of finding a classical weight remains open.

VII. BOUNDARY VALUE PROBLEMS. The first three sets of orthogonal polynomials really satisfy more than differential equations. They are the eigenfunctions for three boundary value problems. since these (singular) problems are not well understood, let us deviate briefly to discuss their essential nature [9].

We consider the differential expression

(46) $$\ell y = [(py')'+qy]/w$$

on $L^2(a,b;w)$ under the assumptions that on (a,b), $p,q,w,1/p$ are continuous and $w > 0$. (46) then generates a minimal operator, a maximal operator, which are Hilbert space adjoints of each other, as well as many others, which are extensions of the minimal operator, restrictions of the maximal opertor. The adjoints of these operators also are extensions of the minimal operator, restrictions of the maximal operator. Occasionally an operator and its adjoint will coincide, and the operator is self-adjoint. We need to know how to describe them, since the classic orthogonal polynomials are the eigenfunctions of self-adjoint operators.

The following, due to Herman Weyl [16], is the basis for discussing singular problems.

VII.1. THEOREM. *Let* $\lambda = \mu+i\nu$, $\nu \neq 0$, *and let* $c \in (a,b)$. *Then there exist independent solutions* u *and* v *of*

(47) $$(py')' + qy = \lambda wy$$

such that $u \in L^2(c,b;w)$ *and* $v \in L^2(a,c;w)$.

There are two situations that can arise at each end. First, u (v) may be the only solution in the L^2 space. If this occurs for a particular choice of λ, then u (v) is the only L^2 solution for all $\nu \neq 0$. Further u (v) is analytic in λ.

Or, second, for *all* λ every solution of (47) may be in the L^2 space.

For technical reasons the first case is Weyl's limit point case. The second is Weyl's limit circle case. Hence there are four situations.

1. a is limit point. b is limit point.
2. a is limit circle. b is limit point.
3. a is limit point. b is limit circle.
4. a is limit circle. b is limit circle.

Each situation must be handled a bit differently.

Boundary conditions are somewhat more complicated than for regular problems. If y is in $L^2(a,b;w)$ and ℓy is also in $L^2(a,b;w)$ and $u \in L^2(c,b;w)$, $v \in L^2(a,c;w)$, then

VII.2. THEOREM.

(48) $$B_u(y) = \lim_{x\to b} p(x)[y(x)u'(x)-y'(x)u(x)]$$

and

(49) $$B_v(y) = \lim_{x\to a} p(x)[y(x)v'(x)-y'(x)v(x)]$$

exist, and are boundary values in the sense of Dunford and Schwartz.

The proof is essentially through Green's formula.

VII.3. THEOREM. *If* b *is in the limit point case, then* $B_u(y) = 0$. *If* b *is in the limit circle case and* $B_u(y) = 0$ *for* $\lambda = \lambda_0$, *then* $B_u(y) = 0$ *for all* λ.

If a *is in the limit point case, then* $B_v(y) = 0$. *If* a *is in the limit circle case and* $B_v(y) = 0$ *for* $\lambda = \lambda_0$, *then* $B_v(y) = 0$ *for all* λ.

VII.4. DEFINITION. *We denote by* D_L *those elements* y *in* $L^2(a,b;w)$ *satisfying*

1. $\ell y \in L^2(a,b;w)$
2. *If* b *is in the limit circle case and* u *is given by Weyl's theorem, then* $B_u(y) = 0$.
3. *If* a *is in the limit circle case and* v *is given by Weyl's theorem, then* $B_v(y) = 0$.

Conditions 2 and 3 are automatic if the limit point case holds.

VII.5 DEFINITION. *We define the operator* L *by setting* $Ly = \ell y$ *for all* y *in* D_L.

VII.6 THEOREM. L *is self-adjoint in* $L^2(a,b;w)$.

As a self-adjoint operator, L possesses a "spectral resolution".

(50) $$Ly = \int_{\Sigma} \lambda dP(\lambda)y$$

where $P(\lambda)$ is a projection valued measure and Σ is the spectrum of L.

Since

$$(51) \qquad Iy = \int_{\Sigma} dP(\lambda)y,$$

(51) is sometimes called a resolution of the identity.

In the case of the three sets of orthogonal polynomials, $P(\lambda)$ is a discrete jumping measure, giveing what is traditionally called an eigenfunction expansion when (51) is written explicitly as a series.

We close the first section by giving explicity the boundary value problems satisfied by the Jacobi, Laguerre and Hermite polynomials.

VIII. THE JACOBI BOUNDARY VALUE PROBLEM. The Jacobi operator is in the limit circle case at 1, if $-1 < A < 1$. It is in the limit circle case at -1, if $-1 < B < 1$. For $A \geq -1$, 1 is limit point. For $B \geq -1$, -1 is limit point. If either is ≤ -1, then the boundary value problem no longer has a discrete spectrum with Jacobi polynomial eigenfunctions. Instead the polynomials lie in an indefinite inner product space.

For $A > -1$, $B > -1$, the Jacobi boundary value problem is

$$(52) \quad Ly = (1-x)^{-A}(1+x)^{-B}((1-x)^{1+A}(1+x)^{1+B}y')' = -n(n+A+B+1)y, \qquad -1 \leq x \leq 1,$$

$$(53) \qquad B_u(y) = \lim_{x\to 1} (1-x)^{1+A}(1+x)^{1+B}y'(x) = 0 \quad , \quad -1 < A < 1,$$

$$(54) \qquad B_v(y) = \lim_{x\to -1} (1-x)^{1+A}(1+x)^{1+B}y'(x) = 0 \quad , \quad -1 < B < 1.$$

The special case of $A = B = 0$ is of more general interest. The Legendre boundary value problem is

$$(55) \qquad Ly = ((1-x^2)y')' = -n(n+1)y, \qquad -1 \leq x \leq 1,$$

$$(56) \qquad B_u(y) = \lim_{x\to 1} (1-x^2)y'(x) = 0,$$

$$(57) \qquad B_v(y) = \lim_{x\to -1} (1-x^2)y'(x) = 0.$$

In the past most people have required that $y(x)$ remain bounded as $x \to \pm 1$ instead of (56) and (57). Such a constraint is, in fact, equivalent to (56) and (57) in this case. It is not always equivalent.

IX. THE LAGUERRE BOUNDARY VALUE PROBLEM. The Laguerre operator is limit point at ∞, and so no boundary condition is required there. It is automatic. At 0 it is limit circle if $-1 < \alpha < 1$, limit point if $\alpha \geq 1$. If $\alpha < -1$, the spectral resolution does not involve the polynomials as eigenfunctions. The polynomials instead lie in an indefinite inner product space.

For $\alpha > -1$, the Laguerre boundary value problem is

(58) $$Ly = x^{-\alpha}e^{x}(x^{\alpha+1}e^{-x}y')' = -ny, \quad 0 \le x < \infty,$$

(59) $$B_{v}(y) = \lim_{x\to 0} x^{\alpha+1}e^{-x}y'(x) = 0.$$

When $\alpha = 0$, the ordinary Laguerre problem is found.

X. THE HERMITE BOUNDARY VALUE PROBLEM. Both $\pm\infty$ are in the limit point case for the Hermite operator. Hence no boundary conditions must be explicitly required. They are automatic. The Hermite boundary value problem is

(60) $$Ly = e^{x^2}(e^{-x^2}y')' = -2ny, \quad -\infty < x < \infty.$$

XI. THE SPECTRAL RESOLUTIONS. In all three of the problems just mentioned, the spectrum of L is discrete with eigenfunctions $\{f_n\}_{n=0}^{\infty}$ the various normalized orthogonal polynomials, and eigenvalues $\{\lambda_n\}_{n=0}^{\infty}$ the various coefficients of y in (52), (58) and (60). L can be expressed by

(61) $$Ly = \sum_{n=0}^{\infty} \lambda_n(y,f_n)f_n.$$

The domain of L consists of all y such that

(62) $$\|Ly\| = \sum_{n=0}^{\infty} \lambda_n^2|(y,f_n)|^2 < \infty.$$

The identitiy can be expressed by

(63) $$Iy = \sum_{n=0}^{\infty} (y,f_n)f_n.$$

(64) $$\|y\|^2 = \sum_{n=0}^{\infty} |(y,f_n)|^2$$

is Parseval's equality. The spectral integrals are infinite series.

ORTHOGONAL POLYNOMIALS SATISFYING DIFFERENTIAL EQUATIONS OF FOURTH, SIXTH AND EIGHTH ORDER.

XII. A ROAD MAP FOR HIGHER ORDER PROBLEMS. In addition to the four sets of orthogonal polynomials mentioned in part 1, there are three new sets of orthogonal polynomials satisfying fourth order differential equations. In addition to these seven there are at least two new sets satisfying sixth order equations. In addition to these nine, there is at least one new set satisflying an eighth order equation.

These sets of orthogonal polynomials are all intimately related. The following charts illustrate the relationships.

2nd Order:	Jacobi	(Legendre)	Laguerre
	↓	↓	↓
4th Order:	Jacobi type	Legendre type	Laguerre Type
		↓	↓
6th Order:		H.L. Krall	Littlejohn
			↓
8th Order:			Littlejohn II

WEIGHTS

Jacobi	⟶	$(1-x)^A(x)^B$,	$0 \le x \le 1$
Jacobi Type	⟶	$(1-x)^\alpha + \frac{1}{M}\,\delta(x)$,	$0 \le x \le 1$
Legendre	⟶	1 ,	$-1 \le x \le 1$
Legendre Type	⟶	$\frac{\alpha}{2} + \frac{1}{2}[\delta(x-1)+\delta(x+1)]$,	$-1 \le x \le 1$
H.L. Krall	⟶	$C + \frac{1}{A}\,\delta(x+1) + \frac{1}{B}\,\delta(x-1)$,	$-1 \le x \le 1$
Koornwinder	⟶	$(1-x)^A(1+x)^B+M\delta(x-1)+N\delta(x+1)$,	$-1 \le x \le 1$
Laguerre	⟶	$x^\alpha e^{-x}$,	$0 \le x < \infty$
Laguerre Type	⟶	$e^{-x} + \frac{1}{R}\,\delta(x)$,	$0 \le x < \infty$
Littlejohn	⟶	$xe^{-x} + \frac{1}{R}\,\delta(x)$,	$0 \le x < \infty$
Littlejohn II	⟶	$x^2e^{-x} + \frac{1}{R}\,\delta(x)$,	$0 \le x < \infty$
xxx	⟶	$x^\alpha e^{-x} + \frac{1}{R}\,\delta(x)$,	$0 \le x < \infty$.

As in the second order case, orthogonal polynomials satisfy an ordinary differential equation of order 4, 6, 8 (or $2n$) if and only if the moments satisfy 2, 3, 4 (or n) recurrence relations. The weights satisfy 2, 3, 4 (or n) differential equations. Rather than burden the reader with their general forms (see [10]), we shall give specific examples.

Further since the problems satisfying sixth order or eighth order equations shed little real light and only complexity of detail, we shall devote our attention to those sets of orthogonal polynomials satisfying fourth order equations.

We remark that the Jacobi, Laguerre, Hermite and Bessel polynomials also satisfy fourth, sixth, ..., order equations, as can be seen by iterating the second order equations. The moments and weights remain unchanged, however. We cite [10] as a reference.

XIII. THE LEGENDRE-TYPE POLYNOMIALS. The Legendre-type polynomials

$\{P_n^{(\alpha)}(x)\}_{n=0}^{\infty}$ satisfy

(65) $$((x^2-1)^2y'' + 4((\alpha(x^2-1)-2)y')' = \lambda_n y$$

where

(66) $$\lambda_n = [8\alpha n+(4\alpha+12)n(n-1)+8n(n-1(n-2)+n(n-1)(n-2)(n-3)],$$

$n = 0,1,\ldots$. They are

(67) $$P_n^{(\alpha)} = \sum_{k=0}^{[n/2]} \frac{(-1)^k(2n-2k)!(\alpha + \frac{1}{2}n(n-1)+2k)}{2^n\, k!(n-k)!(n-2k)!} x^{n-2k},$$

$n = 0,1,\ldots$.

Dehesa, Buendia and Sanchez-Buendia [1] have shown that if μ'_r is the r-th moment of the zeros of $P_n^{(\alpha)}$ (Let the zeros be $\{x_i\}_{i=1}^{n}$. Then $\mu'_r = \frac{1}{n}\sum_{i=1}^{n} x_i^r$), then

(68a) $$\mu'_{2r+1} = 0,$$

(68b) $$\mu'_{2r} = \frac{1}{2^r}\binom{2r}{r} + 0(n^{-1}),$$

$r = 1,2,\ldots,\ n = 0,1,\ldots$. These have a number of physical interpretations.

The moments $\{\mu_n\}_{n=0}^{\infty}$ satisfy

(69) $$(n+1)\mu_n - 2(n-1)\mu_{n-2} + (n-3)\mu_{n-4} = 0 \quad , \quad n \geq 3,$$

(70) $$(n-1+\alpha)(n+1)\mu_n - (n+1+\alpha)(n-1)\mu_{n-2} = 0 \quad , \quad n \geq 1.$$

They are

(71a) $$\mu_{2n+1} = 0,$$

(71b) $$\mu_{2n} = (\alpha+2n+1)(2n+1),$$

$n = 0,1,\ldots$.

Using the moments,

(72) $$w = \sum_{n=0}^{\infty} \frac{(\alpha+2n+1)}{(2n+1)!}\,\delta^{(2n)}(x)$$

is found to be a weight function. If the weight equations, which (72) satisfies,

$$(x^2-1)^2 w' = 0 \tag{73}$$

$$(x^2-1)^2 w''' + 12x(x^2-1)w'' + [(24-4\alpha)x^2+4\alpha]w' = 0 \tag{74}$$

are solved directly,

$$w = \frac{1}{2}[\delta(x-1)+\delta(x+1)] + \frac{\alpha}{2} \quad , \quad -1 \leq x \leq 1, \tag{75}$$

is found. Both (72) and (75) have the same Fourier transform.

The polynomials satisfy a recurrence relation

$$P_{n+1}^{(\alpha)} = \frac{(2n+1)(\alpha + \frac{1}{2} n(n+1))}{(n+1)(\alpha + \frac{1}{2} n(n-1))} x P_n^{(\alpha)} - \frac{n(\alpha + \frac{1}{2} (n+1)(n+2))}{(n+1)(\alpha + \frac{1}{2} n(n-1))} P_{n-1}^{(\alpha)} , \tag{76}$$

and have a generating function

$$(\alpha - x \frac{\partial}{\partial x} + \frac{1}{2} t \frac{\partial^2}{\partial t^2} t)(1-2xt+t^2)^{-\frac{1}{2}} = \sum_{n=0}^{\infty} P_n^{(\alpha)}(x)t^n. \tag{77}$$

Related to (77) is a connection to the Legendre polynomials

$$\left\{\alpha - x \frac{d}{dx} + \frac{n(n+1)}{2}\right\} P_n(x) = P_n^{(\alpha)}(x). \tag{78}$$

This may be inverted to give

$$P_n(x) = \int_x^{\infty} \frac{\left[x^{\alpha + \frac{1}{2} n(n+1)} P_n^{(\alpha)}(\xi)\right]}{\xi^{\alpha + \frac{1}{2} n(n+1)+1}} d\xi . \tag{79}$$

(78) also yields the Rodrigues formula

$$P_n^{(\alpha)}(x) = (\alpha - xD + \frac{1}{2} n(n+1))D^n(x^2-1)^n/2^n n! \; . \tag{80}$$

Rather interestingly, Koornwinder [2] has shown the following

XIII.1. THEOREM. *Let* $\{u_n\}_{n=0}^{\infty}$ and $\{y_n\}_{n=0}^{\infty}$ *be sets of orthogonal polynomials satisfying*

(81) $$u_n = p_n y_n + q_n y_n',$$

(82) $$y_n'' + \alpha_n y_n' + \beta_n y_n = 0,$$

where $p_n, q_n, \alpha_n, \beta_n$ *are rational functions, quotients of polynomials (in* n*) of bounded degree. Then*

(83) $$y_n = r_n u_n + s_n u_n'$$

(84) $$u_n'' + \gamma_n u_n' + \delta_n u_n = 0$$

where r_n, s_n, γ_n, δ_n *are rational functions, quotients of polynomials (in* n*) of bounded degree.*

This can be used to show that if λ_n is given by (66) and

(85) $$\nu_n = n(n+1)(n^4+2n^3-97n^2-98n+192-372\alpha-12\alpha^2)$$

then

(86) $$\begin{aligned}(x^2-1)[(4\alpha^2+4\alpha+\lambda_n)x^2 - (4\alpha^2-4\alpha+\lambda_n)y'' &+ \\ + 2x[(4\alpha^2+4\alpha+\lambda_n)x^2 - (4\alpha^2-12\alpha+\lambda_n)]y' &- \\ - [(\nu_n+4\alpha+96)x^2 - (\nu_n+4\alpha+96-4\lambda_n)]y &= 0,\end{aligned}$$

$n = 0,1,\dots$.

When $\alpha > 0$, (75) generates a Hilbert space H with inner product

(87) $$(y,z) = \int_{-1}^{1} \bar{z}(x)y(x)\,\frac{\alpha}{2}\,dx + \frac{1}{2}\,\bar{z}_1 y_1 + \frac{1}{2}\,\bar{z}_{-1}y_{-1}.$$

The polynomials $\{P_n^{(\alpha)}\}_{n=0}^{\infty}$ form a complete orthogonal set in H. The norm square of $P_n^{(\alpha)}$ is

(88) $$\|P_n^{(\alpha)}\|^2 = \alpha\left[\alpha + \frac{(n-1)n}{2}\right]\left[\alpha + \frac{(n+1)(n+2)}{2}\right]/(2n+1).$$

When $\alpha < 0$ and is not the negative of a triangular number, (87) generates an indefinite inner product space, a Pontrjagin space, in which $\{P_n^{(\alpha)}\}_{n=0}^{\infty}$ is complete. (88) still holds, and the left of (65) generates a self-adjoint operator [12].

It is easy to see that if

$$\frac{-N(N+1)}{2} < \alpha < -\frac{(N-1)N}{2} \tag{89}$$

then

$$\|P_n^{(\alpha)}\|^2 \begin{cases} < 0 , & n = 0,\dots,N-2 \\ > 0 , & n = N-1,N \\ < 0 , & n = N+1. \end{cases} \tag{90}$$

XIV. THE LAGUERRE-TYPE POLYNOMIALS. The Laguerre-type polynomials $\{R_n\}_{n=0}^{\infty}$ satisfy

$$(x^2e^{-x}y'') - (([2R+2]x+2)e^{-x}y')' = \lambda_n e^{-x}y \tag{91}$$

where

$$\lambda_n = (2R+2)n + n(n-1), \tag{92}$$

$n = 0,1,\dots$. They are

$$R_n = \sum_{k=0}^{n} \frac{(-1)^k}{(k+1)!} \binom{n}{k} [k(R+n+1)+R]x^k, \tag{93}$$

$n=0,1,\dots$.

Dehesa, Buendia and Sanchez-Buendia [1] have shown that the r-th moments of the zeros of R_n are given by

$$\mu_r' = \frac{1}{r+1} \binom{2r}{r} n^r + 0(n^{r-1}), \tag{94}$$

$r = 1,2,\dots, n = 0,1,\dots$.

The moments $\{\mu_n\}_{n=0}^{\infty}$ satisfy

$$\mu_n - (n-1)\mu_{n-2} = 0 \quad , \quad n \geq 3, \tag{95}$$

$$(2R+n+1)\mu_n - (2R+[2R+6][n-1]+[n-1][n-2])\mu_{n-1} + (2[n-1][n-2])\mu_{n-2} = 0 \quad , \quad n \geq 1. \tag{96}$$

They are

$$\mu_0 = (R+1)/R, \tag{97a}$$

(97b) $$\mu_n = n!$$

$n = 1,2,\ldots$.

Using the moments,

(98) $$w = \frac{1}{R}\,\delta(x) + \sum_{n=0}^{\infty} (-1)^n \delta^{(n)}(x)$$

is found to be a weight function. If the weight equations, which (98) satisfies,

(99) $$x^2 w' + x^2 w = 0$$

(100) $$x^2 w''' + 6xw'' + [-x^2+(2R+6)x+6]w' + [2Rx+6]w = 0$$

are solved directly,

(101) $$w = \frac{1}{R}\,\delta(x) + e^{-x} \quad , \quad 0 \le x < \infty$$

is found. (98) and (101) have the same Fourier transform.

The polynomials satisfy a recurrence relation

(102) $$R_n = \frac{-[(R+n)(R+n-1)]x+[(2n-1)R^2+4n(n-1)R+(2n-1)n(n-1)]}{n(R+n-1)^2}\,R_{n-1} - \frac{[(n-1)(R+n)^2]}{n(R+n-1)^2}\,R_{n-2},$$

and have the generating function

(103) $$\frac{R(1-t^2)-xt}{(1-t)^3}\exp\left[\frac{-xt}{1-t}\right] = \sum_{n=0}^{\infty} R_n(x)t^n.$$

Related to (103) is a connection to the Laguerre polynomials

(104) $$\left[\frac{d}{dx} + R + n\right] L_n(x) = R_n(x).$$

This may be inverted to give

$$L_n(x) = e^{-(R+n)x}\int_{-\infty}^{x} e^{(R+n)\xi} R_n(\xi)\,d\xi .$$

(104) also yields the Rodrigues formula

(106) $$R_n(x) = \frac{e^x}{n!}\frac{d^n}{dx^n}\left(e^{-x}[(R+n)x^n+nx^{n-1}]\right)$$

If Koornwinder's theorem is applied to the Laguerre and Laguerre type polynomials, a second order differential equation is found [2]. With λ_n given by (92) and

(107) $$\nu_n = (3R^2+45R+42)n + (8n(n-1) - n(n-1)(n-2),$$

(108) $$[(R^2+R+\lambda_n)x^2-Rx]y'' + [-(R^2+R+\lambda_n)x^2+(R^2+2R+\lambda_n)x-2R]y' + [(2R\lambda_n+22\lambda_n-\nu_n)x-\lambda_n]y = 0.$$

$n = 0,1,\ldots$.

When $R > 0$, (101) generates a Hilbert space H with inner product

(109) $$(y,z) = \int_0^\infty \bar{z}(x)y(x)e^{-x}dx + \frac{1}{R}\,\bar{z}_0y_0.$$

The polynomials $\{R_n\}_{n=0}^\infty$ form a complete orthogonal set in H. The norm square of R_n is

(110) $$\|R_n\|^2 = (R+n+1)(R+n).$$

When $R < 0$ and is not a negative integer, (109) generates an indefinite inner product space, a Pontrjagin space, in which $\{R_n\}_{n=0}^\infty$ is complete. (110) still holds, and the left of (91) generates a self-adjoint operator [13].

It is easy to see that if

(111) $$-N-1 < R < -N,$$

then

(112) $$\|R_n\|^2 \begin{cases} > 0, & n = 0,\ldots,N-1, \\ < 0, & n = N \\ > 0, & n = N+1,\ldots . \end{cases}$$

XV. THE JACOBI-TYPE POLYNOMIALS. The Jacobi-type polynomials $\{S_n\}_{n=0}^\infty$ satisfy

(113) $$([(1-x)^{\alpha+4}-2(1-x)^{\alpha+3}+(1-x)^{a+2}]y'')'' + ([(2\alpha+2+M)(1-x)^{\alpha+2}-(2+4+2M)(1-x)^{\alpha+1}]y')'$$

$$= \lambda_n(1-x)^\alpha y,$$

where

(114) $$\lambda n = (\alpha+2)(2\alpha+2+2M)n + (\alpha^2+9\alpha+14+2M)n(n-1) + 2(\alpha+4)n(n-1)(n-2) + n(n-1)(n-2)(n-3),$$

$n = 0,1,\ldots$. They are

(115) $$S_n = \sum_{k=0}^{n} \frac{(-1)^{n-k}\binom{n}{k}(1+\alpha)_{n+k}(k[n+\alpha][n+1]+[k+1]M)x^k}{(k+1)!(1+\alpha)_n},$$

$n = 0,1,\ldots$.

Dehesa, Buendia and Sanchez-Buendia [1] have shown that the r-th moments of the zeros of S_n are given by

$$\mu_r' = \frac{1}{2^{2r}}\binom{2r}{r} + O(n^{-1})$$

$r = 1,2,\ldots,\ n = 0,1,\ldots$.

The moments $\{\mu_n\}_{n=0}^{\infty}$ satisfy

(116) $$(n+\alpha+1)\mu_n - (2n+\alpha)\mu_{n-1} + (n-1)\mu_{n-2} = 0 \quad , \quad n \geq 3,$$

(117) $$(n+\alpha+1)\mu_n - n\mu_{n-1} = 0 \quad , \quad n \geq 2,$$
$$(\alpha+1+M)(\alpha+2)\mu_1 - M\mu_0 = 0.$$

They are

(118a) $$\mu_0 = (\alpha+1+M)/M(\alpha+1)$$

(118b) $$\mu_n = 1/(\alpha+1)_{n+1},$$

$n = 1,2,\ldots$.

Using the moments,

(119) $$w = \frac{1}{M}\delta(x) + \sum_{n=0}^{\infty} (-1)^n\delta^{(n)}(x)/(\alpha+1)_{n+1}$$

is found to be a weight function. If the weight equations, which (119) satisfies,

(120) $$x^2(x-1)^2w' + \alpha x^2(1-x)w = 0,$$

(121) $$x^2(x-1)w''' + (12x^3-18x^2+6x)w'' + \\ + [(-\alpha^2-9\alpha+22)x^2+(6\alpha-24-2M)x+6]w' + [(2M\alpha-12\alpha)x+6\alpha]w = 0,$$

are solved directly,

(122) $$w = \frac{1}{M}\,\delta(x) + (1-x)^\alpha \quad , \quad 0 \le x \le 1$$

is found. (119) and (122) have the same Fourier transform.

The polynomials satisfy a recurrence relation

(123) $$S_n = (A_{n-1}x+B_{n-1})S_{n-1} - C_{n-1}S_{n-2},$$

where

(124) $$A_{n-1} = \frac{(2n+\alpha)(2n+\alpha-1)(n^2+\alpha n+M)}{(n+\alpha)(n)([n-1]^2+\alpha[n-1]+M)},$$

(125) $$B_{n-1} = \frac{(2n+\alpha-1)\;P(n)}{(n+\alpha)(n)(n+\alpha-2)([n-1]^2+\alpha[n-1]+M)^2}$$

with

(126) $$\begin{aligned} P(n) = {} & -2n^6+(-6\alpha+6)n^5+(-6\alpha^2+15\alpha-8-4M)n^4 \\ & + (-2\alpha^3+12\alpha^2-16\alpha-8\alpha M+8M+6)n^3 \\ & + (3\alpha^3-9\alpha^2+9\alpha-4\alpha^2M+12\alpha M-4M-2M^2-2)n^2 \\ & + (-\alpha^3+3\alpha^2-2\alpha+4\alpha^2M-4\alpha M-2\alpha M^2+2M^2)n \\ & + (\alpha\, M^2), \end{aligned}$$

and

(127) $$C_{n-1} = \frac{(n+\alpha-1)(n-1)(2n+\alpha)(n^2+\alpha n+M)^2}{(n+\alpha)(n)(2n+\alpha-2)([n-1]^2+\alpha[n-1]+M)^2}.$$

The generating function function can also be found. If

(128) $$F = ((1-x)\frac{\partial}{\partial x} + t^2\frac{\partial^2}{\partial t^2} + (\alpha+2)t\frac{\partial}{\partial t} + M)$$

and

(129) $$G_0(x,t) = \rho^{-1}(2/[1-t+\rho])^\alpha,$$

where

(130) $$\rho = (1-2[2x-1]t+t^2)^{1/2},$$

is the generating function for the Jacobi polynomials $\{P_n^{(\alpha,0)}\}_{n=0}^{\infty}$, then

(131) $$F \cdot G_0(x,t) = \sum_{n=0}^{\infty} S_n(x)t^n.$$

Related to (128)-(131) is the connection to the Jacobi polynomials

(132) $$S_n = [(1-x)\frac{d}{dx} + (n^2+[\alpha-1]n+M)]P_n^{(\alpha,0)}$$

This may be inverted to yield

(133) $$P_n^{(\alpha,0)}(x) = \int_{-\infty}^{x} \frac{(1-x)^{(n^2+[\alpha-1]n+M)}}{(1-\xi)^{(n^2+[\alpha-1]n+M+1)}} S_n(\xi)d\xi.$$

(132) also give the Rodrigues formula

(134) $$S_n(x) = (-1)^n(1-x)^{-\alpha}\frac{d^n}{dx^n}(n(1-x)^{n+\alpha}x^{n-1}+(n^2+\alpha n+M)(1-x)^{n+\alpha}x^n).$$

The second order differential equation for the Jacobi-type polynomials is, [2],

(135) $$\begin{aligned}&[(M^2+M\alpha+M+\lambda_n)x^3-(M^2+M\alpha+2M+\lambda_n)x^2+Mx]y''\\ &+ [(2M^3+3M\alpha+2M+M^2\alpha+M\alpha^2+2\lambda_n+\alpha\lambda_n)x^2 - (M^2+2M\alpha+4M+\lambda_n)x+2M]y'\\ &+ [\{(-22\alpha+2M+110)\lambda_n-\nu_n\}x+\lambda_n]y = 0\end{aligned}$$

where λ_n is given by (114) and

(136) $$\begin{aligned}\nu_n = &-(42\alpha^3+342\alpha^2+732\alpha+432+6M^2+438M+309M\alpha+3M^2\alpha+45M\alpha^2)n\\ &-(18\alpha^3+270\alpha^2+1188\alpha+1440+3M^2+219M+45M\alpha)n(n-1)\\ &+(\alpha^3-21\alpha^2-274\alpha-696)n(n-1)(n-2)\\ &+(3\alpha^2+15\alpha-12)n(n-1)(n-2)(n-3)\\ &+(18+3\alpha)n(n-1)(n-2)(n-3)(n-4)\\ &+n(n-1)(n-2)(n-3)(n-4)(n-5).\end{aligned}$$

When $\alpha > -1$ and $M > 0$, (122) generates a Hilbert space with inner product

$$(y,z) = \int_0^1 \bar{z}(x)y(x)(1-x)^{\alpha}dx + \frac{1}{M}\bar{z}_0 y_0. \tag{137}$$

The polynomials $\{S_n\}_{n=0}^{\infty}$ form a complete orthogonal set in H. The norm square of S_n is

$$\|S_n\|^2 = \frac{(n^2+\alpha n+M)([n+1]^2+\alpha[n+1]+M)}{(2n+\alpha+1)} \tag{138}$$

If $\alpha < -1$ a Cauchy regularization of the integral in (137) is required. If either $\alpha < -1$ or $M < 0$, (137) generates an indefinite inner product space, a Pontrjagin space, in which $\{S_n\}_{n=0}^{\infty}$ is complete. (138) still holds, and the left side of (113) generates a self-adjoint operator. The sign of (138) varies [14].

XVI. BOUNDARY VALUE PROBLEMS. Finally we turn to the boundary value problems the Legendre, Laguerre and Jacobi-type polynomials satisfy. Just as in the earlier problems, the problems are singular, and are therefore fairly complicated to describe. Further the singular boundary conditions are λ dependent, causing a further difficutly [10], [11].

Let q_0, q_1, q_2, w be sufficiently smooth and let $q_2 \neq 0$, $w > 0$ on (a,b). Let $c \in (a,b)$. Then

$$\ell y = [(q_2 y'')'' + (q_1 y')' + (q_0 y)]/w \tag{139}$$

represents a general formally self-adjoint, fourth order operator on $L^2(a,b;w)$. The equation $\ell y = \lambda y$, $\mathrm{Im}(\lambda) \neq 0$, has either 4, 3, or 2 solutions in $L^2(a,c;w)$ and another 4, 3 or 2 solutions in $L^2(c,b;w)$. The numbers are independent of λ so long as $\mathrm{Im}(\lambda) \neq 0$. Further the L^2 solutions can be made analytic in λ.

For various technical reasons, limiting circles in the solutions constructions, the cases mentioned are called limit -4, limit -3, limit -2. In the limit -4 case, two boundary conditions (from 4) must be chosen in a special manner. In the limit -3 case one boundary condition is automatically equal to 0, one must be chosen from the two remaining. In the limit -2 case, there are two boundary conditions, both automatically 0. These boundary conditions are evaluated as follows. If

$$\begin{aligned}[y,z] = {} & (q_2 y'')'\bar{z} - (q_2 y'')\bar{z}' - (q_2\bar{z}'')y \\ & + (q_2\bar{z}'')y' + (q_1 y')\bar{z} - (q_1\bar{z}')y, \end{aligned} \tag{140}$$

if u satisfies $\ell u = \lambda u$ and is in $L^2(c,b;w)$, then

(141) $$B_u(y) = \lim_{x\to b} [y,u](x)$$

is a boundary condition at b. Similarly, if v satisfies $\ell v = \lambda v$ and is in $L^2(a,c;w)$, then

(142) $$B_v(y) = \lim_{x\to b} [y,v](x)$$

is a boundary condition at a.

For the standard Jacobi, Laguerre and Hermite problems of fourth order, the expression (139), with an appropriate number of boundary conditions, determine a self-adjoint operator in $L^2(a,b;w)$ for which the polynomials are eigenfunctions.

For the Legendre, Laguerre and Jacobi-type polynomials we instead use the boundary conditions to define λ-dependent conditions. The polynomials are then eigenfunctions in a rigged Hilbert space.

Rather than go into full detail, we cite [10] and content ourselves with a description of the three boundary value problems for the Legendre, Laguerre and Jacobi-type polynomials. Problems for polynomials satisfying higher order equations are similar.

XVII. THE LEGENDRE-TYPE BOUNDARY VALUE PROBLEM. The setting is $L^2(-1,1;w)$, where w is given by (87). We assume that $\alpha > 0$. Since both ± 1 are in the limit -3 case, the boundary conditions

(143) $$\lim_{x\to 1} [y,(1-x^2)](x) = 0, \quad \lim_{x\to -1} [y,(1-x^2)](x) = 0$$

automatically. λ dependent boundary conditions

(144) $$\alpha[y,1](1) = \lambda[y,(1-x)](1)/8$$

(145) $$\alpha[y,1](-1) = \lambda[y,(1+x)](-1)/8$$

together with

(146) $$((x^2-1)^2 y'')'' + 4((\alpha(x^2-1)-2)y')' = \lambda y$$

form the Legendre-type boundary value problem.

XVIII. THE LAGUERRE-TYPE BOUNDARY VALUE PROBLEM. The setting is $L^2(0,\infty;w)$, where w is given by (98). ∞ is in the limit -2 case, so there are 2 automatic boundary conditions there, which need not concern us. The limit -3 case holds at 0.

(148) $$\lim_{x\to 0} [y,x^2] = 0$$

automatically. Other boundary forms are

(149) $$\lim_{x\to 0} [y,x] \quad , \quad \lim_{x\to 0} [y,1].$$

The boundary value problem satisfied by the Laguerre type polynomials is

(150) $$R[y,1](0) = (\lambda/2)[y,x](0)$$

together with

(151) $$(x^2 e^{-x} y'') - (([2R+2]x+2)e^{-x}y')' = \lambda e^{-x} y.$$

XIX. THE JACOBI TYPE BOUNDARY VALUE PROBLEM. The setting is $L^2(0,1;w)$, where w is given by (122), $\alpha > -1$, $M > 0$. the limit -3 case holds at $x = 0$. The boundary condition

(152) $$\lim_{x\to 0} [y,x^2] = 0$$

automatically. The two other boundary forms generate the λ-dependent condition

(153) $$-2M[y,1](0) = \lambda[y,(1-x)](0).$$

At $x = 1$, the limit -4 case holds when $-1 < \alpha < 1$. The limit -3 case holds when $1 \leq \alpha < 3$. The limit -2 case holds when $3 \leq \alpha$. Boundary conditions are

(154) $$[y,1](1) = 0 \quad , \quad (\text{automatic if } 3 \leqq \alpha).$$

(155) $$[y,(1-x)](1) = 0 \quad , \quad (\text{automatic if } 1 \leqq \alpha).$$

The Jacobi-type boundary value problem consists of (152) (automatically), (153), (154) and (155) (either required or automatic), together with

(156) $$([(1-x)^{\alpha+4}-2(1-x)^{\alpha+3}+(1-x)^{\alpha+2}]y'')'' + \\ + ([(2\alpha+2+M)(1-x)^{\alpha+2}-(2\alpha+4+2M)(1-x)^{\alpha+1}]y')' \\ = \lambda(1-x)^{\alpha} y.$$

XX. REMARKS. The H.L. Krall polynomials strongly resemble the Legendre-type polynomials. Formulas are a bit more complicated. The differential equation is of 6-th order.

The Littlejohn and Littlejohn II polynomials follow the pattern of the

Laguerre type polynomials. The differential equations are of 6-th and 8th order.

Koornwinder's polynomials genralize the Jacobi, Legendre, Legendre-type, Jacobi-type and H.L. Krall polynomials. A traditional differential equation has not been found for them as this is written, however.

The spectral resolutions of the self-adjoint opertors generated by the boundary value problems are, again, series, as in part 1.

REFERENCES

1. J.S. Dehesa, E. Buendia, M.A. Sanchez-Buendia, "On the polynomial solutions of ordinary differential equations of fourth order", J. Math. Phys. 26(1985), 1547-1552.

2. T.H. Koornwinder, "Orthogonal polynomials with weight function $(1-x)^{\alpha}(1+x)^{\beta}+M\delta(x+1)+N\delta(x-1)$", Canad. Math. Bull. 27(1984), 205-214.

3. A.M. Krall, "Chebyshev sets of polynomials which satisfy an ordinary differential equation", SIAM Rev. 22(1980), 436-441.

4. __________, "Laguerre polynomial expansions in indefinite inner product spaces, "J. Math. Anal. Appl. 70(1979), 267-279.

5. __________, "On boundary values for the Laguerre operator in indefinite inner product spaces", J. Math. Anal. Appl. 85(1982), 406-408.

6. __________, "On the generalized Hermite polynomials $\{H_n^{(\mu)}\}_{n=0}^{\infty}$, $\mu < -1/2$", Indiana U. Math. J. 30(1981), 73-78.

7. __________, "The Bessel polynomial moment problem", Acta. Math. Acad. Sci. Hungar. 38(1981), 105-107.

8. __________, "Orthogonal polynomials satisfying fourth order differential equations", Proc. Roy. Soc. Edin. 87A(1981), 271-288.

9. L.L. Littlejohn and A.M. Krall, "Orthogonal polynomials and singular Sturm-Liouville systems, I", Rocky Mt. J. Math., to appear.

10. __________, "Orthogonal polynomials and singular Sturm-Liouville systems, II", submitted.

11. __________, "On the classification of differential equations having orthogonal polynomial solutions", submitted.

12. A.B. Mingarelli and A.M. Krall, "Legendre-type polynomials under an indefinite inner product", SIAM J. Math. Anal. 14(1983), 399-402.

13. __________, "Laguerre-type polynomials under an indefinite inner product", Acta. Math. Sci. Hungar. 40(1982), 237-239.

14. __________, "Jacobi-type polynomials under an indefinite inner product", Proc. Roy. Soc. Edin. 90A(1981), 147-153.

15. R.D. Morton and A.M. Krall, "Distributional weight functions for orthogonal polynomials", SIAM J. Math. Anal. 9(1978), 604-626.

16. H. Weyl, "Über gewöhnliche Differentialgleichungen mit Singularitäten und die Entwicklungen Willkurlicher Functionen", Math. Ann. 68(1910), 220-269.

ORTHOGONAL POLYNOMIAL SOLUTIONS TO ORDINARY AND PARTIAL DIFFERENTIAL EQUATIONS

L.L. Littlejohn
Utah State University

Introduction

From the viewpoint of boundary value problems of ordinary differential operators, the following problem has become very important: classify all differential equations of the form

$$(*) \qquad \sum_{j=0}^{r} a_j(x)y^{(j)}(x) = \lambda y(x)$$

that have a sequence of orthogonal polynomial solutions. As we shall see, equations of the form (*) will give rise to self adjoint operators in the appropriate Hilbert or Krein space when the correct boundary conditions are determined. Although the theory of self adjoint extensions of formally symmetric differential expressions is well known, examples are sorely needed to illustrate the theory when $r > 2$. This is why equations of the form (*) which have orthogonal polynomial solutions is important: they serve as excellent examples. The reader is urged to consult [11] and [12] to see the connections between orthogonal polynomials and boundary value problems of ordinary differential operators.

This paper is broken down into two chapters. In Chapter One, we discuss the theory of orthogonal polynomial solutions to ordinary differential equations. No proofs are given for these results; instead, we refer the reader to the appropriate references. In Chapter Two, we present a new approach to orthogonal polynomial solutions to second order partial differential equations. Such theory was initially developed by H.L. Krall and I.M. Sheffer [8]; our approach in this paper follows the approach taken by Littlejohn in his study of orthogonal polynomial solutions to ordinary differential equations.

Chapter One Orthogonal Polynomials and Ordinary Differential Equations

§1.1 The Work of Bochner, H.L. Krall and A.M. Krall

It is well known that the classical orthogonal polynomials of Jacobi, Laguerre, and Hermite and the Bessel polynomials all satisfy second order differential equations of the form

$$a_2(x)y''(x) + a_1(x)y'(x) = \lambda y(x).$$

Are there any other orthogonal polynomial solutions to second order differential equations? S. Bochner [3] showed in 1929 that, up to a linear change of variable, that the answer is no - the above four sets are the only ones. See [11] and [17] for an historical account of Bochner's work as well as others. Bochner implicitly poses the following problem:

FIND ALL DIFFERENTIAL EQUATIONS OF THE FORM

(1.1.1) $$\sum_{j=0}^{r} a_j(x)y^{(j)}(x) = \lambda y(x)$$ classification problem

HAVING ORTHOGONAL POLYNOMIAL SOLUTIONS.

It is easy to see that if (1.1.1) has polynomial solutions of degree $0,1,\ldots r$, then it is necessary that $a_j(x)$ be a polynomial of degree $\leq j$, $j = 0,1,\ldots r$. Hence admissible differential equations must have the form

(1.1.2) $$\sum_{i=1}^{r} \sum_{j=0}^{i} \ell_{ij}x^j y^{(i)}(x) = \lambda_m y(x).$$

Furthermore, it is easy to see that $\lambda_m = m\ell_{11} + m(m-1)\ell_{22} + \ldots + m(m-1)\ldots(m-r+1)\ell_{rr}$. In 1938, H.L. Krall [5] proved his 'classification' theorem:

<u>Theorem 1.1.1</u>

Suppose $\Phi_m(x)$, $-\infty < x < \infty$, is a polynomial of degree m, $m = 0,1,\ldots$. Then $\{\Phi_m(x)\}$ is an orthogonal polynomial sequence (OPS) and $\Phi_m(x)$ satisfies (1.1.2) if and only if the moments $\{\mu_m\}$ of the weight function satisfy

$$\text{(i)} \quad \Delta_m =: \begin{vmatrix} \mu_0 & \mu_1 & \cdots & \mu_m \\ \mu_1 & \mu_0 & \cdots & \mu_{m+1} \\ \vdots & & & \\ \mu_m & \mu_{m+1} & \cdots & \mu_{2m} \end{vmatrix} \neq 0 \ , \ m = 0,1,\ldots$$

<u>and</u>

(1.1.3) $$\text{(ii)} \quad S_k(m) =: \sum_{i=2k+1}^{r} \sum_{u=0}^{i} \binom{i-k-1}{k} P(m-2k-1,i-2k-1)\ell_{i,i-u}\mu_{m-u} = 0,$$

$2k+1 \leq r$, $m = 2k+1, 2k+2,\ldots$, where $P(n,k) = n(n-1)\ldots(n-k+1)$. Furthermore, r is necessarily even. That is to say, if r is the smallest order of a differential equation of the form (1.1.2) having $\{\Phi_m(x)\}$ as solutions, then r is even. ■

<u>Definition 1.1.1</u>

The recurrence relations (1.1.3) will be called the <u>moment equations</u> associated with $\{\Phi_m(x)\}$.

Observe since $r = 2n$ for some $n \geq 1$, that there are n moment equations in (1.1.3).

In 1940, H.L. Krall [6] classified all fourth order differential equations having orthogonal polynomial solutions. He discovered three such differential equations having nonclassical orthogonal polynomial solutions. These polynomials were found and studied in detail (including the appropriate boundary value problems) by A.M. Krall [10] in 1981. Krall found the orthogonalizing weights to these polynomials by using a very novel approach based on results he and R.D. Morton [9] established in 1978. Krall first found the moments $\{\mu_m\}$ using the moment equations (1.1.3). He then substituted these moments into

$$w(x) = \sum_{m=0}^{\infty} \frac{(-1)^m \mu_m \delta^{(m)}(x)}{m!} \tag{1.1.4}$$

and showed that w(x) is a weight distribution for the polynomials. By using the Fourier transform, he was able to obtain classical representations of (1.1.4). For example, the Legendre type polynomials [10] satisfy the fourth order equation

$$(x^2-1)^2 y^{(4)} + 8x(x^2-1)y^{(3)} + (4\alpha+12)(x^2-1)y'' + 8\alpha xy' = \lambda_m y$$

The moment equations are readily solved to yield (up to a constant multiple)

$$\mu_{2n} = (\alpha+2n+1)/(2n+1), \quad \mu_{2n+1} = 0.$$

Substituting these values into (1.1.4) yields

$$w(x) = \sum_{n=0}^{\infty} \frac{(\alpha+2n+1)\delta^{(2n)}(x)}{(2n+1)!}.$$

The Fourier transform of w is $\hat{w}(t) = \frac{e^{it}+e^{-it}}{2} + \frac{\alpha \sin t}{t}$ and the inverse Fourier transform of $\hat{w}$ yields $w(x) = \frac{\alpha}{2}\left(H(x+1)-H(x-1) \right) + \frac{1}{2}\delta(x+1) + \frac{1}{2}\delta(x-1)$, where H(x) is the Heaviside function.

In 1981, Littlejohn generalized the Legendre type polynomials when he found the Krall polynomials [14]. Since then, Littlejohn has found two more sets of orthogonal polynomial solutions to differential equations [16],[17]. At the end of Chapter One, we list all known OPS's that satisfy equations of the form (1.1.2). In 1984, T.H. Koornwinder significantly generalized the work of Krall and Littlejohn by explicitly finding the polynomials orthogonal with respect to the weights

$$(1-x)^\alpha(1+x)^\beta + M\delta(x+1) + N\delta(x-1), \quad -1 \le x \le 1$$

$$\text{and} \quad x^\alpha e^{-x} + A\delta(x), \quad x > 0.$$

It is unclear, at this point, if these polynomials will satisfy differential equations of the form (1.1.2) for arbitrary α and β. It appears, in the case of the second weight above, that when $\alpha = n$, n a natural number, then the (minimal) order of the differential equation is 2n+4. This author is unable to find a differential equation of the form (1.1.2) when α is not a positive integer. Much more work needs to be done.

§1.2 Results during the Period 1981-1986

Littlejohn observed that all of the known differential equations having orthogonal polynomial solutions can be made formally self adjoint (symmetric) when multiplied by the appropriate function. He then set out to find conditions for when higher order differential expressions can be made symmetric.

Definition 1.2.1

Let

$$L(y) = \sum_{j=0}^{r} a_j(x)y^{(j)}(x).$$ The Lagrange adjoint of $L(y)$ is

$$L^+(y) = \sum_{j=0}^{r} (-1)^j (a_j(x)y(x))^{(j)}.$$ If $L(y) = L^+(y)$, then L is said to be symmetric.

If $f(x)L(y)$ is symmetric, we shall say that $f(x)$ is a symmetry factor of $L(y)$.

Remarks 1. Every second order differential expression

$$a_2(x)y''(x) + a_1(x)y'(x) + a_0(x)y(x)$$

can be made symmetric when multiplied by $f(x) = \exp(\int^x (a_1(t)/a_2(t))dt)/a_2(x)$.

However, higher order differential expressions cannot always be made symmetric.

2. It is well known that when $a_j(x)$ is real valued, $j = 0,1,\dots r$, and $L(y)$ is symmetric, then r is even.

In [17], Littlejohn and A.M. Krall prove:

Theorem 1.2.1

$f(x)$ is a symmetry factor for $L(y) = \sum_{j=0}^{2n} a_j(x)y^{(j)}(x)$ if and only if $f(x)$ simultaneously satisfies the following equations:

$$(1.2.1)\qquad \sum_{i=2k+1}^{2n} (-1)^i \binom{i-k-1}{k} (a_i(x)f(x))^{(i-2k-1)} = 0,\ k = 0,1,\dots(n-1). \blacksquare$$

Notice that these equations have orders $1,3,\dots(2n-1)$. In particular, the first order equation can be solved to yield

$$f(x) = \exp(\int^x \tfrac{1}{n}(a_{2n-1}(t)/a_{2n}(t)dt)/a_{2n}(x).$$

Definition 1.2.2

The n equations appearing in (1.2.1) are called the symmetry equations of $L(y)$.

Example 1

The symmetry equations attached to the fourth order equation for the Legendre type polynomials are

$$\text{(i)}\quad (x^2-1)^2 f'(x) = 0$$

$$\text{and (ii)}\quad (x^2-1)^2 f^{(3)}(x) + 12x(x^2-1)f''(x) + ((24-4\alpha)x^2+4\alpha)f'(x) = 0.$$

Clearly $f(x) = 1$ is a symmetry factor, i.e. the equation is already symmetric.

Notice what happens if we solve (i) and (ii) distributionally on polynomials:

to solve (i), first divide both sides by $(x^2-1)^2$ to get

$$\Lambda(x) = c_1\delta'(x+1) + c_2\delta'(x-1) + c_3\delta(x+1) + c_4\delta(x-1).$$

Integrate with respect to x and we get

$$\Lambda(x) = c_1\delta(x+1) + c_2\delta(x-1) + c_3H(x+1) + c_4H(x-1) + c_5.$$

If we require that $\Lambda(x) \to 0$ as $|x| \to \infty$, we see that $c_5 = 0$ and $c_3 = -c_4$. Substituting $\Lambda(x)$ into (ii) yields

$$0 = 8\alpha c_1\Phi(-1) - 8\alpha c_2\Phi(1) + c_3\int_{-1}^{1}[(x^2-1)^2\Phi^{(3)} - (4\alpha+12)(x^2-1)\Phi' - 8\alpha x\phi]dx.$$

Integration by parts three times simplifies this equation to

$0 = (8\alpha c_1 - 8c_3)\Phi(-1) + (8c_3 - 8\alpha c_2)\Phi(1)$, for all polynomials Φ. Clearly, we must have $\alpha c_1 = c_3 = \alpha c_2$. Choosing $c_1 = \frac{1}{2}$ yields $\Lambda(x) = \frac{\alpha}{2}(H(x+1)-H(x-1)) + \frac{1}{2}\delta(x+1) + \frac{1}{2}\delta(x-1)$. The weight function for the Legendre type polynomials!

If we assume that the moment equations (1.1.3) have a unique solution once $\mu_0 \neq 0$ is chosen, then Littlejohn and Krall can prove [17]:

Theorem 1.2.2

Suppose $\{\Phi_m(x)\}$ is an OPS and $\Phi_m(x)$ satisfies

$$\sum_{i=1}^{2n} a_i(x)y^{(i)}(x) = \lambda_m y(x), \quad m = 0,1,\ldots$$

Suppose $\Lambda(x)$ is a nontrivial solution to the symmetry equations (1.2.1), found distributionally on polynomials. Furthermore, suppose $\Lambda(x) \to 0$ as $|x| \to \infty$. Then $\Lambda(x)$ is a weight distribution for $\{\Phi_m(x)\}$.■

Corollary 1.2.3

Under the above conditions $< \Lambda(x), a_{2j-1}(x) > = 0$, $j = 1,2,\ldots n$.■

See [15] for a proof of this corollary.

Definition 1.2.3

If the equations (1.2.1) are solved distributionally on polynomials, we shall refer to these equations as the weight equations for $\{\Phi_m(x)\}$.

Example 2

The Bessel polynomials $\{y_n(x,a,b)\}$ satisfy

$$x^2y'' + (ax+b)y' - m(m+a-1)y = 0,$$

where $b \neq 0$ and $a \neq 0,-1,-1,\ldots$. They are orthogonal in the complex plane in the following sense:

$$\int_{\Gamma} y_n(z,a,b)y_m(z,a,b)\rho(z)dz = \frac{(-1)^{n+1}bn!\delta_{nm}}{(2n+a-1)(n+a-2)^{(n-1)}}$$

where Γ is any Jordan curve encircling the origin, $(k)^{(n)} = k(k-1)\ldots(k-n+1)$ and

$$\rho(z) = \frac{1}{2\pi i} \sum_{n=0}^{\infty} \frac{\Gamma(a)\ (-b)^n}{\Gamma(a+n-1) z^n} .$$

From the moment equations (1.1.3), the moments can easily be calculated to yield

$$\mu_m = (-1)^{m+1} b^{m+1} / (a)_m ,$$

where $(a)_m = a(a+1)\ldots(a+m-1)$. Since the moments are all real, a well known theorem of Boas [2] implies that the Bessel polynomials are orthogonal on the real line with respect to some $d\mu(x)$ where $\mu(x)$ has bounded variation on $(-\infty,\infty)$. A suitable $d\mu(x)$ has never been found. The weight equation in this case is

$$x^2\Lambda'(x) + (\ (2-a)x - b)\Lambda(x) = 0.$$

Notice that this equation has an irregular singular point at $x = 0$; the theory of distributional solutions (with test functions polynomials) to such equations is insufficiently developed (of course, the distributional solution to this equation in D' is identically zero). Substituting the moments into (1.1.4) yields the formal weight distribution for the Bessel polynomials

$$w(x) = \sum_{m=0}^{\infty} \frac{b^{m+1}\delta^{(m)}(x)}{(a)_m\ m!} .$$

However, use of the Fourier transform does not simplify this expression. Because a suitable weight function for the Bessel polynomials has not been found, the appropriate boundary value problem associated with the second order differential equation cannot be fully studied.

Example 3

The Jacobi type polynomials satisfy the fourth order equation

$$(x^2-x)^2y^{(4)} + 2x(x-1)[(\alpha+4)x-2]y^{(3)} + x[(\alpha^2+9\alpha+14+2M)x-(6\alpha+12+2M)]y''$$
$$+ [(\alpha+2)(2\alpha+2+2M)x-2M]y' = \lambda_m y.$$

In this case the symmetry equations are

(i) $(x^2-x)^2f' - \alpha x^2(x-1)f = 0$

(ii) $(x^2-x)^2f^{(3)} + x(x-1)(\ (4-2\alpha)x -2)f'' + [(\alpha^2-3\alpha+2+2M)x^2 + (2\alpha-2M)x - 2]f'$

$-2\alpha(Mx+1)f = 0.$

Following the method in example 1, it is not too difficult to see that

$$\Lambda(x) = H(x)(1-x)^{\alpha} + \frac{1}{M}\delta(x)$$

is the general distributional solution to (i) and (ii). It is the weight function for the Jacobi type polynomials (see [10]).

Example 4

So far in this section, we have shown how to construct the weight distribution for an OPS given the differential equation that they satisfy. We now show, by way of example, that this process may be reversed. That is, we show how to construct the differential equation given the weight distribution for the orthogonal polynomials. Indeed, let

(1.2.2) $$w(x) = H(x)x^2e^{-x} + \frac{1}{R}\delta(x).$$

Since the polynomials orthogonal with respect to $H(x)e^{-x} + \frac{1}{R}\delta(x)$ satisfy a fourth order equation and the polynomials orthogonal with respect to $H(x)xe^{-x} + \frac{1}{R}\delta(x)$ satisfy a sixth order equation (see [10] and [16]), we shall assume the polynomials orthogonal with respect to (1.2.2) satisfy an eighth order equation

$$L_8(y) =: \sum_{i=1}^{8} \sum_{j=0}^{i} \ell_{ij} x^j y^{(i)}(x) = \lambda_m y(x).$$

In addition, we shall assume that (i) $x^2e^{-x}L_8(y)$ is symmetric

and (ii) $w(x)$ satisfies the four weight equations associated with $L_8(y)$. These two equations yield many equations for the ℓ_{ij}'s. If we solve these equations, we find that the differential equation is

$$x^4y^{(8)} + (-4x^4+24x^3)y^{(7)} + (6x^4-84x^3+168x^2)y^{(6)}$$

$$+ (-4x^4+108x^3-504x^2+336x)y^{(5)} + (x^4-60x^3+540x^2-840x)y^{(4)} + (12x^3-240x^2+720x)y^{(3)}$$

$$+ (36x^2-(48R+240)x)y'' + ((24+48R)x-144R)y' = (m^4+6m^3+11m^2+(6+18R)m)y.$$

The orthogonal polynomial solutions are $\Phi_m(x) = 6R\, L_m^{2,\ 1/2R}(x)$, where $L_m^{\alpha,R}(x)$ is the generalized Laguerre type polynomial, found by Koornwinder [4], which are orthogonal on $(0,\infty)$ with respect to $x^\alpha e^{-x} + R\delta(x)$. The reader is also referred to [17] where properties of the polynomials are discussed as well as the appropriate self adjoint boundary value problem is established in the Hilbert space $L^2(0,\infty,\ x^2e^{-x}+ (1/R)\delta(x))$.

What is the connection between the moment equations (1.1.3) and the weight equations (1.2.1)? To see what this relationship is, suppose $w(x)$ is a nontrivial distributional solution to the equations (1.2.1), found on polynomials Φ, such that $w(x) \to 0$ as $|x| \to \infty$. Then, for $m \geq 2k+1$, we have

$$0 = \left\langle \sum_{i=2k+1}^{2n} (-1)^i \binom{i-k-1}{k} (a_i w)^{(i-2k-1)},\ x^{m-2k-1} \right\rangle$$

$$= \langle w(x), \sum_{i=2k+1}^{2n} \binom{i-k-1}{k} a_i(x)\, D^{(i-2k-1)}(x^{m-2k-1}) \rangle \qquad (D = \frac{d}{dx})$$

$$= \langle w(x), \sum_{i=2k+1}^{2n} \sum_{j=0}^{i} \binom{i-k-1}{k} P(m-2k-1, i-2k-1)\ell_{ij} x^{m-i+j} \rangle$$

$$= \langle w(x), \sum_{i=2k+1}^{2n} \sum_{j=0}^{i} \binom{i-k-1}{k} P(m-2k-1, i-2k-1)\ell_{i,i-u} x^{m-u} \rangle$$

$$= S_k(m).$$

That is to say, <u>if we solve the weight equations on the monomials x^{m-2k-1}, $m \geq 2k+1$, we generate the moment equations (1.1.3)</u>.

<u>Example 5</u>

The weight equation for the Hermite polynomials is

$$w' + 2xw = 0.$$

Hence $0 = \langle w'+2xw, x^m \rangle = \langle w, -mx^{m-1} + 2x^{m+1} \rangle = -m\mu_{m-1} + 2\mu_{m+1}$. This recurrence relation is easily solved to yield the moments for the Hermite polynomials:

$$\mu_{2m} = \frac{(2m)!\mu_0}{4^m m!}, \quad \mu_{2m+1} = 0, \quad m = 0,1,2,\ldots .$$

The following theorem connects the approach taken by Krall and Morton to the approach taken by Littlejohn (see [17]):

<u>Theorem 1.2.4</u>

Suppose $\{\Phi_m(x)\}$ is an OPS and $\Phi_m(x)$ satisfies

$$\sum_{j=1}^{2n} a_j(x) y^{(j)}(x) = \lambda_m y(x), \quad m = 0,1,\ldots$$

Let $w(x)$ be given by (1.1.4) where $\{\mu_m\}$ is the sequence of moments associated with $\{\Phi_m(x)\}$. Then $w(x)$ formally satisfies the weight equations (1.2.1).■

Closely related to the problem of classifying all orthogonal polynomial solutions to differential equations of the form (1.1.1) is the problem of finding all OPS's that satisfy second order differential equations of the form

(1.2.3) $$A(x,m)\Phi''_m(x) + B(x,m)\Phi'_m(x) + C(x,m)\Phi_m(x) = 0.$$

Observe here that the coefficients of the derivatives are functions of x and the indexing parameter, unlike the situation with our original classification problem. All of the known OPS's that satisfy equations of the form (1.1.1) also satisfy equations of the form (1.2.3). However, the converse is not true. In fact, Atkinson and Everitt [1] have established the following important and general theorem:

Theorem 1.2.5

Suppose $\{\phi_m(x)\}$ is an OPS with respect to $d\mu(x) = d\mu_1(x) + d\mu_2(x)$ where $d\mu_1(x) = w(x)dx$ and $\mu_2(x)$ is a nondecreasing step function with a finite number of jumps. Suppose $w(x)$ is nonnegative and continuous on some interval (a,b), where $-\infty \le a < b \le \infty$ and suppose $w(x)$ satisfies a first order equation

$$\sigma_1 w' + \sigma_0 w = 0,$$

where σ_1 and σ_2 are polynomials. Then $\phi_m(x)$, $m = 0,1\ldots$, satisfies a differential equation of the form (1.2.3) with polynomial coefficients.■

As a consequence of this theorem and Theorem 1.2.2, it is easy to prove the following theorem (see [17]):

Theorem 1.2.6

Suppose $\mu(x)$ and $w(x)$ satisfy the conditions of Theorem 1.2.5. Suppose $\{\phi_m(x)\}$ is an OPS with respect to $\Lambda(x)dx = d\mu(x)$ and suppose $\phi_m(x)$ satisfies a differential equation of the form (1.1.1), $m = 0,1,2\ldots$. If $\Lambda(x)$ satisfies the weight equations, then $\{\phi_m(x)\}$ satisfies an equation of the form (1.2.3).■

We mentioned in the introduction the intimacy between symmetric differential equations and orthogonal polynomials. We now make this connection precise with the following theorem (see [17]).

Theorem 1.2.7

Suppose $\{\phi_m(x)\}$ is an OPS with respect to $\Lambda(x)dx = d\mu(x) = f(x)dx + d\mu_1(x)$ on an interval I of the real line, where $f(x) \ge 0$, $x \in I$, is continuous and $\mu_1(x)$ is a nondecreasing step function with a finite number of jumps.

(a) If I is compact and $\{\phi_m(x)\}$ satisfies the differential equation

$$L_{2n}(y) =: \sum_{i=1}^{2n} a_i(x) y^{(i)}(x) = \lambda_m y(x) \tag{1.2.4}$$

and if $f \in C^{2n-1}(I)$, then $f(x)L_{2n}(y)$ is symmetric.

(b) If $\Lambda(x)$ simultaneously satisfies the equations

$$\sum_{i=2k+1}^{2n} (-1)^i \binom{i-k-1}{k} (a_i\Lambda)^{(i-2k-1)} = 0,\ k = 0,1,\ldots (n-1),$$

distributionally on polynomials, where $a_i(x)$ is a polynomial of degree $\le i$, $i = 1,2,\ldots 2n$ and suppose $\Lambda(x) \to 0$ as $|x| \to \infty$. Then $\{\phi_m(x)\}$ satisfies (1.2.4) and $f(x)L_{2n}(y)$ is symmetric.■

Notice that this theorem can be interpreted, in a sense, as a restatement of H.L. Krall's 1938 classification theorem (Theorem 1.1.1). On the one hand, it is a weaker version of Theorem 1.1.1 and on the other hand, it is slightly stronger since symmetry is established.

We now list all known OPS's that satisfy differential equations of the form (1.1.1).

OPS	ORDER OF DIFFERENTIAL EQUATION	WEIGHT	INTERVAL
Bessel	2	?	$(0,\infty)$
Jacobi	2	$(1-x)^{\alpha}(1+x)^{\beta}$	$(-1,1)$
Laguerre	2	$x^{\alpha}e^{-x}$	$(0,\infty)$
Hermite	2	e^{-x^2}	$(-\infty,\infty)$
Laguerre type	4	$e^{-x}+ \frac{1}{R}\delta(x)$	$(0,\infty)$
Jacobi type	4	$(1-x)^{\alpha}+ \frac{1}{M}\delta(x)$	$(0,1)$
Legendre type	4	$\frac{\alpha}{2} + \frac{1}{2}\delta(x+1) + \frac{1}{2}\delta(x-1)$	$(-1,1)$
Krall	6	$C + \frac{1}{A}\delta(x+1) + \frac{1}{B}\delta(x-1)$	$(-1,1)$
Generalized Laguerre type	6	$xe^{-x} + \frac{1}{R}\delta(x)$	$(0,\infty)$
Generalized Laguerre type	8	$x^2e^{-x} + \frac{1}{R}\delta(x)$	$(0,\infty)$

A few words are in order regarding the classification problem. Will we solve it? The answer is quite possibly. Indeed, it is quite possible that Koornwinder has found the most general formulas for the orthogonal polynomials that satisfy differential equations of the form (1.1.1). This author feels that it would be a significant result if someone could either prove or disprove this statement. If indeed all of the orthogonal polynomials have been found, it still remains a fact that the differential equations are important to differential equators and to functional analysts: afterall, they give mathematics a new class of self adjoint operators.

Chapter Two Orthogonal Polynomials and Second Order Partial Differential Equations

§2.1 Orthogonal Polynomials in Two Variables

How much of the theory developed and stated in §1 carries over to the problem of classifying all orthogonal polynomial solutions to partial differential equations? We shall consider this question in this chapter with particular emphasis on second order partial differential equations. The classification of all second order PDE's having orthogonal polynomial solutions was achieved by H.L. Krall and I.M. Sheffer [8] in 1967. Besides developing some general theory of orthogonal polynomials in two variables, they showed that there are nine different sets of orthogonal polynomials (up to a linear change of variable) that satisfy a second order partial differential equation. Before discussing their results and our approach to this classification problem, we briefly mention some general properties of orthogonal polynomials in two variables.

Notation Let P denote the set of all polynomials $\Phi(x,y)$ with real coefficients. Let $V_k = \text{span}\{x^k, x^{k-1}y, x^{k-2}y^2, \ldots y^k\}$ so V_k is a (k+1) dimensional vector space.

Definition 2.1.1

Suppose $L: P \to R$ is a (real) linear functional. Then L is called an orthogonalizing weight functional if there exists a sequence of polynomials $\{\Phi_{mn}(x,y)\}$, m,n = 0,1,... such that

(i) $\Phi_{mn}(x,y)$ is a polynomial of degree exactly m+n, m,n = 0,1,2...

(ii) $\{\Phi_{mn}(x,y)\}_{m+n=k}$ is a basis for V_k , k = 0,1,2,...

(iii) $L(\Phi_{mn}(x,y)\Phi_{kl}(x,y)) = 0$ whenever $(m,n) \neq (k,l)$

(iv) $L(\Phi^2_{mn}(x,y)) \neq 0$, m,n = 0,1,2...

The sequence $\{\Phi_{mn}(x,y)\}$ is called an orthogonal polynomial sequence in the classical sense with respect to L. If $\{\Phi_{mn}(x,y)\}$ is a sequence of polynomials satisfying all of the above properties except possibly (iv), then $\{\Phi_{mn}(x,y)\}$ is called a weak orthogonal polynomial sequence in the classical sense. The real number $\mu_{mn} =: L(x^m y^n)$ is called the mn-th moment of L.

Remark

Proving that $\{\Phi_{mn}(x,y)\}$ satisfies conditions (i), (ii), (iii) and (iv) can be quite difficult. In particular, it could be quite tedious showing

(a) $L(\Phi_{mn}(x,y)\Phi_{kl}(x,y)) = 0$ when m+n = k+l but $(m,n) \neq (k,l)$

or (b) $L(\Phi^2_{mn}(x,y)) \neq 0$

especially if a weight function is not readily available for the polynomials. Krall and Sheffer begin their paper with a considerably less restrictive definition of an orthogonal polynomial sequence $\{P_{mn}(x,y)\}$ and show that it is possible, in some cases, to obtain an orthogonal polynomial sequence in the classical sense $\{\Phi_{mn}(x,y)\}$ from $\{P_{mn}(x,y)\}$ by taking appropriate linear combinations. In particular, it is quite possible that the sequence $\{P_{mn}(x,y)\}$ does not satisfy the conditions (iii) and (iv) in the above definition but, of course, $\{\Phi_{mn}(x,y)\}$ does.

We make this clear with Krall and Sheffer's definition:

Definition 2.1.2

Suppose $L:P \to R$ is a linear functional. If there exists a unique monic set $\{P_{mn}(x,y)\}$, m,n = 0,1,2... where

$$P_{mn}(x,y) = x^m y^n + \text{lower terms}$$

such that $L(P_{mn}(x,y)\Psi(x,y)) = 0$ for all polynomials Ψ of degree $\le m+n-1$, then $\{P_{mn}(x,y)\}$ is called a monic orthogonal set relative to L. If the sequence $\{P_{mn}(x,y)\}$ cannot be determined uniquely, we shall say that $\{P_{mn}(x,y)\}$ is a monic weak orthogonal set relative to L.

Using this definition, Krall and Sheffer established the following results:

Theorem 2.1.1

Suppose $L: P \to R$ is a linear functional with moments $\mu_{mn} = L(x^m y^n)$, m,n = 0,1,2,...

Let Δ_n be the determinant of order $\frac{1}{2}(n+1)(n+2)$ where the first row consists of

$$\mu_{00}, \mu_{10}, \mu_{01}, \mu_{20}, \mu_{11}, \mu_{02}, \dots \mu_{n0}, \mu_{n-1,1}, \dots \mu_{0n}$$

and whose subsequent rows are obtained by adding the integers (i,j) to the above subscript pairs, where (i,j) runs successively through the values

$$(i,j) = (1,0), (0,1), (2,0), (1,1), (0,2), \dots (n,0), (n-1,1), \dots (0,n).$$

Then a monic orthogonal set $\{P_{mn}(x,y)\}$ exists if and only if $\Delta_n \neq 0$, n = 0,1,2,... .

In fact,

$$P_{mn}(x,y) = \frac{\Delta_{m+n;m,n}(x,y)}{\Delta_{m+n-1}} \qquad (P_{00}(x,y) = \mu_{00})$$

where $\Delta_{m+n;m,n}(x,y)$ is the determinant of order $1+\frac{1}{2}(m+n)(m+n+1)$ obtained from Δ_{m+n-1} as follows:

(a) Adjoin a column on the extreme right of Δ_{m+n-1} of which the first element is μ_{mn} and whose subsequent elements are gotten by adding to the index pair (m,n) the respective pairs (1,0), (0,1),... (m+n-1,0), (m+n-2,1),... (0,m+n-1).

(b) Then adjoin a new bottom row whose elements are

$$1, x, y, \dots, x^{m+n-1}, x^{m+n-2}y, \dots y^{m+n-1}, x^m y^n. \blacksquare$$

For example,

$$\Delta_0 = \mu_{00}\ ,\quad \Delta_1 = \begin{vmatrix} \mu_{00} & \mu_{10} & \mu_{01} \\ \mu_{10} & \mu_{20} & \mu_{11} \\ \mu_{01} & \mu_{11} & \mu_{02} \end{vmatrix},\quad \Delta_2 = \begin{vmatrix} \mu_{00} & \mu_{10} & \mu_{01} & \mu_{20} & \mu_{11} & \mu_{02} \\ \mu_{10} & \mu_{20} & \mu_{11} & \mu_{30} & \mu_{21} & \mu_{12} \\ \mu_{01} & \mu_{11} & \mu_{02} & \mu_{21} & \mu_{12} & \mu_{03} \\ \mu_{20} & \mu_{30} & \mu_{21} & \mu_{40} & \mu_{31} & \mu_{22} \\ \mu_{11} & \mu_{21} & \mu_{12} & \mu_{31} & \mu_{22} & \mu_{13} \\ \mu_{02} & \mu_{12} & \mu_{03} & \mu_{22} & \mu_{13} & \mu_{04} \end{vmatrix}$$

and

$$P_{10}(x,y) = \frac{\begin{vmatrix} \mu_{00} & \mu_{10} \\ 1 & x \end{vmatrix}}{\mu_{00}}\ ,\qquad P_{01}(x,y) = \frac{\begin{vmatrix} \mu_{00} & \mu_{01} \\ 1 & y \end{vmatrix}}{\mu_{00}}$$

$$P_{20}(x,y) = \frac{\begin{vmatrix} \mu_{00} & \mu_{10} & \mu_{01} & \mu_{20} \\ \mu_{10} & \mu_{20} & \mu_{11} & \mu_{30} \\ \mu_{01} & \mu_{11} & \mu_{02} & \mu_{21} \\ 1 & x & y & x^2 \end{vmatrix}}{\Delta_1}\ .$$

Theorem 2.1.2

Suppose $\{P_{mn}(x,y)\}$ is a monic orthogonal set. For each $n = 0,1,2\ldots$, V_n has a basis $\{\Phi_{n0}(x,y),\ \Phi_{n-1,1}(x,y),\ldots\Phi_{0n}(x,y)\}$ such that $\{\Phi_{mn}(x,y)\}$ is an orthogonal polynomial sequence in the classical sense.■

Observe that each $\Phi_{n-j,j}$ is a linear combination of $P_{n0},\ P_{n-1,1},\ \ldots\ P_{0n}$.

Theorem 2.1.3

(a) Suppose $\{P_{mn}(x,y)\}$ is a monic orthogonal set with respect to the weight functional L. Let

$$\beta_{ij}^{kl} = L(P_{kl}P_{ij}) \qquad i,j,k,l = 0,1,2,\ldots$$

and

$$\delta_n = \begin{vmatrix} \beta_{n0}^{n0} & \beta_{n-1,1}^{n0} & \cdots & \beta_{0n}^{n0} \\ \beta_{n0}^{n-1,1} & \beta_{n-1,1}^{n-1,1} & \cdots & \beta_{0n}^{n-1,1} \\ \vdots & & & \\ \beta_{n0}^{0n} & \beta_{n-1,1}^{0,n} & \cdots & \beta_{0n}^{0n} \end{vmatrix}$$

Then (i) $\beta_{ij}^{kl} = \beta_{kl}^{ij}$

(ii) $\beta_{ij}^{kl} = 0$ if $i+j \neq k+l$

(iii) $\delta_n \neq 0,\ n = 0,1,2\ldots$

(b) Suppose $\{P_{mn}(x,y)\}$ is at least a weak monic orthogonal set and suppose β_{ij}^{kl} and δ_n are as above. If $\delta_n \neq 0$ for $n = 0,1,2,\ldots$, then $\{P_{mn}(x,y)\}$ is a monic orthogonal set.■

§2.2 Orthogonal Polynomials and Second Order Partial Differential Equations

Definition 2.2.1

A partial differential equation

$$(2.2.1) \qquad L_r(u) =: \sum_{p+q=1}^{r} Q_{pq}(x,y)D_x^pD_y^q u + \lambda u = 0$$

is called admissible if there exists an infinite sequence $\{\lambda_n\}$ such that for $\lambda = \lambda_n$, there are no non-zero polynomial solutions of degree $< n$ and there are precisely $(n+1)$ linearly independent solutions of degree n.

Remark

Notice that we are not assuming any orthogonality requirements on the polynomial solutions of $L_r(u) = 0$. Of course, the condition of admissibility allows for the possibility of orthogonality. From the definition above, one might wonder if when $\lambda = \lambda_n$, there are polynomial solutions of degree $> n$. This cannot happen because of

Theorem 2.2.1

If (2.2.1) is admissible, then it has no non-zero polynomial solutions for $\lambda \neq \lambda_0$, $\lambda_1, \ldots$ and for $\lambda = \lambda_n$, the only polynomial solutions are those in V_n.■

Theorem 2.2.2

Equation (2.2.1) is admissible if and only if

(i) $\lambda_0 = 0$ and $\lambda_m \neq \lambda_n$ for $m \neq n$.

(ii) (2.2.1) has the form:

$$L_r(u) = \sum_{s=p+q=1}^{r} \{ \binom{s}{q} a_s x^p y^q + M_{pq}(x,y)\} D_x^p D_y^q u + \lambda u = 0$$

where $\deg M_{pq} < p+q$

(iii) $\lambda_n = -n! \sum_{k=1}^{n} \frac{a_k}{(n-k)!}$, $n = 1,2,\ldots$ where $a_n = 0$ if $n > r$.■

In particular, for second order partial differential equations, we have:

Theorem 2.2.3

The most general admissible second order partial differential equation is given by:

$$(ax^2 + d_1x + e_1y + f_1)u_{xx} + (2axy + d_2x + e_2y + f_2)u_{xy}$$

(2.2.2) $$+ (ay^2 + d_3x + e_3y + f_3)u_{yy} + (gx + h)u_x + (gy + h_2)u_y$$

$$+ \lambda u = 0,$$

where $\lambda_n = -n(n-1)a - gn$ and $g + na \neq 0$, $n = 0,1,2...$ (The condition $g + na \neq 0$ for $n = 0,1,2...$ is easily seen to be equivalent to $\lambda_n \neq \lambda_m$ when $n \neq m$.)■

Since $g \neq 0$ and we are allowing for linear changes of variable, we can introduce the transformations $x = x^* - h_1/g$, $y = y^* - h_2/g$ and rewrite (2.2.2) as

(2.2.3) $$L_2(u) =: (ax^2 + d_1x + e_1y + f_1)u_{xx} + (2axy + d_2x + e_2y + f_2)u_{xy} + (ay^2 + d_3x + e_3y + f_3)u_{yy} + gxu_x + gyu_y + \lambda u = 0,$$

where we still have $\lambda_n = -n(n-1)a - gn$ and $g + na \neq 0$, $n = 0,1,2...$. In the analysis that follows shortly, most of the differential equations will be put in the form (2.2.3) but, for obvious reasons, some will still be in the form (2.2.2).

Krall and Sheffer posed the following classification problem: Classify, up to a linear change of variables, all weak orthogonal polynomial sets and all orthogonal polynomial sets to equations of the form (2.2.3) (or (2.2.2)).

Krall and Sheffer established the following 'classification' theorem:

Theorem 2.2.4

Suppose

$$P_{mn}(x,y) = x^m y^n + \text{lower order terms}, \quad m,n = 0,1,2,...$$

Define

$$A_{mn} =: (m+n)[a(m+n-1) + g]\mu_{mn} + m[d_1(m-1) + e_2n]\mu_{m-1,n}$$
$$+ n[d_2m + e_3(n-1)]\mu_{m,n-1} + f_1m(m-1)\mu_{m-2,n} + f_2mn\mu_{m-1,n-1}$$
$$+ f_3n(n-1)\mu_{m,n-2} + e_1m(m-1)\mu_{m-2,n+1} + d_3n(n-1)\mu_{m+1,n-2}$$

$$B_{mn} =: 2[a(m+n) + g]\mu_{m,n+1} + me_2\mu_{m-1,n+1} + [md_2 + 2ne_3]\mu_{mn}$$
$$+ mf_2\mu_{m-1,n} + 2nf_3\mu_{m,n-1} + 2nd_3\mu_{m+1,n-1}$$

$$C_{mn} =: 2[a(m+n) + g]\mu_{m+1,n} + [2md_1 + ne_2]\mu_{mn} + nd_2\mu_{m+1,n-1}$$
$$+ 2mf_1\mu_{m-1,n} + nf_2\mu_{m,n-1} + 2me_1\mu_{m-1,n+1} \qquad m,n = 0,1,2...$$

(a) Suppose $P_{mn}(x,y)$ satisfies (2.2.3) for $m,n = 0,1,2....$ If $\{P_{mn}(x,y)\}$ is at least a weak monic orthogonal polynomial set with respect to the weight functional L, then the moments $\mu_{mn} = L(x^m y^n)$ satisfy the recurrence relations $A_{mn} = B_{mn} = C_{mn} = 0$, $m,n = 0,1,2\ldots$

(b) Suppose $\{P_{mn}(x,y)\}$ is a monic polynomial set that satisfies (2.2.3), $m,n = 0,1,\ldots$ Suppose $\{\mu_{mn}\}$ is a double sequence of real numbers satisfying the recurrence relations $A_{mn} = B_{mn} = C_{mn} = 0$, $m,n = 0,1,2\ldots$ where $\mu_{00} \neq 0$. Then $\{P_{mn}(x,y)\}$ is at least a weak orthogonal polynomial set with respect to the weight functional T defined by $T(x^m y^n) = \mu_{mn}$.■

<u>Definition 2.2.2</u>

The recurrence relations $A_{mn} = B_{mn} = C_{mn} = 0$, $m,n = 0,1,2\ldots$ are called the <u>moment equations</u> of (2.2.3).

Theorem 2.2.4 can be simplified since $2A_{mn} = nB_{m,n-1} + mC_{m-1,n}$. Consequently,

<u>Theorem 2.2.5</u>

If any two of the relations $A_{mn} = 0$, $B_{mn} = 0$, $C_{mn} = 0$ hold for $m,n = 0,1,2\ldots$, then so does the third.

Using Theorem 2.2.4, Krall and Sheffer classified all second order partial differential equations having at least a weak orthogonal polynomial set. We list their findings.

1. $xu_{xx} + yu_{yy} + (1+\alpha-x)u_x + (1+\beta-y)u_y + nu = 0$

The polynomial solutions are $\Phi_{mn}(x,y) = L_{n-m}^{(\alpha)}(x)L_m^{(\beta)}(y)$, $n = 0,1,\ldots$, $m = 0,1,\ldots n$, where $\{L_n^{(\alpha)}(x)\}$ is the Laguerre polynomial sequence, orthogonal on $(0,\infty)$ with respect to $x^\alpha e^{-x}$. The polynomials $\{\Phi_{mn}(x,y)\}$ are orthogonal on $0 < x,y < \infty$ with respect to $\rho(x,y) = x^\alpha y^\beta e^{-x-y}$. Clearly, they form an OPS in the classical sense.

2. $u_{xx} + yu_{yy} - xu_x + (1+\alpha-y)u_y + nu = 0$

In this case, $\Phi_{mn}(x,y) = H_{n-m}(x)L_m^{(\alpha)}(y)$ satisfies this PDE where $\{H_k(x)\}$ is the Hermite polynomial sequence. $\{\Phi_{mn}(x,y)\}$ is an OPS in the classical sense with respect to the weight $\rho(x,y) = y^\alpha e^{-y-\frac{1}{2}x^2}$ on $-\infty < x < \infty$, $0 < y < \infty$.

3. $xu_{xx} + u_{yy} + (1+\alpha-x)u_x - yu_y + nu = 0$

Here, $\Phi_{mn}(x,y) = L_{n-m}^{(\alpha)}(x)H_m(y)$ satisfies this PDE, $n = 0,1,\ldots$, $m = 0,1,\ldots n$. $\{\Phi_{mn}(x,y)\}$ is an OPS in the classical sense with respect to the weight

$$\rho(x,y) = x^\alpha e^{-x-\frac{1}{2}y^2} \text{ on } 0 < x < \infty,\ -\infty < y < \infty.$$

4. $u_{xx} + u_{yy} - xu_x - yu_y + nu = 0$

Here, $\phi_{mn}(x,y) = H_{n-m}(x)H_m(y)$ satisfy this second order equation, $n = 0,1,\ldots$ $m = 0,1,\ldots n$. This sequence is also an OPS in the classical sense with respect to the weight function $\rho(x,y) = e^{-\frac{1}{2}(x^2+y^2)}$.

These first four examples are spe~i3l cases of the following:

Theorem 2.2.6

Let $\{P_n(x)\}$, $\{Q_n(x)\}$ be OPS's (not necessarily different) corresponding to the weight functionals L_1 and L_2 respectively. Define

$$R_{mn}(x,y) = P_{n-m}(x)Q_m(x)$$

$n = 0,1,\ldots$, $m = 0,1,\ldots n$. Then $\{R_{mn}(x,y)\}$ is an OPS in the classical sense with respect to the functional L defined by

$$L(x^n y^m) = L_1(x^n)L_2(y^m)$$

The sequence $\{R_{mn}(x,y)\}$ is called the product orthogonal polynomial sequence.

5. $3yu_{xx} + 2u_{xy} - xu_x - yu_y + nu = 0$

Krall and Sheffer showed that this equation does have an orthogonal polynomial sequence in the classical sense. Their weight function is unknown. The moment equations are easily solved to yield

$$\mu_{m+3k,m} = \frac{(m+3k)!}{m!}, \quad \mu_{m+3k+1,m} = \mu_{m+3k+2,m} = 0, \qquad m,k = 0,1,2\ldots$$

$$\mu_{m,m+k} = 0, \quad m = 0,1,\ldots, \quad k = 1,2,\ldots$$

This example illustrates why Krall and Sheffer took their approach to orthogonality (Definition 2.1.2). Indeed, notice that for $n = 2$, three independent polynomial solutions to the partial differential equation are

$$\phi_1(x,y) = x^2-6y, \quad \phi_2(x,y) = xy-1, \quad \phi_3(x,y) = y^2.$$

However, even though these three polynomials are monic polynomials and elements of a monic orthogonal set, they are not members of an orthogonal sequence in the classical sense. It is a nontrivial task to replace a monic orthogonal set by an orthogonal polynomial sequence in the classical sense. For example, we point out that if

$$\Psi_i(x,y) = \alpha_i(x^2-6y) + \beta_i(xy-1) + \gamma_i y^2, \qquad i = 1,2,3$$

are elements of an OPS in the classical sense, the coefficients α_i, β_i, γ_i must satisfy the nonlinear equations:

$$4\alpha_1\gamma_1 + \beta_1^2 \neq 0$$

$$4\alpha_2\gamma_2 + \beta_2^2 \neq 0$$

$$4\alpha_3\gamma_3 + \beta_3^3 \neq 0$$

$$2\alpha_1\gamma_2 + 2\alpha_2\gamma_1 + \beta_1\beta_2 = 0$$

$$2\alpha_1\gamma_3 + 2\alpha_3\gamma_1 + \beta_1\beta_3 = 0$$

$$2\alpha_2\gamma_3 + 2\alpha_3\gamma_2 + \beta_2\beta_3 = 0$$

In general, for any monic orthogonal polynomial set, to replace the (n+1) monic polynomials of degree n, there are (n+1)(n+2)/2 nonlinear equations to solve!

6. $(x^2-x)u_{xx} + 2xyu_{xy} + (y^2-y)u_{yy} + \{(\alpha+\beta+\gamma+3)x - (1+\alpha)\}u_x$
$+ \{(\alpha+\beta+\gamma+3)y - (1+\beta)\}u_y + \lambda u = 0 \qquad (\alpha,\beta,\gamma > -1)$

This equation does have an OPS of solutions in the classical sense. They are called the triangle polynomials since they are orthogonal with respect to

$$\rho(x,y) = x^\alpha y^\beta (1-x-y)^\gamma$$

in the triangular region $x \geq 0$, $y \geq 0$, $x+y \leq 1$.

7. $(x^2-1)u_{xx} + 2xyu_{xy} + (y^2-1)u_{yy} + gxu_x + gyu_y + \lambda u = 0$

The orthogonal polynomials that satisfy this equation are called the circle polynomials because their weight function is

$$\rho(x,y) = (1-x^2-y^2)^{(g-3)/2}$$

on the closed unit circle $x^2+y^2 \leq 1$. They are an OPS in the classical sense. They were first studied by Didon and Hermite and later, in more generality, by Appel and Kampé de Fériet; see [8] for further historical references.

8. $x^2u_{xx} + 2xyu_{xy} + (y^2-y)u_{yy} + g(x-1)u_x + g(y-\gamma)u_y + \lambda u = 0$

This equation has at least a weak monic orthogonal sequence of solutions. It is unknown if there is a monic orthogonal sequence of solutions. The computation of the determinants δ_n seems to be formidable. The weight function for these polynomials is unknown. The moments can easily be calculated to be

$$\mu_{k0} = \frac{g^{k-1}\mu_{00}}{(g+1)_{k-1}}, \quad k = 1,2,\ldots$$

$$\mu_{kj} = \frac{\gamma g^k (g\gamma+1)_{j-1}\mu_{00}}{(g+1)_{k+j-1}} \qquad k = 0,1,2\ldots,\ j = 1,2,\ldots$$

9. $(x^2+y)u_{xx} + 2xyu_{xy} + y^2u_{yy} + gxu_x + g(y-1)u_y + \lambda u = 0$

Krall and Sheffer show that this equation also has at least a weak monic orthogonal set - as in example 8, it is unknown whether or not the polynomials form a monic orthogonal set. The weight function is also unknown for this polynomial sequence. The moments are readily calculated to yield

$$\mu_{0k} = \frac{g^k \mu_{00}}{(g)_k} \qquad k = 0,1,2\ldots$$

$$\mu_{2j,k} = \frac{(-1)^j (2j-1)!\, g^{k+j} \mu_{00}}{2^{j-1}(j-1)!\,(g)_{k+2j}} \qquad j = 1,2,\ldots,\ k = 0,1,\ldots$$

$$\mu_{2j-1,k} = 0 \quad , \quad j = 1,2,\ldots,\ k = 0,1,\ldots .$$

§2.3 Another Approach to Krall and Sheffer's Results

Definition 2.3.1

Let $L(u) = A(x,y)u_{xx} + 2B(x,y)u_{xy} + C(x,y)u_{yy} + D(x,y)u_x + E(x,y)u_y + F(x,y)u$

The Lagrange adjoint of $L(u)$ is the expression

$$L^+(u) = (A(x,y)u)_{xx} + (2B(x,y)u)_{xy} + (C(x,y)u)_{yy} - (D(x,y)u)_x - (E(x,y)u)_y + F(x,y)u.$$

We say that L is symmetric if $L(u) = L^+(u)$. A function $\rho(x,y)$ is called a symmetry factor for $L(u)$ if $\rho(x,y)L(u)$ is symmetric.

It is easy to see that $\rho(x,y)$ is a symmetry factor for $L(u)$ above if and only if ρ simultaneously satisfies the following equations:

(2.3.1) $\quad (\rho A)_x + (\rho B)_y - \rho D = 0$

(2.3.2) $\quad (\rho B)_x + (\rho C)_y - \rho E = 0$

(2.3.3) $\quad (\rho A)_{xx} + (2B\rho)_{xy} + (\rho C)_{yy} - (\rho D)_x - (\rho E)_y = 0.$

Observe that (2.3.3) is superfluous: we can obtain (2.3.3) by differentiating (2.3.1) with respect to x, (2.3.2) with respect to y and then adding. We list equation (2.3.3) because it will arise later in our approach to Krall and Sheffer's results. Assuming the coefficients of $L(u)$ are sufficiently differentiable and $B^2 - AC \neq 0$, we can solve for ρ. If $B^2 - AC = 0$, ie $L(u)$ is parabolic, then ρ may or may not exist. For example, consider the equation

$$x^2u_{xx} + 2xyu_{xy} + y^2u_{yy} + 4xu_x + 5yu_y + u = 0 \qquad (x,y) \in R^2$$

It is easy to check that no symmetry factor can exist for this equation. Now consider the equation

$$x^2u_{xx} + 2xyu_{xy} + y^2u_{yy} + 3xu_x + 3yu_y + u = 0 \qquad (x,y) \in R^2\backslash(0,0)$$

Equations (2.3.1) and (2.3.2) are

$$x(x\rho_x + y\rho_y) = 0$$

$$y(x\rho_x + y\rho_y) = 0$$

The general solution to these equations is $\rho(x,y) = f(\theta)$ where $f(\theta) \in C^1(R)$ and $f(\theta)$ has period 2π.

We remark that all of the partial differential equations listed in §2.2 that have orthogonal polynomial solutions where the weight function is known, are elliptic equations in the region of orthogonality.

Suppose $\{\phi_{mn}(x,y)$ is at least a weak monic orthogonal polynomial sequence of solutions to (2.2.3). Assume that $L_2(u)$ can be made symmetric. Suppose that $\Lambda(x,y)$ is a nonzero weak solution to (2.3.1) and (2.3.2) on polynomials. By this we mean

(2.3.4) $$< (\Lambda A)_x + (\Lambda B)_y - \Lambda D,\ x^my^n > = 0$$

(2.3.5) $$< (\Lambda B)_x + (\Lambda C)_y - \Lambda E,\ x^my^n > = 0\ , \quad m,n = 0,1,\ldots$$

In addition, suppose $\Lambda(x,y) \to 0$ as $|x|,|y| \to \infty$. Set $\mu_{mn} = < \Lambda,\ x^my^n >$

Then

$$0 = < (\Lambda A)_x + (\Lambda B)_y - \Lambda D,\ x^my^n >$$

$$= -< \Lambda A, \frac{\partial}{\partial x}(x^my^n)> - < \Lambda B, \frac{\partial}{\partial y}(x^my^n)> - < \Lambda D,\ x^my^n>$$

$$= -< \Lambda,\ mx^{m-1}y^n(ax^2+d_1x+e_1y+f_1> - < \Lambda,\ nx^my^{n-1}(axy+\frac{d_2x}{2} + \frac{e_2y}{2} + \frac{f_2}{2}) >$$

$$-< \Lambda,\ gx^{m+1}y^n>$$

$$= \frac{-1}{2}\ C_{mn}$$

Similarly, $0 = < (\Lambda B)_x + (\Lambda C)_y - \Lambda E,\ x^my^n > = \frac{-1}{2}B_{mn}$. Incidentally, if we substitute Λ into (2.3.3), we find that

$$0 = A_{mn} = < (\Lambda A)_{xx} + (2\Lambda B)_{xy} + (\Lambda C)_{yy} - (\Lambda D)_x - (\Lambda E)_y\ ,\ x^my^n >$$

$$= < (\Lambda A)_{xx} + (\Lambda B)_{xy} - (\Lambda D)_x\ ,\ x^my^n > + < (\Lambda B)_{xy} + (\Lambda C)_{yy} - (\Lambda E)_y\ ,\ x^my^n >$$

$$= - < (\ A)_x + (\ B)_y - D,\ mx^{m-1}y^n > - < (\Lambda B)_x + (\Lambda C)_y - \Lambda E,\ nx^my^{n-1} >$$

$$= \frac{m}{2}\ C_{m-1,n} + \frac{n}{2}\ B_{m,n-1}$$, which is in agreement with Theorem 2.2.5.

Summarizing, we have:

Theorem 2.3.1

Suppose $\{\Phi_{mn}(x,y)\}$ is at least a weak monic orthogonal set of solutions to (2.2.3). Suppose $L_2(u)$ can be made symmetric and there exists a nonzero weak solution Λ to equations (2.3.4) and (2.3.5) satisfying $\Lambda(x,y) \to 0$ as $|x|, |y| \to \infty$. Assuming the moment equations have a unique solution once $\mu_{oo} \neq 0$ is specified, Λ is an orthogonalizing weight functional for $\{\Phi_{mn}(x,y)\}$.■

Definition 2.3.2

Equations (2.3.1) and (2.3.2) are called the symmetry equations of L(u) while equations (2.3.4) and (2.3.5) are called the weight equations for $\{\Phi_{mn}(x,y)\}$.

The next result gives us a formal weight distribution for $\{\Phi_{mn}(x,y)\}$. Notice how the formula below is similar to the formula in equation (1.1.4).

Theorem 2.3.2

Suppose $\{\Phi_{mn}(x,y)\}$ is at least a weak monic orthogonal polynomial sequence with moment sequence $\{\mu_{mn}\}$. Define

$$(2.3.6) \qquad w(x,y) = \sum_{m=0}^{\infty}\sum_{n=0}^{\infty} \frac{(-1)^{m+n}\mu_{mn}\,\delta^{(m)}(x)\delta^{(n)}(y)}{m!n!}$$

Then w(x,y) is a formal orthogonalizing weight functional for $\{\Phi_{mn}(x,y)\}$.

Proof

It suffices to show $< w(x,y),\ x^k y^l > = \mu_{kl}$, $k,l = 0,1,\ldots$

Formally,

$$< w(x,y),\ x^k y^l > = \sum_{m=0}^{\infty}\sum_{n=0}^{\infty} \frac{(-1)^{m+n}\mu_{mn}}{m!n!} < \delta^{(m)}(x)\delta^{(n)}(y),\ x^k y^l >$$

By definition,

$$< \delta^{(m)}(x)\delta^{(n)}(y),\ x^k y^n > = < \delta^{(m)}(x),\ x^k > < \delta^{(n)}(y),\ y^l >$$
$$= (-1)^{k+l} k!l!\delta_{mk}\delta_{nl}$$
$$= \begin{cases} (-1)^{k+l}k!l! & \text{if } m=k \text{ and } n=l \\ 0 & \text{otherwise} \end{cases}$$

Substitution of this above yields the result.■

In Theorem 1.2.4 we mentioned that w(x), given in (1.1.4), satisfies the weight equations (1.2.1). Can we expect a similar result in this situation? The answer is affirmative:

Theorem 2.3.3

Suppose $\{\Phi_{mn}(x,y)\}$ is at least a weak orthogonal polynomial sequence of solutions to (2.2.3). Suppose $\{\mu_{mn}\}$ is the moment sequence associated with $\{\Phi_{mn}(x,y)\}$. Then $w(x,y)$, given in (2.3.6), formally satisfies the weight equations (2.3.4) and (2.3.5) on polynomials.

Proof

$$- \langle (wA)_x + (wB)_y - wD, x^m y^n \rangle$$

$$= \langle w, mx^{m-1}y^n(ax^2+d_1x + e_1y + f_1) + nx^m y^{n-1}(axy + \tfrac{d_2x}{2} + \tfrac{e_2y}{2} + \tfrac{f_2}{2}) + gx^{m+1}y^n \rangle$$

Since $\langle w, x^m y^n \rangle = \mu_{mn}$, we find that

$- \langle (wA)_x + (wB)_y - wD, x^m y^n \rangle = -\tfrac{1}{2} C_{mn} = 0$, by Theorem 2.2.4.

Similarly, it is easy to check that w satisfies (2.3.5).■

Of course, we give meaning to $w(x,y)$ as follows: the Fourier transform of $w(x,y)$ is

$$F(w(x,y)) = \sum_{m=0}^{\infty}\sum_{n=0}^{\infty} \frac{(-1)^{m+n}\mu_{mn}(it)^m(is)^n}{m!n!}$$

Suppose $F(w(x,y))$ converges pointwise to some function $f(t,s)$ in some region of R^2. If $F^{-1}(f(t,s))$ exists, where F^{-1} is the inverse Fourier transform, we <u>define</u>

$$\sum_{m=0}^{\infty}\sum_{n=0}^{\infty} \frac{(-1)^{m+n}\mu_{mn}\delta^{(m)}(x)\delta^{(n)}(y)}{m!n!} = F^{-1}(f(t,s))(x,y).$$

We illustrate an example of this below. We now consider some of the partial differential equations found by Krall and Sheffer and show how our methods apply.

1. $xu_{xx} + yu_{yy} + (1+\alpha-x)u_x + (1+\beta-y)u_y + nu = 0$, $\alpha,\beta > -1$

The symmetry equations in this PDE are

$$\text{(i)}\quad x\rho_x + (x-\alpha)\rho = 0$$

$$\text{(ii)}\quad y\rho_y + (y-\beta)\rho = 0$$

These equations may be rewritten as

$$\text{(a)}'\quad x^{\alpha+1}e^{-x}(x^{-\alpha}e^x\rho)_x = 0$$

$$\text{(b)}'\quad y^{\beta+1}e^{-y}(y^{-\beta}e^y\rho)_y = 0$$

Clearly, their classical solution if $\rho(x,y) = x^\alpha y^\beta e^{-x-y}$. To find the distributional solution to (a)', first divide by $x^{\alpha+1}e^{-x}$:

$$(x^{-\alpha}e^{x}w)_x = \beta_1(y)\delta(x) + \beta_2(y)\delta'(x) + \ldots + \beta_{[\alpha]}(y)\delta^{[\alpha]}(x)$$

where $[\alpha]$ is the greatest integer part of α. Integrating with respect to x yields

$$x^{-\alpha}e^{x}w(x,y) = \beta_1(y)H(x) + \beta_2(y)\delta(x) + \ldots + \beta_{[\alpha]}(y)\delta^{[\alpha-1]}(x) + k(y)$$

so that

$$w(x,y) = \beta_1(y)H(x)x^{\alpha}e^{-x} + k(y)x^{\alpha}e^{-x} \tag{2.3.7}$$

Similarly, the distributional solution to (b)' is

$$w(x,y) = \gamma_1(x)H(y)y^{\beta}e^{-y} + p(x)y^{\beta}e^{-y} \tag{2.3.8}$$

The simultaneous distributional solution to (a)' and (b)' is

$$w(x,y) = c_1H(x)H(y)x^{\alpha}y^{\beta}e^{-x-y} + c_2x^{\alpha}y^{\beta}e^{-x-y}.$$

Since we are requiring that $w(x,y) \to 0$ as $|x|, |y| \to \infty$, we see that our weak solution to (a)' and (b)' is

$$\Lambda(x,y) = c_1H(x)H(y)x^{\alpha}y^{\beta}e^{-x-y}$$

Of course, this is the weight function for the OPS satisfying the above second order PDE. Alternatively, we can compute a weight function via formula (2.3.6). The moment equations are

$$\mu_{m+1,n} - (m+\alpha+1)\mu_{mn} = 0$$

and

$$\mu_{m,n+1} - (n+\beta+1)\mu_{mn} = 0.$$

They are easily solved to yield $\mu_{mn} = \Gamma(m+\alpha+1)\Gamma(n+\beta+1)\mu_{00}$

Substituting these into (2.3.6) yields

$$w(x,y) = \sum_{m=0}^{\infty}\sum_{n=0}^{\infty}\frac{(-1)^{m+n}\Gamma(m+\alpha+1)\Gamma(n+\beta+1)\mu_{00}\delta^{(m)}(x)\delta^{(n)}(y)}{m!n!}$$

The Fourier transform of this is

$$F(w(x,y)) = \frac{1}{2\pi}\sum_{m=0}^{\infty}\sum_{n=0}^{\infty}\frac{(-1)^{m+n}\Gamma(m+\alpha+1)\Gamma(n+\beta+1)\mu_{00}(it)^{m}(is)^{n}}{m!n!}$$

$$= \frac{\Gamma(\alpha+1)\Gamma(\beta+1)\mu_{00}}{2\pi(1+it)^{\alpha+1}(1+is)^{\beta+1}}$$

The inverse Fourier transform of this is well known to be $w(x,y) = H(x)H(y)x^{\alpha}y^{\beta}e^{-x-y}$.

2. $(x^2-x)u_{xx} + 2xyu_{xy} + (y^2-y)u_{yy} + \{(\alpha+\beta+\gamma+3)x - (1+\alpha)\}u_x$

$+ \{(\alpha+\beta+\gamma+3)y - (1+\beta)\}u_y + \lambda u = 0$

Here the symmetry equations are

$$\text{(a)}\quad (x^2-x)\rho_x + xy\rho_y + ((-\alpha-\beta-\gamma)x + \alpha)\rho = 0$$

$$\text{(b)}\quad xy\rho_x + (y^2-y)\rho_y + ((-\alpha-\beta-\gamma)y + \beta)\rho = 0$$

It is easy to check that $\rho(x,y) = x^\alpha y^\beta(1-x-y)^\gamma$ is the simultaneous solution to (a) and (b). To find the weak solution, we combine (a) and (b) and obtain:

$$\text{(a)}'\quad y(1-x-y)^{\gamma+1}x^{\alpha+1}((1-x-y)^{-\gamma}x^{-\alpha}w)_x = 0$$

and $$\text{(b)}'\quad x(1-x-y)^{\gamma+1}y^{\beta+1}((1-x-y)^{-\gamma}y^{-\beta}w)_y = 0$$

Following the method in example 1, we find that the simultaneous weak solution to (a)' and (b)' is

$$w(x,y) = H(x)H(y)H(1-x-y)x^\alpha y^\beta(1-x-y)^\gamma.$$

3. Our next example concerns one of the equations that Krall and Sheffer found where the weight function is unknown for the corresponding orthogonal polynomial solutions. We shed some light on a probable reason why a suitable representation of this weight is unavailable.

$$\text{(2.3.9)}\quad x^2u_{xx} + 2xyu_{xy} + (y^2-y)u_{yy} + (gx-g)u_x + (gy-g\gamma)u_y + \lambda u = 0$$

In this case, the symmetry equations may be written as:

$$\text{(a)}\quad xy^{g\gamma}e^{-gy/x}(y^{1-g\gamma}e^{gy/x}\rho)_y = 0$$

$$\text{(b)}\quad x^{g-g\gamma}e^{(-gy+g)/x}(x^{2+g\gamma-g}e^{(gy-g)/x}\rho)_x = 0$$

The simultaneous solution to (a) and (b) for a symmetry factor yields

$$\rho(x,y) = e^{(g-gy)/x}x^{g-2-g\gamma}y^{g\gamma-1}.$$

Because of the essential singularity at $x = 0$, it is unclear what the simultaneous weak solution is to (a) and (b). Of course, if we substitute the moments (found in §2.2) into (2.3.6), we do get the following formal weight distribution:

$$w(x,y) = \sum_{m=0}^{\infty}\frac{(-1)^m g^{m-1}\delta^{(m)}(x)\delta(y)}{(g+1)_{m-1}\,m!} + \sum_{n=1}^{\infty}\sum_{m=0}^{\infty}\frac{(-1)^{m+n}\gamma g^m(g\gamma+1)_{n-1}\delta^{(m)}(x)\delta^{(n)}(y)}{(g+1)_{m+n-1}\,m!n!}$$

Unfortunately, use of the Fourier transform does not seem to simplify this formal expression.

It is easy to check that $P_n^{(g-g\gamma-1,g\gamma-1)}(2y-1)$ is a solution to (2.3.9) where $P_n^{(g-g\gamma-1,g\gamma-1)}(2y-1)$ is the Jacobi polynomial with parameters $\alpha = g-g\gamma-1$ and $\beta = g\gamma-1$. Similarly, $y_n(x,g,-g)$ is also a solution to (2.3.9) where $y_n(x,g,-g)$ is the generalized Bessel polynomial. (Unfortunately, the product polynomials

$$\Phi_{mn}(x,y) = y_{n-m}(x,g,-g)P_m^{(g-g\gamma-1,g\gamma-1)}(2y-1)$$

only satisfy (2.3.9) when m = 0 or m = n.) The appearance of the Bessel polynomials is a good reason why a suitable weight for the orthogonal polynomial solutions to (2.3.9) has never been found.

4. This next example exhibits behavior similar to the example above.

(2.3.10) $\quad (x^2+y)u_{xx} + 2xyu_{xy} + y^2u_{yy} + gxu_x + g(y-1)u_y + \lambda u = 0$

Here, the symmetry equations may be combined and rewritten as:

$$\text{(a)}\quad y^g e^{(gx^2+2gy)/2y^2}\left(y^{3-g} e^{(-gx^2-2gy)/gy^2}\rho\right)_y = 0$$

$$\text{and}\quad \text{(b)}\quad y^2 e^{gx^2/2y^2}\left(e^{-gx^2/2y^2}\rho\right)_x = 0$$

Their classical solution is

$$\rho(x,y) = y^{g-3}\, e^{(gx^2+2gy)/2y^2}$$

Again, because of the essential singularity at y = 0, the weak solution to these equations is unknown. The formal weight distribution is

$$w(x,y) = \sum_{m=0}^{\infty}\sum_{n=0}^{\infty} \frac{(-1)^{m+n} g^{m+n} \delta^{(2m)}(x)\delta^{(n)}(y)}{2^m m!\,n!\,(g)_{2m+n}}$$

Again, the Fourier transform fails to simplify this double series. It is interesting to note that $y_n(y,g,-g)$ is a solution of (2.3.10). As in example 3, the appearance of Bessel polynomial solutions could explain why a classical weight function is unknown for the orthogonal polynomials.

5. $\quad (x^2-1)u_{xx} + 2xyu_{xy} + (y^2-1)u_{yy} + gxu_x + gyu_y + \lambda u = 0$

The symmetry equations for PDE can be rewritten as:

$$\text{(a)}\quad (1-x^2-y^2)^{(g-1)/2}\left((1-x^2-y^2)^{(3-g)/2}\rho\right)_y = 0$$

and

$$\text{(b)} \quad (1-x^2-y^2)^{(g-1)/2}\left((1-x^2-y^2)^{(3-g)/2}\,\rho \right)_x = 0,$$

from which we immediately get the symmetry factor $\rho(x,y) = (1-x^2-y^2)^{(g-3)/2}$. Following the method given in example 1, we find that the distributional solution to (a) and (b) is

$$w(x,y) = H(1-x^2-y^2)(1-x^2-y^2)^{(g-3)/2}.$$

6. Our last example could be the most intriguing of all our examples.

$$(2.3.11) \qquad 3yu_{xx} + 2xyu_{xy} - xu_x - yu_y + nu = 0$$

The symmetry equations are

$$\text{(a)} \quad 3y\rho_x + \rho_y + x\rho = 0$$

and

$$\text{(b)} \quad \rho_x + y\rho = 0$$

They are easily solved to yield the symmetry factor $\rho(x,y) = e^{-xy+y^3}$. It appears also that $\rho(x,y)$ is also the weak solution to these symmetry equations! What is the region of orthogonality then? If we substitute the moments found in §2.2, we arrive at the formal weight:

$$w(x,y) = \sum_{m=0}^{\infty}\sum_{n=0}^{\infty} \frac{(-1)^n \delta^{(m+3n)}(x)\,\delta^{(m)}(y)}{m!\,n!}$$

The Fourier transform of $w(x,y)$ is

$$F(w(x,y))(t,s) = e^{-st}(\cos t^3 + i\sin t^3)$$

This author does not know the inverse Fourier transform of this function.

In closing, much work needs to be done on orthogonal polynomial solutions to partial differential equations. In particular, much is still to be learned about the polynomial solutions of the equations appearing in examples 3,4 and 6 above.

REFERENCES

1. F.V. ATKINSON AND W.N. EVERITT, "Orthogonal polynomials which satisfy second order differential equations", in E.B. Christoffel, ed. P.L. Butzer and F. Fehér, Birkhäuser, Basel, 1981, 173-181.
2. R.P. BOAS, JR., "The Stieltjes moment problem for functions of bounded variation", Bull. Amer. Math. Soc., 45, 1939, 399-404.
3. S. BOCHNER, "Über Sturm-Liouvillesche Polynomsysteme", Math. Z. 29, 1929, 730-736.

4. T.H. KOORNWINDER, "Orthogonal polynomials with weight function $(1-x)^{\alpha}(1+x)^{\beta} + M\delta(x+1) + N\delta(x-1)$", Canad. Math. Bull., 27(2), 1984, 205-214.

5. H.L. KRALL, "Certain differential equations for Tchebycheff polynomials", Duke Math. J., 4, 1938, 705-718.

6. H.L. KRALL, "On orthogonal polynomials satisfying a certain fourth order equation", The Pennsylvania State College Studies, No. 6, The Pennsylvania State College, State College, Pa., 1940.

7. H.L. KRALL and O. FRINK, "A new class of orthogonal polynomials: the Bessel polynomials", Trans. Amer. Math. Soc., 65, 1949, 100-115.

8. H.L. KRALL and I.M. SHEFFER, "Orthogonal polynomials in two variables", Ann. Mat. Pura Appl., 4, 1967, 325-376.

9. A.M. KRALL and R.D. MORTON, "Distributional weight functions for orthogonal polynomials", SIAM J. Math. Anal., 9, 1978, 604-626.

10. A.M. KRALL, "Orthogonal polynomials satisfying fourth order differential equations", Proc. Roy. Soc. Edinburgh Sect. A, 87, 1981, 271-288.

11. A.M. KRALL and L.L. LITTLEJOHN, "Orthogonal polynomials and singular Sturm-Liouville systems I", Rocky Mountain J. Math., 16(3), 1986, 435-479.

12. A.M. KRALL and L.L. LITTLEJOHN. "Orthogonal polynomials and singular Sturm-Liouville systems II", submitted.

13. A.M. KRALL and L.L. LITTLEJOHN, "Sturm-Liouville operators and orthogonal polynomials", submitted.

14. L.L. LITTLEJOHN, "The Krall polynomials: a new class of orthogonal polynomials", Quaestiones Math., 5, 1982, 255-265.

15. L.L. LITTLEJOHN, "On the classification of differential equations having orthogonal polynomial solutions", Ann. Mat. Pura Appl., 4, 1984, 35-53.

16. L.L. LITTLEJOHN, "An application of a new theorem on orthogonal polynomials and differential equations", Quaestiones Math., 10, 1986, 49-61.

17. L.L. LITTLEJOHN and A.M. KRALL, "On the classification of differential equations having orthogonal polynomial solutions II", Ann. Mat. Pura Appl., to appear.

18. L.L. LITTLEJOHN and A.M. KRALL, "Necessary and sufficient conditions for the existence of symmetry factors for real ordinary differential expressions", submitted.

Department of Mathematics,
Utah State University,
Logan, Utah, 84322-4125,
U.S.A.

Rational approximations, orthogonal polynomials and equilibrium distributions

G. López. Math. Dept., Hav. Univ., Havana
E.A. Rakhmanov, Steklov Math. Inst., Moscow

A number of questions of rational approximation and orthogonal polynomials may be reduced to some general problems concerning the equilibrium distributions of a "system of charges" on a "system of conductors" in the presence of exterior fields. The corresponding concept of equilibrium for "vector-potentials" is considered in the paper of A.A. Goncar and E.A. Rakhmanov [9]. The most simple case of a "single charge" was discused in their paper [8]. Here we consider several problems of approximation theory mainly connected with the equilibrium distributions for a single potential. We note that we do not pretend to give a complete review of all the contributions obtained in this direction, we limit ourselves to those results due to A.A.Gončar, G.López and E.A.Rakhmanov.

We begin with the following simple but important example which contains almost all the essential features.

1. Rational interpolation of Markov-type functions.

Let F be a weight (integrable almost everywhere positive function) on the segment $I = [-1,1]$ and

$$\hat{F}(z) = \int_I \frac{F(x)\,dx}{z-x}, \qquad z \in \mathcal{D} = \overline{\mathbb{C}} \setminus I \tag{1}$$

Let E be another segment on the real line, that doesn't intersect I, on E is given a triangular table of interpolation points

$$\Omega = \{ \beta_{i,n}, \; i = 1,2,\dots,2n, \quad n \in \mathbb{N} \}$$

We can construct a sequence of rational functions

$$R_n(z) = \frac{P_{n-1}}{Q_n}(z), \qquad P_{n-1} \in \mathbb{P}_{n-1}, \; Q_n \in \mathbb{P}_n$$

($\mathbb{P}_n$ is the set of algebraical polynomials of degree at most n) that interpolate to $\hat{F}$ with respect to table Ω according to the following formula

$$\frac{Q_n \hat{F} - P_{n-1}}{\omega_{2n}}(z) \in \mathcal{H}(E), \qquad \omega_{2n}(z) = \prod_{i=1}^{2n} (z-\beta_{i,n}) \tag{2}$$

(where $\mathcal{H}(E)$ is the set of analytic functions on E). Now, suppose we wish to study the convergence properties of the sequence $\{R_n\}$. The first thing which we encounter is a set of orthogonal polynomials with respect to varying weights.

1.1 Orthogonal polynomials with respect to varying weights.

Standard operations show that the conditions of interpolation (2) for function $\hat{F}$ are equivalent to the following relations of orthogonality for the denominators Q_n

$$\int_I Q_n(x)\, x^j \, \frac{F(x)}{\omega_{2n}(x)}\, dx = 0, \qquad j = 0, 1, \dots, n-1 \tag{3}$$

If we also consider the interpolation formula of Hermite

$$\hat{F}(z) - R_n(z) = \frac{\omega_{2n}(z)}{Q_n^2(z)} \int_I \frac{Q_n^2(x)}{\omega_{2n}(x)} \frac{F(x)\,dx}{z-x}, \qquad z \in \mathcal{D} \tag{4}$$

then we can see that the study of rational interpolation of Markov -type functions is reduced to the study of the asymptotics of the polynomials Q_n . This asymptotics may be written in terms of some equilibrium distribution.

1.2 Equilibrium in the field of a fixed negative charge.

In the following, all measures are considered to be finite and positive. S_μ and $\|\mu\|$ denote the support and the norm of measure μ . By $V^\mu(z) = \int \log|z-t|^{-1} d\mu(t)$ we denote the logarithmic potential of μ . A charge is the difference of two measures.

Let ω be a fixed unit measure on the segment E . Then there existe a unique unit measure $\mu = \mu(\omega)$ on the segment I such that

$$V^\mu(x) - V^\omega(x) \equiv w = \text{const}, \qquad x \in I \tag{5}$$

We call this, the equilibrium condition in the exterior field $\varphi(x) = -V^\omega(x)$. This terminology is due to the well known electrostatic interpretation. The measure is interpreted as the density of the distribution of the electric charge, whose different parts interact according to the law of the logarithmic potential. Then, $V^\mu(x) - V^\omega(x)$-is the full potential and (5) is the usual physical condition of equilibrium in the exterior field.

On the other hand, equality (5) simply means that the measure μ is the balayage of ω from $\mathcal{D}$ on I.

We note inmediately an importante feature of the equilibrium distribution of a positive unit charge in the exterior field of a negative unit charge. The support of the equilibrium distribution is the whole segment (already this is not so in a field generated by a positive charge, the general case is considered in section 2). Because of this, the equilibrium condition is simplified and it is possible to find measure μ and its potential V^{μ} explicitly for an arbitrary ω

$$d\mu(x) = \left\{\frac{1}{\pi}\int\frac{\sqrt{t^2-1}}{|x-t|}\,d\omega(t)\right\}\frac{dx}{\sqrt{1-x^2}}$$

$$w = \int g(t,\infty)\,d\omega(t)$$

$$V^{\mu}(z) = w + V^{\omega}(z) + \int g(z,t)\,d\omega(t)$$

(where $g(z,t)$ is Green's function for the region $\mathcal{D}$ with singularity at point $t\in\mathcal{D}$).

1.3 Asymptotics for polynomials orthogonal with respect to varying weights and convergence of rational interpolants.

Let P be an arbitrary polynomial. We denote

$$\chi(P) = \sum_{P(\xi)=0}\delta(\xi)$$

the associated measure ($\delta(\xi)$ is the unit measure with support at point ξ); note that $\|\chi(P)\| = \deg P$

THEOREM. Suppose that the sequence of measures $\frac{1}{2n}\chi(\omega_{2n})$ is weakly convergent to measure ω on some subsequence of indexes $\Lambda\subset\mathbb{N}$. Then for this subsequence Λ we have

$$\frac{1}{n}\chi(Q_n) \to \mu = \mu(\omega), \quad n\in\Lambda, \ n\to\infty$$

where μ is the equilibrium measure in the field $\varphi(x) = -V^{\omega}(x)$ (see (5)), and

$$\lim_{\substack{n\to\infty\\ n\in\Lambda}} |\hat{F}(z) - R_n(z)|^{\frac{1}{2n}} = \exp\left\{-\int g(z,t)\,d\omega(t)\right\} < 1, \quad z\in\mathcal{D}$$

Corollary. The sequence $\{R_n\}_{n\in\mathbb{N}}$ uniformly converges to $\hat{F}$ on each compact subset of $\mathcal{D}$ for every table Ω on $E\subset\mathbb{R}\setminus I$.

We note that the assertion of the corollary is valid for arbitrary non-negative functions F on I (not only for weights). The classical Markov Theorem is obtained when $\beta_{i,n} = +\infty, \forall i,n$. For other details see A.A.Gončar [3], A.A.Gončar and G.López [6].

So, we have seen that the problem of convergence of multi-point Pade approximants for Markov-type functions is reduced to the study of orthogonal polynomials with respect to sequences of weight functions. The asymptotics of these polynomials can be described in terms of a certain equilibrium distribution. In the following, we consider more examples which develop these ideas.

2. Equilibrium in an arbitrary field and logarithmic (weak) asymptotics of orthogonal polynomials.

This section is based on the paper of A.A.Gončar and E.A.Rakhmanov [8].

Various problems lead to exterior fields which cannot be represented as potentials of charges and so we must introduce a more general class $\mathcal{F}$ of exterior fields. Let

$$\mathcal{F} = \{ \varphi : \mathbb{R} \to (-\infty, +\infty] : \text{(i) and (ii) are satisfied}$$

(i) $\varphi(x)$ is a lower semicontinuous and weak approximatively continuous at every point $x_0 \in \mathbb{R}$

$$\underline{\lim}_{x \to x_0} \varphi(x) = \lim_{x \to x_0,\, x \in e(x_0)} \varphi(x) = \varphi(x_0)$$

where $e(x_0)$ is a certain set (depending also on φ) which has a positive density at point x_0

(ii) $$\underline{\lim}_{|x| \to \infty} \frac{\varphi(x)}{\log|x|} = +\infty$$

Let $\mathcal{M}_n = \{ \mu : S_\mu \subset \mathbb{R}, \|\mu\| = n \}$ be the set of measures on $\mathbb{R}$ with norm $n > 0$.

Lemma. Suppose that $n > 0$, $\varphi \in \mathcal{F}$, then there exists a unique measure $\mu \in \mathcal{M}_n$ which satisfies the following "equilibrium condition".

$$V^{\mu}(x)+\varphi(x)\quad \begin{matrix} \equiv w = \text{const}, & x\in S_{\mu} \\ \geq w \quad , & x \in \mathbb{R} \end{matrix} \qquad (6)$$

We call $\mu=\mu_{n,\varphi}$ the equilibrium measure in the field φ (with norm n) ; $w = w_{n,\varphi}$ is the corresponding extremal constant. We underline the fact that the equilibrium condition uniquely determines the pair (μ, w) .

The equilibrium measure satisfies a series of important characteristical properties; we write down some of them without further comment.

(a) The main extremal property is

$$w = \min_{x\in\mathbb{R}} \left(V^{\mu}+\varphi\right)(x) = \max_{\sigma\in\mathcal{M}_n} \min_{x\in\mathbb{R}} \left(V^{\sigma}+\varphi\right)(x). \qquad (7)$$

(this explains why we call w the extremal constant).

(b) The extremal growth of a normed potential is given by

$$w - V^{\mu}(z) \geq \min_{x\in\mathbb{R}} \left(V^{\sigma}+\varphi\right)(x) - V^{\sigma}(z)$$

for all $\sigma\in\mathcal{M}_n$ and $z\in\mathbb{C}$ (if equality takes place at least in one point $z\in\mathbb{C}\setminus S_{\mu}$ then $\sigma=\mu$).

(c) The equilibrium measure minimizes the energy integral in the exterior field; if

$$\mathcal{E}_{\varphi}(\sigma) = \iint \log\frac{1}{|x-y|}\,d\sigma(x)\,d\sigma(y) + 2\int\varphi(x)\,d\sigma(x)$$

then

$$\mathcal{E}_{\varphi}(\mu) = \min_{\sigma\in\mathcal{M}_n} \mathcal{E}_{\varphi}(\sigma)$$

(d) The characteristic property of the support of the equilibrium measure is: let $\sigma\in\mathcal{M}_n$, if $x\in\mathbb{R}$ and

$$V^{\sigma}(x)+\varphi(x) = \min_{x\in\mathbb{R}} \left(V^{\sigma}+\varphi\right)(x) \quad \text{then} \quad x\in S_{\mu}$$

These statements are proved similar to the way that it is done for the classical case ($\varphi\equiv 0$, $x\in E$, $\varphi\equiv+\infty$, $x\in\mathbb{R}\setminus E$ where E is a compact set). The main difficulty arises when trying to find explicitly the equilibrium measure and the extremal constant; this is due mainly to the fact that the support of the equilibrium measure is not known in advance. Let us discuss in more detail the following important example.

2.1 The case of a single segment.

Let $\varphi \in \mathcal{F}$, $n > 0$ and suppose it is apriori known that the support of the equilibrium measure $\mu_n = \mu_{n,\varphi}$ is a single segment

$$S_{\mu_n} = [x_n, y_n] \qquad (x_n = x_{n,\varphi},\ y_n = y_{n,\varphi}).$$

Then from the equilibrium condition (8) we can get the equality

$$V^{\mu_n}(z) = W_n - u_{n,\varphi}(z) - n\, g_n(z) \tag{8}$$

where

$$g_n(z) = \operatorname{Re} \int_{x_n}^{z} \frac{d\zeta}{\sqrt{(\zeta - x_n)(\zeta - y_n)}}, \quad z \in \mathcal{D}_n = \mathbb{C} \setminus [x_n, y_n]$$

is Green's function for the domain $\mathcal{D}_n = \mathbb{C} \setminus [x_n, y_n]$ and

$$u_{n,\varphi}(z) = \operatorname{Re} \left\{ \sqrt{(z - x_n)(z - y_n)}\ \frac{1}{\pi} \int_{x_n}^{y_n} \frac{\varphi(t)}{z - t} \frac{dt}{\sqrt{(t - x_n)(y_n - t)}} \right\}, \quad z \in \mathcal{D}_n$$

is the harmonic function in $\mathcal{D}_n$ whose boundary values are $u^{\pm}_{n,\varphi}(x) = \varphi(x)$, $x \in \partial \mathcal{D}_n$ (the existence of the integral is supposed).

Now, the analysis of the behavior of $V^{\mu_n}(x)$ for $x \to x_n - 0$ and $x \to y_n + 0$ gives the system of equations which allows to determine the extreme points x_n, y_n of the support of the measure. The system can be written, for example, as follows

$$\begin{cases} \dfrac{1}{\pi} \displaystyle\int_{x_n}^{y_n} \frac{2t - x_n - y_n}{\sqrt{(t - x_n)(y_n - t)}} \varphi'(t)\, dt = \frac{1}{2\pi} \int_{x_n}^{y_n} \sqrt{(t - x_n)(y_n - t)}\, \varphi''(t)\, dt - 2n \\[2ex] \dfrac{1}{\pi} \displaystyle\int_{x_n}^{y_n} \frac{\varphi'(t)\, dt}{\sqrt{(t - x_n)(y_n - t)}} = 0 \end{cases} \tag{9}$$

Here we must suppose that all these integrals are finite. There are interesting cases when this is not so; for example, when we have equilibrium on a finite segment $[a, b]$ ($\varphi(x) \equiv +\infty, x \in \mathbb{R} \setminus [a, b]$). Such situations require small modifications which we will not discuss.

For the density of the equilibrium measure we have the expression

$$d\mu_n(t) = \left\{ \sqrt{(t - x_n)(y_n - t)}\ \frac{1}{\pi} \int \frac{\varphi'(t) - \varphi'(\tau)}{t - \tau} \frac{d\tau}{\sqrt{(\tau - x_n)(y_n - \tau)}} \right\} dt, \quad t \in [x_n, y_n] \tag{10}$$

For instance, if $\varphi \in C^{1+\varepsilon}$ then this integral exist in the usual sense for each point t . In more general situations, the integral must be understood in the sense of the principal value.

2.2 A condition for S_{μ_n} to be a segment.

As was pointed out above it is difficult to give a simple condition on φ, n to guarantee that S_{μ_n} be a segment.

In some cases, this can be obtained on the basis of the general properties of the field. For instance, if φ is convex then it is easy to check that S_{μ_n} is a segment for each $n>0$.

In a series of cases, the following criteria happens to be useful. It is a straightforward consequence of the statement on the uniqueness of the measure satisfying the equilibrium condition.

Suppose that system (9) can be solved for given φ, n , the integral in (10) exists for almost all $t\in[x_n, y_n]$ and defines a non-negative function from L_1 , moreover for $x\in \mathbb{R}\setminus[x_n, y_n]$ we have $u_{n,\varphi}(x)+n g_n(x) \leq \varphi(x)$. Then (10) defines the distribution of the equilibrium measure.

In the next section we give an example where this criteria allows to conclude that S_{μ_n} is a segment and to obtain μ_n in explicit form.

2.3 The case of power fields: $\varphi(x)=\frac{1}{2}|x|^{\lambda}$, $\lambda>0$

In this case all the formulas become very simple. The end points can be expressed as follows : $S_{\mu_n}=[-x_n, x_n]$, where

$$x_n=\{n/A(\lambda)\}^{\frac{1}{\lambda}}, \qquad A(\lambda)=\frac{1}{\sqrt{\pi}}\Gamma\left(\frac{\lambda+1}{2}\right)/\Gamma\left(\frac{\lambda}{2}\right) \tag{11}$$

The equilibrium measures for various n can be obtained from each other by means of contraction and normalization; let

$$a_\lambda(x)=\frac{\lambda}{\pi}|x|^{\lambda-1}\int_{|x|}^{1} t^{-\lambda}(1-t^2)^{-\frac{1}{2}}dt, \qquad x\in[-1,1] \tag{12}$$

then

$$d\mu_n(x)=\frac{n}{x_n}a_\lambda\left(\frac{x}{x_n}\right), \qquad x\in[-x_n, x_n]$$

We note that in this case S_{μ_n} is a segment for every $n>0$ and $\lambda>0$. For $\lambda\geq 1$ it is a consequence of the convexity of function $\varphi(x)$. For $\lambda\in(0,1)$ this fact can be obtained from the criteria of section 2.2.

2.4 Weak (logarithmic) asymptotics for the extremal polynomials with respect to varying norms.

We begin with a general description concerning the polynomials

$Q_n(x) = x^n + \dots$ orthogonal with respect to varying weights F_n

$$\int Q_n(x)\, x^j F_n(x)\, dx = 0, \qquad j = 0, 1, \dots, n-1.$$

It happens that if the sequence $F_n^{1/n}(x)$, in a certain sense, converges to the function $\exp\{-2\varphi(x)\}$ then the sequence of normalized associated measures $\frac{1}{n}\chi(Q_n)$ weakly converges to $\mu_{1,\varphi}$, the unit equilibrium measure in the field φ.

Now, we will specify the concept of convergence in the class of exterior fields which we will need. We denote

$$\varphi_n \xrightarrow{\mathcal{F}} \varphi \qquad \text{if} \quad \varphi_n, \varphi \in \mathcal{F} \quad \text{and:}$$

(i) for any segment $I \subset \mathbb{R}$ we have $\varphi_n(x) \to \varphi(x)$ in measure on I and

$$\lim_{n\to\infty} \min_{x\in I} \varphi_n(x) = \min_{x\in I} \varphi(x)$$

(ii) also $$\underline{\lim}_{|x|\to\infty, n\to\infty} \frac{\varphi(x)}{\log|x|} = +\infty$$

THEOREM. Suppose that $p \in (0, +\infty]$ and a sequence of weights $F_n(x)$ are given. The sequence of polynomials $Q_n(x) \in \mathbb{P}_n^{(1)}$ is defined by the sequence of extremal relations

$$N_n = \| Q_n F_n \|_{L_p(\mathbb{R})} = \min_{q\in\mathbb{P}_n^{(1)}} \| q F_n \|_{L_p(\mathbb{R})}$$

If $F_n(x) = F(x) e^{-n\varphi_n(x)}$ where $F(x) > 0$ a.e. on $\mathbb{R}$, $F(x) \to \mathrm{const} > 0$ as $|x| \to \infty$ and $\varphi_n \xrightarrow{\mathcal{F}} \varphi$

then

$$N_n^{\frac{1}{n}} \to e^{-W_{1,\varphi}}, \qquad \frac{1}{n}\chi(Q_n) \to \mu_{1,\varphi}$$

In potential theoretic terms the statement about the weak convergence of the associated measures can be expressed as follows

$$\log|Q_n(z)| \sim -n V^{\mu_{1,\varphi}}(z) = V^{\mu_{n,n\varphi}}(z), \qquad z \in \mathbb{C}\setminus\mathbb{R}$$

We can write also the following asymptotic estimate for an arbirary sequence of polynomials $q_n \in \mathbb{P}_n$, $n \in \mathbb{N}$

$$\overline{\lim_{n\to\infty}} \left(\frac{|q_n(z)|}{\| q_n F_n \|} \right)^{\frac{1}{n}} \leq \exp\left\{ w_{1,\varphi} - V^{\mu_{1,\varphi}}(z) \right\}, \quad z \in \mathbb{C}\setminus\mathbb{R}$$

This theorem extends and complements several interesting results obtained in papers of H.Stahl, E.A.Rakhmanov, J.Ullman, R.S.Varga, E.B.Saff, H.N.Mhaskar, D.Lubinski, P.Nevai, J.Dehesa and others (see e.g. [19], [21], [26], [29], [31], [34]).

The most important cases are $p=\infty$ (weighted Tchebyshev polynomials) and $p=2$ (orthogonal polynomials). The last case has important applications in various problems connected with the "vector equilibrium" (Hermite - Pade approximation, rational approximation for orthogonal expensions and so on).

3. Orthogonal polynomials on the real axis.

The general approach to the problem of asymptotic of orthogonal polynomials on the real axis was outlined in [25],[26], [27] where the weak forms of asymptotics are mainly discussed. Here, we present some of those results (in a bit more general form) and some applications of them in approximation theory. At this moment we will not comment the basis of our approach; this will be done in section 5 where the problem of strong asymptotics is discussed

Let φ be a convex function on $\mathbb{R}$ such that $|\varphi'(x)| \Rightarrow +\infty$ as $|x| \to \pm\infty$, $\mu_n = \mu_{n,\varphi}$, $w_n = w_{n,\varphi}$ are the equilibrium measures and the corresponding extremal constants for the field φ Let

$$S_{\mu_n} = [x_n, y_n]$$

be the support of the equilibrium measure (see (9)).

The main result consists in the following: the parameters of the equilibrium measures adequately describe the logarithmic asymptotics for the polynomials $Q_n(z) = \kappa_n z^n + \dots$ orthonormal with respect to an arbitrary weight of the form

$$F(x) = e^{-2f(x)}, \qquad f(x) \sim \varphi(x), \quad |x| \to \infty$$

(the condition $f \in L_{loc}(\mathbb{R})$ is also assumed).

More precisely, let $\beta_{i,n}$ be the zeros of Q_n

$$X_n = \min_{1\le i\le n} \beta_{i,n}, \quad Y_n = \max_{1\le i\le n} \beta_{i,n}, \quad \chi_n = \chi(Q_n) = \sum_{i=1}^{n} \delta(\beta_{i,n})$$

In order to describe the zero distribution of the orthogonal polynomials it is convenient to use the contraction of the associated measures and the equilibrium measures obtained by means of the linear transformation

$$\ell_n(t) = \frac{1}{2}(x_n+y_n) + \frac{1}{2}t(y_n-x_n) \; : \; [-1,1] \to S_{\mu_n}$$

THEOREM. Under the above conditions on the weight the corresponding orthogonal polynomials $Q_n(z) = \kappa_n z^n + \dots$ satisfy:

1° $\frac{1}{n}\chi_n(\ell_n) - \frac{1}{n}\mu_n(\ell_n) \to 0$,

2° $X_n \sim x_n$, $Y_n \sim y_n$,

3° $\log \kappa_n \sim w_n$

4° $\log|Q_n(z)| \sim w_n - V^{\mu_n}(z)$, $z \in \mathbb{C}\setminus\mathbb{R}$

Hence, the form of the exterior logarithmic asymptotics of the orthonormal polynomials depends in the given case on the logarithmic asymptotics of the weight as $|x| \to \infty$. In connection with assertion 3 we note that a slightly stronger result is true $\lim_{n\to\infty} \frac{1}{n}(\log\kappa_n - w_n) = 0$.

We have formulated the theorem in terms directly connected with the equilibrium measure and its potential because in such form the results are true under much more general conditions than those stated.

Making use of the formulas for the equilibrium measure given in section 2 the statements of the theorem can be expressed more explicitly. Under certain additional conditions further simplifications can be achieved. This takes place, for instance, in the symmetrical case.

3.1 The symmetrical case ($\varphi(x) = \varphi(-x)$).

In this situation, the support of the equilibrium measure is a symmetrical segment $S_{\mu_n} = [-x_n, x_n]$ and x_n is determined by the equation

$$\frac{2}{\pi}\int_0^{x_n} \frac{x\varphi'(x)}{\sqrt{x_n^2-x^2}}\,dx = n$$

Formulas 3 and 4 of the previous theorem can be expressed in the form

$$\lim_{n\to\infty}\left\{\frac{1}{n}\log \kappa_n + \log x_{n,\varphi} - \frac{2}{\pi n}\int_0^{x_n}\frac{\varphi(t)\,dt}{\sqrt{x_n^2-t^2}}\right\} = \log 2\,,$$

$$\log|Q_n(z)| \sim D_{n,\varphi}\cdot|\operatorname{Im} z|,\quad z\in\mathbb{C}\setminus\mathbb{R}$$

$$D_{n,\varphi} = x_n\cdot\frac{2}{\pi}\int_1^{x_n}\frac{\varphi'(t)\,dt}{t\sqrt{x_n^2-t^2}}\,.$$

Moreover, we have $\ell_n(t) = x_n t$, but further simplifications for the asymptotic distribution of the zeros cannot be obtained; the sequence $\frac{1}{n}\chi_n(\ell_n)$, in general, is not weakly convergent (as in the general case).

3.2 The power case ($\varphi(x) = \frac{1}{2}|x|^\lambda$)

Here, we have

1° $\quad \frac{1}{n}\chi_n(\ell_n) \to a_\lambda(x)\,dx$

2° $\quad -X_n = Y_n \sim x_n = \left\{\frac{n}{A(\lambda)}\right\}^{\frac{1}{\lambda}}$

3° $\quad \lim_{n\to\infty}\left\{\frac{1}{n}\log\kappa_n + \frac{1}{\lambda}\log\frac{n}{e\,A(\lambda)}\right\} = \log 2$

4° $\quad \log|Q_n(z)| \sim \frac{\lambda}{\lambda-1}A(\lambda)^{\frac{1}{\lambda}}\,n^{1-\frac{1}{\lambda}}\,|\operatorname{Im} z|,\quad z\in\mathbb{C}\setminus\mathbb{R}$

Several results on the asymptotic distribution of the zeros of orthogonal polynomials for particular weight functions were obtained earlier by P.Erdos, G.Freud, P.Nevai, J.Dehesa, J.Ullman. In particular, for weights $|x|^\alpha e^{-|x|^\lambda}$, $\lambda=4,6$ G.Freud found the asymptotics of the greatest zero, [1]. For the same weights J.Ullman [34] obtained the previous formula 1° for the limit distribution of the zeros. Using some euristical arguments, he also obtained this formula for arbitrary λ . Hence, the distributions $a_\lambda(x)dx$ have been called Ullman distributions

3.3 Speed of convergence of Pade approximants for Stieltjes type functions.

Let $F(x)$ be a non negative function on $\mathbb{R}_+ = [0, +\infty)$ and

$$\hat{F}(z) = \int_0^\infty \frac{F(x)\,dx}{z-x}, \qquad z \in \mathcal{D} = \mathbb{C} \setminus \mathbb{R}_+ \tag{13}$$

If all moments of F are finite then function $\hat{F}$ can be put in correspondence with the following asymptotic expansion at point $z = \infty$ (as $z \to \infty$, $|\arg z| \geq r > 0$)

$$\hat{F}(z) \sim \sum_{n=0}^{\infty} \frac{c_n}{z^{n+1}}, \qquad c_n = \int_0^\infty x^n F(x)\,dx \tag{14}$$

Fundamental results on the convergence of Pade approximants $[n/n]_{\hat{F}}$ (continued fractions) for such series were obtained by T.Stieltjes (for the more general case of a positive measure in place of $F dx$) In particular, he proved that the sequence $[n/n]_{\hat{F}}$ converges (to $\hat{F}$) on each compact subset of $\mathcal{D}$ if the moment problem for measure $F(x)dx$ is determined. In other words, if there exists only one function of type (13) with the given asymptotic expansion (14), then this function can be obtained through the coefficients of the series by means of the Pade approximants. Later, T.Carleman showed that a sufficient condition for the moment problem to be determined is that $\sum_{n=0}^{\infty} c_n^{-1/2n} = \infty$

The question on the speed of convergence of $[n/n]_{\hat{F}} \to \hat{F}$ as in the compact (Markov) case reduces to the study of the asymptotics of the orthogonal polynomials $Q_n(z) = K_n z^n + \ldots$ on $\mathbb{R}_+$ with respect to the weight F (see section 1). In essence, here fine results are not needed. For example, in order to obtain a rough picture of the dependance of the speed of convergence to zero of the remainder

$$\varepsilon_{n,F}(z) = \left| \hat{F}(z) - [n/n]_{\hat{F}}(z) \right|$$

as $n \to \infty$ with respect to the decreasing speed of $F(x)$ at infinity it is sufficient to find the exponential asymptotic order of $\varepsilon_{n,F}$ as $n \to \infty$; for this it is needed to study a "very weak" double logarithmic asymptotics of the polynomials $Q_n(z)$. Nevertheless, the question remained open until 1982 when in the papers [25], [26] it was obtained a rather genersl answer (precisely, the attempt of solving this problemled the author to different results on the asymptotics of orthogonal polynomials).

THEOREM. Suppose $\lim\limits_{x\to\infty} \log\log \frac{1}{F(x)} / \log x = \lambda > \frac{1}{2}$

then $\lim\limits_{n\to\infty} \log\log |Q_n(z)| / \log n = 1 - \frac{1}{2\lambda}, \quad z \in \mathcal{D}$

Corollary. If $F(x) = \exp\{-x^{\lambda(x)}\}, \quad \lambda(x) \to \lambda, \quad x \to \infty$

then $\mathcal{E}_{n,F}(z) = \exp\{-n^{\lambda_n(z)}\}, \quad \lambda_n(z) \rightrightarrows 1 - \frac{1}{2\lambda}, \quad z \in \mathcal{D}$

We note that in the given case it is not supposed that F is a weight ($F > 0$ a.e.) on $\mathbb{R}_+$. Such condition (or some other on $\operatorname{supp}(F\,dx)$) becomes necessary if it is required to indicate the dependance of the decreasing speed of $\mathcal{E}_{n,F}(z)$ on the point $z \in \mathcal{D}$, [27].

THEOREM. Suppose $F(x) = e^{-f(x)}$ is a weight on $\mathbb{R}_+$,

$x^{-\frac{1}{2}} f(x) \in L_{loc}(\mathbb{R}_+)$ and there exists a function $\varphi \in C^1(\mathbb{R})$ such that $\varphi(x^2)$ is convex, $x^{-1/2}\varphi(x) \to \infty$ as $x \to \infty$ and $f(x) \sim \varphi(x)$ as $x \to \infty$. Then

$$\log \mathcal{E}_{n,F}(z) \sim -\frac{2}{\pi}\sqrt{y_n} \int_1^{y_n} \frac{\varphi'(t)\,dt}{\sqrt{t(y_n-t)}} \cdot \operatorname{Im}\sqrt{-z}, \qquad z \in \mathcal{D}$$

($n \to \infty$), where y_n can be determined as the root of the equation

$$\frac{1}{\pi}\int_0^{y_n} \sqrt{\frac{x}{y_n - x}}\,\varphi'(x)\,dx = n.$$

Results related to the more general case when $F(x)$ is given on the whole real axis (so called Hamburger series) can be obtained directly from the formulas in section 3.

3.4 Rational and polynomial approximation connected with orthogonal polynomials.

In this connection we can make the following remark. The theory of general orthogonal polynomials arised in the last century in the context of the theory of continued fractions. The polynomials orthogonal with respect to "an arbitrary weight" appeared in the papers of P.L.Tchebyshev as the denominators of the partial fractions of certain type of continued fractions. And preciesely it is the theory of rational approximation (in particular, continued fractions) that essentially require the study of orthogonal polynomials with respect to arbitrary weights. Indeed, in this case the system of orthogonal polynomials is generated by the approximated function and the study of a particular system of orthogonal polynomials allows to investigate the convergence of the rational approximants to that particular function.

A completely different situation takes place for polynomial approximations connected with orthogonal polynomials (orthogonal expansions, Lagrange interpolation). We note that precisely this type of approximation schemes connected with orthogonal polynomials had the leading role in the middle of our century. In this case, it is not essentially required to study the general orthogonal polynomials. Indeed, the system of orthogonal polynomials become now the tool of approximation and a particular system of polynomials can be used to approximate a large class of functions. For example, to study the polynomial approximation for the class of functions analytic on the segment (in the sense of the order of approximation) it is sufficient to use the Tchebyshev system of polynomials whose zeros also give a good extremal interpolation table.

This remark does not mean that the authors claim against polynomial approximators. As a proof we will discuss in short one problem in this direction.

3.5 Extremal tables of interpolation points for analytic functions on a band.

Let $(\mathcal{B}, \rho)$ be a function space with a metric $\rho(*, *)$. We will suppose that all the functions $f \in \mathcal{B}$ are defined and continuous on $\mathbb{R}$ and $\mathbb{P}_n \subset \mathcal{B}$ for each $n \in \mathbb{N}$. Suppose, $\Omega = \{\alpha_{i,n} \in \mathbb{R}\ ;\ i = 1, 2, \ldots, n+1,\ n \in \mathbb{N}\}$ is a triangular table of interpolation points and

$$q_{n,f}(x) \in \mathbb{P}_n \quad : \quad q_{n,f}(\alpha_{i,n}) = f(\alpha_{i,n}), \quad i = 1, 2, \ldots, n+1$$

the corresponding interpolating polynomials to $f \in \mathcal{B}$ (for simplicity we will assume that $\alpha_{i,n} \neq \alpha_{j,n},\ i \neq j$).

DEFINITION. We say that table Ω is estremal for class $\mathcal{B}$ if for any function $f \in \mathcal{B}$ we have

$$\log \rho(f, q_{n,f}) \sim \log \min_{q \in \mathbb{P}_n} (f, q), \qquad n \to \infty$$

(in other words, the interpolating polynomials give the best order of approximation).

This concept is essentially contained in Walsh's monograph where in particular, interpolation of analytic function on the segment $I=[-1,1]$ is considered. For the uniform metric on I, the table of interpolation points consisting of the zeros of orthogonal polynomials $Q_n(x)$ on I with respect to an arbitrary weight F is extremal. Moreover, this is true for any reasonable metric on I (e.g. for $L_p(\sigma)$, where σ is a measure on I such that $\sigma'>0$ a.e.). In essence, this means that for the class $\mathcal{H}(I)$ all metrics are identical. The approximation speed of functions $f\in\mathcal{H}(I)$ is determined by the index $r(f)$ of the maximal ellipse where f is analytic.

Let us briefly consider the case of weighted polynomial approximation on $\mathbb{R}_+$ of analytic functions on a band. Here, the most important characteristic becomes the index (width) of the maximal symmetrical band where f is analytic:

$$r(f)=\sup\{r: f\in\mathcal{H}(I_r)\} \quad \text{where} \quad I_r=\{z\in\mathbb{C}: |\mathrm{Im}\, z|<r\}$$

Another important characteristic is the growth of $|f(z)|$ as $|\mathrm{Re}\, z|\to\infty$. We will limit ourselves to the most simple "bounded" case. More specifically, we will consider the following class of functions

$$\mathcal{H}_r=\{f\in\mathcal{H}(\mathbb{R}): r(f)=r\in(0,+\infty) \text{ and } \sup_{z\in I_s}|f(z)|<\infty \ \forall s<r\}$$

There are different possible metrics on $\mathbb{R}$. Possibly, the most important ones are generated by the norms

$$\|f\|_{\lambda,p}=\Big\{\int_{\mathbb{R}}|f(x)|^p F_\lambda(x)\,dx\Big\}^{\frac{1}{\lambda}}, \qquad F_\lambda(x)=e^{-|x|^\lambda}$$

Different values of $p\in(0,+\infty]$ generate essentially the same norm For definiteness, we will consider the norm $\|f\|_\lambda=\|f\|_{\lambda,2}$ (this allows to consider Fourier series with respect to orthogonal polynomials). The next result, due to E.A.Rakhmanov was presented at the International Conference on Approximation Theory, Kiev'83.

THEOREM.

Let $\lambda>1$, $r\in(0,+\infty)$, then the following statements hold:

1. The table of zeros of the orthonormal polynomials $Q_n(z)$ with with respect to the weights $F_\lambda(x)$ is extremal for the class $\mathcal{H}_r$,

$r > 0$ with norm $\| * \|_\lambda$.

2. For any function $f \in \mathcal{H}_r$ we have the formula

$$\overline{\lim_{n\to\infty}} \frac{1}{n^{1-1/\lambda}} \log \left(\min_{q \in \mathbb{P}_n} \| f - q \|_\lambda \right) = -c_\lambda r$$

3. An analogous formula takes place in terms of the orthogonal expansion of f with respect to the system $\{Q_n\}$; if

$$f(z) = \sum_{n=0}^{\infty} f_n Q_n(z) , \qquad f_n = \int f(x) Q_n(x) F_\lambda(x) dx$$

then $$\overline{\lim_{n\to\infty}} \frac{1}{n^{1-1/\lambda}} \log |f_n| = -c_\lambda r .$$

In different problems an important role is played by the following explicit table (near to the table of zeros of Tchebyshev polynomials with respect to the weight $F_\lambda(x)$). Let

$d\mu_n(x) = \frac{n}{x_n} a_\lambda\left(\frac{x}{x_n}\right) dx$ be the equilibrium measure in the exterior field $\frac{1}{2}|x|^\lambda$. We define the points $x^*_{i,n} = x^*_{i,n}(\lambda)$ which are uniformly distributed with respect to the measure $d\mu_n$

$$\int_{-x_n}^{x^*_{1,n}} d\mu_n = \int_{x^*_{n,n}}^{x_n} d\mu_n = \frac{1}{2} , \qquad \int_{x^*_{i,n}}^{x^*_{i+1,n}} d\mu_n = 1 , \qquad i = 1,2,\dots,n-1 \tag{15}$$

Now, the statements of the previous theorem can be complemented with the following:

4. The table $\Omega^* = \{x^*_{i,n}\}$ is extremal for $\mathcal{H}_r$ with the norm $\| * \|_\lambda$.

In section 5 we will return to such tables in connection with the relative approximation by polynomials.

4. The asymptotics of the ratio of orthogonal polynomials and the convergence of Pade approximants for Markov and Stieltjes type meromorphic functions.

In section 1 we considered the question of the rational approximation of Markov functions

$$\hat{F}(z) = \int_I \frac{F(x)dx}{z-x} , \qquad z \in \mathcal{D} = \overline{\mathbb{C}} \setminus I$$

This result was complemented by E.A.Rakhmanov in [22] and [23].

THEOREM. For any weight F the corresponding orthogonal polynomials $Q_n(x)=x^n+\ldots$ satisfy formula (16).

In connection with this result see also [37] and [28]

The study of the convergence of Pade approximants for Markov and Stieltjes type meromorphic functions was continued by G.Lopez. First, he considered the convergence of multipoint Pade approximants for Markov type meromorphic functions.

Suppose, as in section 1, that F is a weight on $I=[-1,1]$,

$$\Omega = \{ \beta_{i,n} , \quad i = 1,2,\ldots,2n , \quad n \in \mathbb{N} \}, \qquad \omega_{2n}(z) = \prod_{i=1}^{2n}(z-\beta_{i,n})$$

is a table of interpolation points on a segment $E \subset \mathbb{R} \setminus I = \mathcal{D}$;

$$f(z) = \hat{F}(z) + r(z)$$

where r is a rational function whose set of poles $\mathcal{P}_r$ is contained $\mathcal{D} \setminus E$ and

$$R_n(z) = P_{n-1}(z)/Q_n^*(z) \; : \qquad \frac{Q_n^* f - P_{n-1}}{\omega_{2n}}(z) \in \mathcal{H}(E)$$

is the sequence of multipoint Pade approximants with respect to the table Ω.

As in the case when $r=0$ an important role is played by the polynomials $Q_{m,n}(x)=x^m+\ldots$ which are orthogonal to the weight

$$F_n(x) = \frac{F(x)}{\omega_{2n}(x)} \; : \qquad \int_I Q_{m,n}(x)\cdot x^j F_n(x)\, dx = 0, \qquad j=0,1,\ldots,m-1$$

THEOREM. The following statements are true:

1°. For any integer j we have

$$\lim_{n\to\infty} \frac{Q_{n+j+1,n}}{Q_{n+j,n}}(z) = \Psi(z) , \qquad z \in \mathcal{D}$$

2°. $$\lim_{n\to\infty} \frac{Q_n^*(z)}{Q_{n,n}(z)} = \frac{1}{\Psi^m(z)} \prod_{j=1}^{m} \frac{[\Psi(z)-\Psi(\beta_j)]^2}{(z-\beta_j)}$$

3°. As $n \to \infty$ each pole of f attracts as many poles of R_n as its order of multiplicity; the rest of the poles of R_n tend to I

In this section we will study analogous problems for meromorphic functions (with a finite number of poles) of Markov type

$$f(z) = \hat{F}(z) + r(z)$$

where F is a non-negative function on the segment $I = [-1,1]$, r is a rational function whose set of poles $\mathcal{P}_r$ belongs to $\mathcal{D}$.

This question was developed in a series of papers of A.A.Gončar, E.A.Rakhmanov and G.López (see [2], [12], [13], [15], [16], [22], [23] and [24]).

The study of this problem started with the paper [2] of A.A. Gončar where he proved the following result:

THEOREM. Suppose that $F \geq 0$ is a non-negative function on $x \in I$ such that the corresponding orthogonal polynomials $Q_n(x) = x^n + \dots$ satisfy the following asymptotic formula:

$$\lim_{n\to\infty} \frac{Q_{n+1}}{Q_n}(z) = \Psi(z) = \frac{1}{2}\left(z + \sqrt{z^2-1}\right), \qquad z \in \mathcal{D} \tag{16}$$

Then for the diagonal sequence of Pade apprroximants $[n/n]_f$ with respect to f we have:

1°. As $n\to\infty$ each pole of f in $\mathcal{D}$ attracts as many poles of $[n/n]_f$ as its order of multiplicity; the rest of the poles of $[n/n]_f$ tend to I (in other words, for any open set $U \Subset \mathcal{D}$, $\partial U \cap \mathcal{P}_r = \emptyset$ the functions $f(z)$ and $[n/n]_f(z)$ have for all sufficiently large n the same number of poles).

2. The sequence $[n/n]_f(z)$ converges to f uniformly inside (on each compact subset) of

These statements were obtained in [2] as a consequence of the following formula of comparative asymptotics for the denominators $Q_n^*(z) = z^n + \dots$ of the Pade approximants $[n/n]_f(z)$ with respect to the initial orthogonal polynomials.

$$\lim_{n\to\infty} \frac{Q_n^*}{Q_n}(z) = \Psi^m(z) \prod_{j=1}^{m} \frac{[\Psi(z) - \Psi(\beta_j)]^2}{(z-\beta_j)}$$

where $\beta_1, \dots, \beta_m$ are the poles of r

We note that the polynomials $Q_n^*(z)$ are the (non-Hermite) orthogonal polynomials with respect to the generalized function $d\nu = F dx + d\eta$ where η is the generalized function defined by relation $\int (\xi - z)^{-1} d\eta(\xi) = r(z)$.

4° The sequence $\{R_n\}$ converges to f uniformly inside of $\mathcal{D}\setminus\mathcal{P}_r$

Hence, the behavior of multipoint Pade approximants in this case is identical to the behavior of the classical ones (compare with Goncar's Theorem)

For details see [13]. We note that in that paper statements 2 and 3 were obtained for extremal interpolations tables: $1/2n\ \chi(\omega_{2n}) \to \mu_E$ where (μ_E, μ_I) is the (usual) equilibrium distribution for the condenser (E, I) (in this case we have $1/n\ \chi(Q_{n,n}) \to \mu_I$ and the sequence $\{R_n\}$ gives the same order of convergence on E as the best rational approximation). Nevertheless, statements 2 and 3 are true also in the general case.

Suppose that the segment E and I have the common point $x=1$. If the certain number of interpolation points tend to 1 (but are different from 1) then we are in a situation which in a certain sense lies between the Markov case (the table of interpolation points is compactly contained in the region of analyticity of the function) and the pure Stieltjes case (all the interpolation points agree with the point $x=1$).

In the intermediate case a condition on the average speed of convergence of $\beta_{i,n}$ to 1 can be given which garantees that the sequence $\{R_n\}$ behaves as in the Markov case.

THEOREM. Suppose that table Ω is such that

$$\lim_{n\to\infty} \sum_{i=1}^{2n} (\beta_{i,n}-1)^{1/2} = +\infty$$

then for any weight F on I and any rational function r: $\mathcal{P}_r \subset \mathbb{C}\setminus I$ statements 1° - 3° of the previous theorem are true.

If $\lim\limits_{n\to\infty} \sum\limits_{i=1}^{2n} (\beta_{i,n}-1)^{\frac{1}{2}} < \infty$ then we are essentialy in the pure Stieltjes situation where additional conditions are required on the weight. We will briefly discuss the pure Stieltjes case in its usual form when interpolation is carried out at point

Suppose that F is a weight on $\mathbb{R}_+$; that is $F>0$ a.e. and all the moments $c_n = \int_{\mathbb{R}_+} x^n F(x)\,dx$ are finite so $\hat{F}(z)$ can be put in correspondence with an asymptotic expansion on powers of $\frac{1}{z}$

$$\hat{F}(z) = \int_0^\infty \frac{F(x)\,dx}{z-x} \sim \sum_{n=0}^{\infty} \frac{c_n}{z^{n+1}}$$

Let r be a rational function with poles in $\mathcal{D} = \mathbb{C}\setminus\mathbb{R}_+$ and

$$f(z) = \hat{F}(z) + r(z)$$

(for the sake of simplicity we will suppose that r is analytic at the point $z=-1$; this can be done without loss of generality since a convenient change of variables translates the problem to an equivalent one where the new function has the desired property). If we consider the convergence of the classical Pade approximants associated to f then problems arise connected with the fact that the sequence Q_{n+1}/Q_n even if it converges (after a convenient normalization) then this limit is constant. Nevertheless, it turns out that under a certain natural condition on the weight a formula of comparative asymptotics can be obtained for the denominator Q_n^* of the Pade approximants. Recently, G.Lopez [16] proved the following.

THEOREM. Suppose that the weight F satisfies Carleman's condition $\sum_{n=0}^{\infty} C_n^{-1/2n} = +\infty$. Then

1. As $n \to \infty$ each pole of f "attracts" as many zeros of as it's multiplicity. The rest of the zeros of $[n/n]f$ tend uniformly to $\mathbb{R}_+$ (in particular, $Q_n^*(-1) \neq 0$ for all large enough n).
2. If Q_n and Q_n^* are normalized so that $Q_n(-1) = Q_n^*(-1) = (-1)^n$ then we have the following comparative asymptotics

$$\lim_{n\to\infty} \frac{Q_n^*}{Q_n}(z) = \frac{(z+1)^m}{\psi\left(\frac{z-1}{z+1}\right)} \prod_{i=1}^{m} \frac{\left[\psi\left(\frac{z-1}{z+1}\right) - \psi\left(\frac{\beta_i-1}{\beta_i+1}\right)\right]^2 (1+\beta_i)}{z-\beta_i} \tag{17}$$

where $\beta_1, \ldots, \beta_m$ are poles of r and $z \in \mathcal{D}$.

3. The sequence of the diagonal Pade approximants $[n/n]_f(z)$ converges to $f(z)$ uniformly on each compact subset of $\mathcal{D} \setminus \mathcal{P}_r$

We note that an analogous result holds if we consider a function of the form
$$f(z) = r(z) + \int \frac{S(t)F(t)}{z-t}\,dt$$
where S is a rational function whose zeros and poles are in $\mathcal{D}$. Standard approximation techniques (see, for example, [28]) allow to extend (17) to more general cases. Also, a similar theorem is true for meromorphic functions of Hamburger type.

In the case when r has real coefficients statements 1 and 3 also hold for any function $F(x) \geq 0$ that satisfies Carleman's condition although statement 2 in general is not true (see [24] and [11]).

5. Strong asymptotics.

So far we have considered the equilibrium distribution for the study of weak form of asymptotics. Here, we will expose a few results and conjectures of E.A.Rakhmanov concerning the problem of strong asymptotics. In order to explain our approach to this problem it is useful to compare it with the classical one.

5.1 The classical approach

Let F be a weight on the segment $I=[-1,1]$ satisfying the Szego condition $\int_{-1}^{1}(1-x^2)^{-\frac{1}{2}}\log F(x)\,dx > -\infty$ and

$$D_F(z) = \exp\left\{-\sqrt{z^2-1}\ \frac{1}{2\pi}\int_{-1}^{1}\frac{\log F(x)}{z-x}\ \frac{dx}{\sqrt{1-x^2}}\right\},\quad z\in\mathcal{D}=\bar{\mathbb{C}}\setminus I$$

is the corresponding Szego function (we have for its boundary value $|D_F^+(x)| = |D_F^-(x)| \overset{a.e.}{=} F(x),\ x\in I$).

The starting point of the well known approach to the problem of the strong asymptotics is the following extremal property of the orthogonal polynomials

$$m_n = \int_I Q_n^2(x)\,F(x)\,dx = \min_{q\in\mathbb{P}_n^{(1)}}\int_I |q(x)|^2 F(x)\,dx$$

Now, the idea consists of the following. Let us consider the same problem in a more general class of functions. To be more precise, instead of the class of polynomials $\mathbb{P}_n^{(1)}$ we consider the space $\mathbb{P}_n^{(1)}\oplus\mathcal{H}_{2,F}^{(0)}$ where

$$\mathcal{H}_{2,F}^{(0)} = \{ f\in\mathcal{H}(\mathcal{D}):\quad fD_F\in\mathcal{H}_2(\mathcal{D}),\ f(\infty)=0\}$$

is the weighted Hardy space ($\mathcal{H}_2(\mathcal{D})$ is the usual Hardy space for the domain $\mathcal{D}$). Let

$$m_n^* = \min_{q\in\mathbb{P}_n^{(1)}\oplus\mathcal{H}_{2,F}^{(0)}}\int_{\partial\mathcal{D}}|q(x)|^2F(x)|dx| = \int_{\partial\mathcal{D}}|Q_n^*(x)|^2F(x)|dx|$$

Here, $Q_n^*\in\mathbb{P}_n^{(1)}\oplus\mathcal{H}_{2,F}^{(0)}$ is the extremal function (we note that in this case we integrate on the boundary of $\mathcal{D}$ in the sense of Caratheodory $\partial\mathcal{D} = I^+\cup I^-$).

Now, one of the central results on the asymptotics of the orthogonal polynomials can be expressed by the two following theorems.

THEOREM A. For any Szego weight $F(x)$ we have

1° $\quad m_n \sim m_n^*$

2° $\quad Q_n(z) \sim Q_n^*(z), \quad z \in \mathcal{D}$

3° $\quad \frac{1}{m_n}\left\{Q_n - (Q_n^*)^+ - (Q_n^*)^-\right\} \xrightarrow{L_{2,F}} 0$

(if F satisfies the Dini-Lipschits condition then the last relation takes place uniformly on every segment inside I).

THEOREM B. We have the following equalities:

1° $\quad m_n^* = \pi^{-1} \cdot 2^{-2n-1} \cdot D_F(\infty)^{-1}$

2° $\quad Q_n^*(z) = 2^{-n-\frac{1}{2}}(z^2-1)^{-\frac{1}{4}}\left(z+\sqrt{z^2-1}\right)^{n+\frac{1}{2}}$

This way of expressing this classical result may seem artifitial Nevertheless, it has one important advantage. Theorem A turns out to be true in a much more general setting than that which is being considered. For example, the segment I can be substituted by a set of curves and contours in the complex plane. The expressions for m_n^* and Q_n^* in the general situation become much more complicated. For details see H.Widom [36].

5.2 Polynomials, orthogonal on the real axis.

Let $F(x) = \exp\{-2\varphi(x)\}$ be a weight on $\mathbb{R}$ and $Q_n(x) = x^n + \ldots$ the corresponding orthogonal polynomial of degree n

$$m_n = \int_{\mathbb{R}} Q_n^2(x) F(x)\,dx = \min_{q \in \mathbb{P}_n^{(1)}} \int_{\mathbb{R}} |q(x)|^2 F(x)\,dx \qquad (18)$$

The approach considered above, connected with the extension of Hilbert spaces is not valid now because it is not possible to find a convenient Hilbert space. Nevertheless, we can apply a similar idea making use of potential theoretic methods. Let

$$\mathcal{M}_n^* = \{\mu : S_\mu \subset \mathbb{R},\ \|\mu\| = n\}$$

$$\mathcal{M}_n = \left\{\mu = \sum_{j=1}^{n} \delta(x_j) : x_1, \ldots, x_n \in \mathbb{R}\right\} \subset \mathcal{M}_n^*$$

Then it is clear that $\{|q| : q \in \mathbb{P}_n^{(1)}\} = \{e^{-V^\sigma} : \sigma \in \mathcal{M}_n^*\}$ and the extremal problem ( ) may be reexpressed in the form

$$m_n = \int_{\mathbb{R}} \exp\{-2(V^{\chi_n}+\varphi)(x)\}dx = \min_{\sigma \in \mathcal{M}_n} \int_{\mathbb{R}} \exp\{-2(V^\sigma+\varphi)(x)\}dx ,$$

$\chi_n = \chi(Q_n)$ is the associated measure corresponding to Q_n .

Let us consider the same problem for a more general class of measures:

$$m_n^* = \int_{\mathbb{R}} \exp\{-2(V^{\chi_n^*}+\varphi)(x)\}dx = \min_{\sigma \in \mathcal{M}_n^*} \int_{\mathbb{R}} \exp\{-2(V^\sigma+\varphi)(x)\}dx$$

The corresponding extremal measure χ_n^* may be used for the analysis of the orthogonal polynomials. But in this situation it does not seem reasonable to introduce and investigate one more new concept. It is simpler to pass from the L_2 metric to the uniform one: let

$$e^{-2W_n} = \max_{x \in \mathbb{R}} e^{-2(V^{\mu_n}+\varphi)(x)} = \min_{\sigma \in \mathcal{M}_n^*} \max_{x \in \mathbb{R}} e^{-2(V^\sigma+\varphi)(x)}$$

We arrive again to the concept of equilibrium measure (compare with (7)).

Suppose that the weight $F(x) = e^{-2\varphi(x)}$ decreases sufficiently regularly as $x \to \pm\infty$. so that for large n takes place the case of a "single segment" for the corresponding equilibrium measure $\mu_{n,\varphi}$ (for example, if φ is convex). Then for large n's the polynomials $Q_n(z)$ are close to the polynomials orthogonal on the finite segment $S_{\mu_{n,\varphi}}$ (with the same weight). In this case the difference between the extremal functions for the L_2 and the L_∞ norms are described by the factor $(z-x_n)^{-\frac{1}{4}}(z-y_n)^{-\frac{1}{4}}$ Considering what has been said we naturally arrive to the conjecture that for the orthogonal polynomials $Q_n(z)$ the following asymptotical formulas hold:

$$m_n \sim \frac{1}{2\pi} \exp\{-2W_{n+\frac{1}{2}}\} \tag{19}$$

$$Q_n(z) \sim [(z-x_n)(z-y_n)]^{-\frac{1}{4}} \exp\{-V^{\mu_{n+1/2}}(z)\}, \quad z \in \mathbb{C}\setminus\mathbb{R}$$

$$Q_n(x) = 2[(x-x_n)(y_n-x)]^{-\frac{1}{4}} e^{-W_{n+\frac{1}{2}}} F(x)^{-\frac{1}{2}} \left\{\varepsilon_n(x) + \cos\int^x d\mu_{n+\frac{1}{2}}(t)\right\}$$

$$\lim_{n\to\infty} \max_{|x| \leq (1-\varepsilon)x_n} |\varepsilon_n(x)| = 0 \qquad \forall \varepsilon > 0$$

where $$\mathcal{V}^{\mu_n}(z) = \int \log \frac{1}{(z-x)}\, d\mu_n(x)$$

is the complex potential of measure μ_n ($V^{\mu_n} = \mathrm{Re}\, \mathcal{V}^{\mu_n}$).
We will not precise the general conditions on the weight. Let's consider one important case where these formulas are true.

THEOREM. Let $F(x) = e^{-|x|^\lambda}$, $\lambda > 0$ $\quad (\varphi(x) = \frac{1}{2}|x|^\lambda)$
then

$$Q_n(z) \sim (z^2 - x_n^2)^{-\frac{1}{4}} \exp\left\{-W_{n+\frac{1}{2}} + \frac{1}{2} z^\lambda - (n+\tfrac{1}{2}) \int_{x_n}^{z} \left(\frac{z^\lambda}{\xi^\lambda} - 1\right) \frac{d\xi}{\sqrt{\xi^2 - x_n^2}}\right\}, \quad z \in \mathbb{C} \setminus \mathbb{R}$$

$$Q_n(x) = (x_n^2 - x^2)^{-\frac{1}{4}} \exp\left\{\frac{1}{2} x^\lambda - W_{n+\frac{1}{2}}\right\} \left\{\varepsilon_n(x) + \cos\left(n \int^{x/x_n} a_\lambda(t)\, dt\right)\right\}$$

We note that there exists a classical approach to the problem of strong asymptotics when $F(x) = e^{-x^{2m}}$ where m is a natural number based on a differential equation satisfied by the orthogonal polynomials. Using this approach P.Nevai [18] obtained the above result for $\lambda = 4$. Later R.Sheen solved the problem for $\lambda = 6$. Based on these particular cases, P.Nevai formulated a correct conjecture for an arbitrary $\lambda > 0$. A series of other interesting results using this method were obtained by A.Magnus, P.Nevai and other authors.

Recently D.Lubinski and E.B.Saff announced to have proved formula (19) under rather general conditions (in terms of the leading coefficient of the extremal polynomials).

There exists another way of expressing the results connected with "the single segment case", (we limit ourselves to the exterior asymptotics).

THEOREM. Let $F(x) = e^{-|x|^\lambda}$, then for the corresponding orthonormal polynomials $Q_n(z)$ we have

$$Q_n(z) \sim (2\pi)^{-\frac{1}{2}} D_{n,F}(z) \left(\frac{z}{x_n} + \sqrt{\frac{z^2}{x_n^2} - 1}\right)^{n+\frac{1}{2}}, \quad z \in \mathbb{C} \setminus \mathbb{R}$$

where

$$D_{n,F}(z) = \exp\left\{-\sqrt{(z^2 - x_n^2)}\, \frac{1}{2\pi} \int_{-x_n}^{x_n} \frac{\log F(x)}{z - x} \frac{dx}{\sqrt{x_n^2 - x^2}}\right\}$$

is Szego's function for the weight F on the segment $S_{\mu_n} = [-x_n, x_n]$, $\mu_n = \mu_{n,\varphi}$, $x_n = x_{n,\varphi}$, $\varphi(x) = \frac{1}{2}|x|^\lambda$

This result illustrates fairly well the statement that in the considered case the orthonormal polynomials behave asymptotically as polynomials orthonormal on a segment that varies with n.

The formulation in terms of equilibrium measures has the advantage that it is not connected with the supposal of existance of Szego's function. This is important in the case of orthogonality on a compact set.

5.3 The non-Szego compact case.

We introduce here two conjectures related with the exterior asymptotitics of orthogonal polynomials on the circle and on a segment. Let us only consider continuous weights.

Let $F(z)$ be a continuous weight ($F(z)>0$ a.e.) on the unit circle $\Gamma=\{z:|z|=1\}$ and $\mu_n=\mu_{n,\varphi}$ the equilibrium measure on Γ with norm n in the field $\varphi(x)=1/2\log[1/F(z)]$; as in the case of the real axis this measure is uniquely determined by the relations $S_{\mu_n}\subset\Gamma$, $\|\mu_n\|=n$ and

$$V^{\mu_n}(z)+\varphi(z)=w_n=\min_{z\in\Gamma}\,(V^{\mu_n}+\varphi)(z)\,,\quad z\in S_{\mu_n}$$

CONJECTURE 1. For the orthonormal polynomials $Q_n(z)=k_n z^n+\ldots$

$$\frac{1}{2\pi}\int_\Gamma Q_n(z)\,\overline{Q_m(z)}\,F(z)\,|dz|=\delta_{n,m}\,,\quad n,m=0,1,2,\ldots$$

we have

$$k_n\sim e^{w_n}\,,\qquad Q_n(z)\sim\exp\left\{w_n-V^{\mu_n}(z)\right\},\quad |z|>1.$$

CONJECTURE 2. For the orthonormal polynomials $Q_n(z)=k_n z^n+\ldots$ with respect to the continuous weight $F(x)=e^{-2\varphi(x)}$ on the segment $I=[-1,1]$ we have

$$k_n\sim\frac{1}{\sqrt{2\pi}}e^{w_{n+\frac12}}\,,\quad Q_n(z)\sim\frac{1}{\sqrt{2\pi}}(z^2-x_n^2)^{-\frac14}\exp\left\{w_{n+\frac12}-V^{\mu_{n+\frac12}}(z)\right\}$$

where $\mu_n=\mu_{n,\varphi}$, $w_n=w_{n,\varphi}$ are the equilibrium measure and the extremal constant for the field φ .

For some simple cases these statements have been proved but the explicit form of the potentials in the right sides seem too complicated. For instance, the formulas for the case when $\varphi(x)$ is a convex function on $I=[-1,1]$ and $\varphi(x)\to+\infty$ as $x\to\pm1$, in such a way that Szego's condition is not satisfied results to be more complicated than when $\varphi(x)$ is a convex function on $\mathbb{R}$.

For the investigation of strong asymptotics of orthogonal polynomials on the whole real axis a key role is played by the problem of the strong asymptotics of polynomials orthogonal with respect to varying weights. At the same time, the polynomials orthogonal with respect to varying weights, as we saw earlier, have their own particular interest.

We will introduce here one result in this direction, considering for a change, the case of orthogonality on the circle. The polynomials orthogonal on the unit circle have an important advantage, this is directly connected with the relative polynomial approximation.

We consider a simple case which essentially contains all the important features. This is when we have $F_n(z) = e^{-2n\varphi(z)}$, $z \in \Gamma$ and $\varphi \in C^{1+\varepsilon}(\Gamma)$, $\varepsilon > 0$

THEOREM. If

$$m_\varphi = \min_{\theta \in [0,2\pi]} \frac{1}{\pi^2} \int_0^{2\pi} \operatorname{ctg}(\theta - \tau)\, d\varphi(e^{i\tau}) > -1$$

then for the orthonormal polynomials $Q_n(z) = K_n z^n + \dots$ on Γ with respect to the weights $F_n(z)$ we have the classical Szego formula

$$Q_n(z) \sim D_{F_n}(z)\, z^n \quad , \quad |z| \geq 1 \tag{20}$$

where

$$D_{F_n}(z) = \exp\left\{ -\frac{1}{4\pi} \int_\Gamma \frac{z+\xi}{z-\xi} \log F_n(\xi)\, |d\xi| \right\} \quad , |z| \geq 1.$$

is the exterior Szego function for the weight F_n.

If $m_\varphi < -1$ then formula (20) is not true even for $|z| > 1$ (the case when $m_\varphi = 1$ needs additional conditions).

We note that much more simple is the case when the weights have the form $F_n(z) = F(z)/|\omega_n(z)|^2$ where $\omega_n \in \mathbb{P}_n$ are polynomials- whose zeros $\beta_{i,n}$ lie in the unit circle and satisfy $\lim_{n\to\infty} \sum_{i=1}^{n} (1 - |\beta_{i,n}|) = +\infty$. The statement of the theorem is true for any such sequence of weights (see [14]). It is usefull to compare that situation with the following $F_n(z) = |z - z_0|^{2n}$, $|z_0| < 1$ which seems to be very simple. Nevertheless, in this case the statement is true only if $|z_0| \leq \frac{1}{3}$ If $|z_0| > \frac{1}{3}$ then the correspoding orthogonal polynomials behave as if they were orthogonal on some arc of the circle.

Let us return to the theorem. The same condition $m_\varphi > -1$ is sufficient for the possibility of relative trigonometrical approximation of functions $F_n(z)$. More precisely, for $m_\varphi > -1$ there exists a sequence of trigonometrical polynomials $t_n(\theta)$, $\deg t_n = n$ such that

$$\lim_{n\to\infty} \| t_n(\theta) F_n(e^{i\theta}) - 1 \| = 0$$

For $m_\varphi < -1$ we can assert that sucn sequence does not exist (if $m_\varphi = 1$ aditional conditions are necessary).

Of course, these statements have a valuable theoretic potential basis. The condition $m_\varphi > -1$ means that the unit equilibrium measure $\mu_{1,\varphi}$ in the field φ on Γ has a positive density on all the points of Γ ($m_\varphi < -1$ means that $S_{\mu_{1,\varphi}}$ does not coincide with Γ).

5.5 A method for constructive relative approximation by polynomials.

Above (sections 5.1 and 5.2) we showed the leading idea of the theoretic potential method in connection with the problem of strong asymptotics. In order to carry out this idea it is necessary to have a constructive method of relative polynomial approximation. In fact, any general scheme for the investigation of the asymptotics of orthogonal polynomials requires some method of relative approximation. Classically, this problem is usually solved with the aid of generalized Faber polynomials (see, e.g. [ ]) which give a fairly good absolute approximation ($\| f - p \| < \varepsilon$). Nevertheless, in essence, extremal (in particular, orthogonal) polynomials are connected with relative approximation ($\| p/f - 1 \| < \varepsilon$). In the situations in which we are interested the difference between these forms of apprroximation becomes essential, and so the construction of Faber polynomials does not work.

A natural method of relative polynomial approximation is directly based on the discrete approximation of the equilibrium measure. In the case of approximation on the real axis this construction is carried out in two steps. First, we construct polynomials which are close to the weighted Tchebyshev ones.

For example, let $\mu_n = \mu_{n,\varphi}$ be the equilibrium measure with norm n in the field $\varphi(x) = \frac{1}{2}|x|^\lambda$ and $x_{i,n}^* = x_{i,n}^*(\lambda)$ ($i = 1,2,\dots,n$) be points on the real axis which are uniformly distributed with respect to $d\mu_n$ (see (15)).

The polynomials $T_n^*(x) = \prod_{i=1}^{n} (x - x_{i,n}^*)$ with zeros at these points can be expressed on the segment $S_{\mu_n} = [-x_n, x_n]$ in the form

$$T_n(x) = A_n(x)\cos\Phi_n(x) \qquad (21)$$

where $\Phi_n(x) = \pi\int_{-x_n}^{x} d\mu_n(t)$, so that $T_n(x)$ and $\cos\Phi_n(x)$ have the same zeros on $[-x_n, x_n]$ and the expression (21) is simply the definiton of a positive function $A_n(x)$

Now, it turns out that for any sequence $\varepsilon_n \to 0$ such that $n\varepsilon_n \to \infty$ as $n \to \infty$ we have

$$\lim_{n\to\infty} \max_{|x|\le(1-\varepsilon_n)x_n} \left| \frac{A_n(x)}{2e^{w_n}F(x)} - 1 \right| = 0$$

Near the end points $\pm x_n$ the ratio $\alpha_n = A_n/2e^{w_n}F$ somemewhat faints (in particular, $\alpha_n(\pm x_n) \to 2^{-3/4}$). In general, the approximation is worst at those points where the density of the equilibrium measure is small. This is an inconvenience of the method and it generates certain difficulties, although they are not essential. In particular, making use of the polynomials we can obtain the asymptotics of Tchebyshev's constant corresponding to the weight $F(x) = \exp\{-\frac{1}{2}|x|^\lambda\}$

$$\tau_n = \min_{q\in\mathbb{P}_n^{(1)}} \max_{x\in\mathbb{R}} |q(x)F(x)| \sim 2e^{-w_n}$$

In order to obtain a nearly best relative approximation of the weight it is convenient to introduce also the polynomials $S^*_{n-1}(x) = \prod_{i=1}^{n-1}(x - y^*_{i,n})$ where $\int_{y^*_{i,n}}^{y^*_{i+1,n}} d\mu_n(t) = 1$, $i = 0,1,\dots,n$; $y_{0,n} = -x_n$, $y_{n,n} = x_n$

Suppose $P^*_{2n}(x) = (T^*_n)^2(x) + (x_n^2 - x^2)(S^*_{n-1})^2(x)$ then the polynomials $P^*_{2n}(x)$ give a rather good relative approximation to $F(x)^{-1}$.

6. Rational apprroximation to e^{-x} on $\mathbb{R}_+$

To conclude this review we consider one problem which leads to a slightly more general type of equilibrium. The results of this section are due to A.A.Goncar and E.A.Rakhmanov.

Suppose that

$$\rho_n = \min_{r\in\mathcal{R}_n} \max_{x\in\mathbb{R}_+} |e^{-x} - r(x)|, \qquad \mathcal{R}_n = \left\{\frac{p}{q} : p, q \in \mathbb{P}_n\right\}$$

Many papers have been dedicated to the research of the asymptotical estimate of this value. Due to limitations of space, it is impossible to name all the authors which have contributed to the solution of this problem. Nevertheless, we wish to remark that A. Magnus making use of an ingenious euristhical method was able to indicate the correct value of the limit $V = \lim_{n\to\infty} \rho_n^{1/n}$ (see [20]).

The final result was obtained by A.A. Gončar and E.A. Rakhmanov [5]

THEOREM.

The following limit exists $V = \lim_{n\to\infty} \rho_n^{1/n}$ which coincides with the root of the equation $f(x) = 1/8$ where

$$f(x) = \sum_{n=1}^{\infty} a_n x^n, \qquad a_n = \Big| \sum_{d \mid n} (-1)^d d \Big|$$

(the sum is taken over all the divisors of n including 1 and n).

In the process of the proof of this result a rational function r_n^* was constructed whose order of approximation coincides with that of the best rational approximation. This function has an interesting asymptotic behavior (in fact, the weak asymptotics of r_n is the same as that of the function of best rational approximation).

Denote $\beta_{1,n}, \dots, \beta_{n,n}$ the poles of r_n^* and $\omega_{1,n}, \dots, \omega_{2n+1,n}$ the interpolation points (according to Tchebyshev's Theorem there exists at least $2n+1$ interpolation points). As $n \to \infty$ the poles and the interpolation points tend to infinity (as n). So, to obtain the limit distribution it is necessary to carry out the contraction. Suppose that

$$\omega_n = \frac{1}{2n+1} \sum_{i=1}^{2n+1} \delta\left(\frac{\omega_{i,n}}{n}\right), \qquad \beta_n = \frac{1}{n} \sum_{i=1}^{n} \delta\left(\frac{\beta_{i,n}}{n}\right)$$

It turns out that the weak limits of the sequences $\{\omega_n\}, \{\beta_n\}$ exist and can be described in terms of a certain equilibrium distribution.

Let γ be a Jordan curve in the domain $\mathcal{D} = \mathbb{C} \setminus \mathbb{R}_+$ which separates $\mathbb{R}_+$ from the values $x \to -\infty$ and such that $\lim_{z\to\infty,\, z\in\gamma} \operatorname{Re} z / \log|z| = +\infty$

Let us place on $\mathbb{R}_+$ a unit negative charge, and on a unit positive charge. Suppose that the charge contained in γ is under the action of the exterior field $\varphi(z) = 1/2 \operatorname{Re} z$. Suppose that $\mu_1, -\mu_2$ is the equilibrium distribution corresponding to this scheme of interaction. Formally, the pair of measures μ_1, μ_2 is determined by the following system of relations

$$W_1(z) = V^{\mu_1}(z) - V^{\mu_2}(z) + \tfrac{1}{2} \operatorname{Re} z = \min_{z\in\gamma} W_1(z), \quad z \in S_{\mu_1}$$

$$W_2(z) = V^{\mu_2}(z) - V^{\mu_1}(z) = \min_{z\in\mathbb{R}_+} W_2(z), \quad z \in S_{\mu_2}$$

(in fact $S_{\mu_2} = \mathbb{R}_+$

There exists a curve γ^* that satifies the following stationary condition

$$\frac{\partial}{\partial n_1} W_1(z) = \frac{\partial}{\partial n_2} W_1(z) , \quad z \in S_{\mu_1^*} , \quad \mu_1^* = \mu_1(\gamma^*)$$

(n_1 and n_2 are the right and left normals to S_{μ_1}). The curve γ^* is uniquely determined on that part which coincides with $S_{\mu_1^*}$. The measures μ_1^*, μ_2^* are uniquely determined by these conditions (of course, μ_1^*, μ_2^* also depends on γ^*).

The stationary condition means that the charge μ_1 remains in equilibrium (unstable) even if we remove the "conductor" γ^* . The stationary condition (as the equilibrium one) can be expressed in terms of the energy.

Now, for the poles and interpolation points of r_n^* we have

$$\beta_n \to \mu_1^* \quad , \qquad \omega_n \to \mu_2^*$$

In the proof of these statements an important role is played by the study of the polynomials which satisfy orthogonal relations of type

$$\int_\gamma Q_n^*(z)\, z^j \, \frac{e^{-z}}{\omega_{2n}(z)}\, dz = 0 \quad , \qquad j = 0,1,\dots, n-1$$

where ω_{2n} are polynomials whose zeros lie on $\mathbb{R}_+$ and γ is a curve which separates $\mathbb{R}_+$ from $x \to -\infty$

The basis of the method which allows to study the aysmptotics of (non-Hermite) orthogonal polynomials on curves in the complex plane with respect to analytic weights is contained in the excelent papers of Herbert Stahl [32], [33]. Stahl's method, in the presence of an exterior field, was developed by A.A.Goncar and E.A.Rakhmanov This improvement gives the possibility to widen the field of applications and to obtain an approach to the solution of a new class of problems.

BIBLIOGRAPHY

1. G.Freud, On the greatest zero of an orthogonal polynomial, 1. Acta Sci. Math. 34 (1973), 91-97.

2. A.A.Gončar, On convergence of Pade approximants for some classes of meromorphic functions, Mat. Sb. 97 (139), 1975, 4 =Math. USSR Sb. 26 (1975), 555-575.

3. A.A.Gončar, On the speed of rational approximation of some analytic functions, Mat. Sb. 105(147), 1978, 2 = Math. USSR Sb. 34 (1978), 2.

4. A.A.Gončar, On the speed of convergence of rational approximants for analytic functions, Trudi MIAN 166 (1984), 52-60.

5. A.A.Gončar, Rational approximation of analytic functions. Proceedings of the ICM, Berkeley'86.

6. A.A.Gončar G.López, On Markov's theorem for multipoint Pade approximants, Mat. Sb. 105 (147), 1978, 513-524 = Math. USSR Sb. 34 (1978), 449-459.

7. A.A.Gončar E.A.Rakhmanov, On the convergence of simultanuous Pade approximants for a system of Markov-type functions, Trudi MIAN 157 (1981), 31-48 = Proc. of the Steklov Inst of Math. 1983 issue 3.

8. A.A.Gončar E.A.Rakhmanov, Equilibrium measure and the distribution of zeros of extremal polynomials, Mat. Sb. 125(167) 1984, 1 =Math. USSR Sb. 53 (1986), 1.

9. A.A.Gončar E.A.Rakhmanov, On the problem of equilibrium for vector-potentials, Uspehi Mat. Nauk, 40(1985),155-156.

10. G.López, On the convergence of multipoint Pade approximants for Stieltjes type functions, Dokl. AN SSSR., 239 (1978), 793-796.

11. G.López, Conditions of convergence of multipoint Pade approximants for Stieltjes type functions, Mat. Sb. 107(149), 1978, 69-83 = Math. USSR Sb. 35 (1979), 363-376.

12. G.López, On the convergence of Pade approximants for Stieltjes type meromorphic fuctions, Mat. Sb. 111(143), 1980, 308-316 =Math. USSR Sb. 38 (1981), 281-288.

13. G.López, On the asymptotics of the ratio of orthogonal polynomials and the convergence of multipoint Pade approximants, Mat. Sb. 128(170), 1985, 2 =Math. USSR Sb. 56 (1987).

14. G.López, On Szego's Theorem for polynomials orthogonal with respect to varying measures, Proceedings of the International Seminar on Orthogonal Polynomials and its Applications, Segovia'86.

15. G.López, Asymptotics of polynomials orthogonal with respect to

varying measures, (submitted to Const. Approx. Theory)

16. G.López, On the convergence of Pade approximants for Stieltjes type meromorphic functions II, (submitted to Mat. Sb.)

17. G.López J.Illan, Sobre la convergencia de los aproximantes multipuntuales de Pade para funciones meromorfas de tipo Stieltjes, Rev. Ciencias Mat., v.III (1981), 43-66.

18. P.Nevai, Asymptotics for orthogonal polynomials associated with $\exp\{-x^4\}$, SIAM. J. Math. Analysis 15(1984), 1177-1187.

19. P.Nevai and J.S.Dehesa, On asymptotic average properties of zeros of orthogonal polynomialsSIAM. J. Math. Analysis, 10(1979), 1184-1192.

20. A.P.Magnus, CFGT determination of Varga's constant "1/9", Institut Mathematique U.C.L., B-1348, Belgium (1986).

21. H.N.Mhaskar and E.B.Saff, Were does the sup norm of orthogonal polynomials live? (A generalization of incomplete polynomials), Constr. Appr., 1(1985), 71- 91.

22. E.A.Rakhmanov, On the asymptotic of the ratio of orthogonal polynomials, Mat. Sb. 103(145), 1977, 237-252 = Math. USSR Sb. 32 (1977), 199-213.

23. E.A.Rakhmanov, On the aysmptotic of the ratio of ortogonal polynomial,II, Mat. Sb. 118(160), 1982, 104-117 = Math. USSR Sb.,46 (1983), 105-117.

24. E.A.Rakhmanov, On the convergence of diagonal Pade approximants, Mat. Sb. 104(146), 1977, 271-291 =Math. USSR Sb. 27 (1977).

25. E.A.Rakhmanov, On the asymptotic properties of orthogonal polynomials on the real axis, Dokladi AN SSSR, 261 (1981), 282-284.

26. E.A.Rakhmanov, On the asymptotic properties of orthogonal polynomials on the real axis, Mat. Sb., 119(161), 1982, 163-203 = Math. USSR Sb. 47 (1984), 155-193.

27. E.A.Rakhmanov, Asymptotic properties of orthogonal polynomials, Thesis, Steklov Math. Inst., Moscow, 1983.

28. E.A.Rakhmanov, On the asymptotic properties of orthogonal polynomials on the circle with respect to non-Szego weights Mat. Sb., 130(172), 1986, 151-169.

29. E.B.Saff, I.L.Ullman, R.S.Varga, Incomplete polynomials: an electrostatic approach. - Approximation theory III (E.W.Cheney ed.), 1980, Academic press, 769-782.

30. R.Sheen, Plancherel-Rotach-type asymptotics for orthogonal polynomials associated with $\exp(-x^6/6)$, J. Approx. Theory 50 (1987), n° 3, 232-293.

31. H.Stahl, Doctoral thesis, Technical Universitat, Berlin, 1976.

32. H.Stahl, Orthogonal polynomials with complex weight functons, I, Constr. Approx., Const. Appr. 2 (1986), 225-240.

33. H.Stahl, Orthogonal polynomials with complex weight functions, II, Constr. Approx., Const. Appr. 2 (1986), 241-252.

34. J.L.Ullman, Orthogonal polynomials associated with an infinite interval, Michigan Math. J., 27 (1980), 353-367

35. J.S.Walsh, Interpolation and approximation by rational functions in the complex domain, published by the American Mathematical Society, 1960.

36. H.Widom, Extremal polynomials associated woth a sistem of curves in the complex plane, Adv. Math.,3, no2, 1969.

37. A.Mate, P.Nevai and V.Totic, On the asymptotics of the leading coefficients of orthogonal polynomials, Constr. Approx. I, 1985.

FACTORIZATION OF SECOND ORDER DIFFERENCE EQUATIONS AND ITS APPLICATION TO ORTHOGONAL POLYNOMIALS

Attila Máté
Department of Mathematics
Brooklyn College of the
City University of New York
Brooklyn, New York 11210, USA

Paul Nevai
Department of Mathematics
The Ohio State University
Columbus, Ohio 43210, USA

ABSTRACT. It is shown that a second order recurrence expression with coefficients having bounded variation, written as a second degree polynomial of the forward shift operator, can be factored as the product of two first order expressions. This result is used to obtain asymptotics over the complex plane for a class of polynomials orthonormal over the real line.

CONTENTS

1. Introduction

Consider the second order recurrence equation

$$f(n+2) + \alpha(n)f(n+1) + \beta(n)f(n) = h(n) \quad (-\infty < n < \infty) \tag{1.1}$$

with f as the unknown function, such that the coefficients α and β are of bounded variation (i.e., (2.4) below holds). Writing A and B for the limits of the sequences $\alpha(n)$ and $\beta(n)$, respectively, the equation

$$t^2 + At + B = 0 \tag{1.2}$$

This material is based upon work supported by the National Science Foundation under Grant Nos. DMS-8601184 (first author) and DMS-8419525 (second author), and by the PSC-CUNY Research Award Program of the City University of New York under Grant No. 6-66429 (first author).

is called the characteristic equation of (1.1). Provided the roots of (2.1) have different absolute values, we will be able to express the left-hand side of (1.1) as an operator product. That is, introducing the forward shift operator E, the left-hand side of (1.1) can be written as

$$((E^2 + \alpha E + \beta)f)(n)$$

(E will be described in some detail in the next section.) This will be factored as

$$((E - \varsigma_2)(E - \varsigma_1)f)(n)$$

where ς_1 and ς_2 are functions on integers (it is shown in (2.9) below how to multiply out this product). Thus (1.1) can be written as

$$(E - \varsigma_2)(E - \varsigma_1)f = h,$$

which can be solved explicitly as

$$f = (E - \varsigma_1)^{-1}(E - \varsigma_2)^{-1}h;$$

the inverses on the right can be written as infinite sums, and they will be convergent if h is well-behaved.

Now, polynomials orthogonal on the real line satisfy a recurrence equation analogous to (1.1):

$$a_{n+2}p_{n+2}(x) + (b_{n+1} - x)p_{n+1} + a_{n+1}p_n(x) = 0 \quad (n \geq -1) \tag{1.3}$$

(see (4.3) below). If the coefficient sequences here are of bounded variation, then we can use the above method to obtain an asymptotic expression for p_n outside the real line (see Theorems 4.1, 5.1 and 5.2 below). The results we obtain go partly beyond Theorem 1 of Máté-Nevai-Totik [5, p. 232] in that our asymptotic estimate holds on unbounded sets as well.

While equation (1.3) is a homogeneous equation, it is only valid for $n \geq -1$. Before the method outlined above can be applied, it has to be extended to an inhomogeneous equation valid for $-\infty < n < \infty$.

PART I. RECURRENCE EQUATIONS
2. The Factorization Theorem

In this section $\alpha, \beta, \ldots, f, g, \ldots$ will denote functions on integers and E will denote the forward shift operator, that is

$$E^k f(n) = (E^k f)(n) = f(n + k) \qquad (-\infty < k, n < \infty).$$

We will write terms involving products of several functions, indicating the argument only once on the right. That is, e.g.,

$$fg(n) = (fg)(n) = f(n) \cdot g(n).$$

Such terms may be interlaced with integral powers of E. An occurrence of E in the term will affect all functions to the right in the same term. That is, e.g.,

$$fEg(n) = f(n)g(n+1),$$
$$f_1E^2f_2E^{-3}f_3f_4E^4f_5(n) = f_1(n)f_2(n+2)f_3(n-1)f_4(n-1)f_5(n+3)$$

(e.g. to obtain $f_4(n-1)$, observe that f_4 is affected by powers of E to the left of it, i.e., by $E^2E^{-3} = E^{-1}$), and

$$\begin{aligned}(E\alpha)^3(n) &= E\alpha E\alpha E\alpha(n) \\ &= \alpha(n+1)\alpha(n+2)\alpha(n+3).\end{aligned} \tag{2.1}$$

Using this notation, we can now state the Factorization Theorem:

THEOREM 2.1. *Consider the recurrence polynomial*

$$(E^2 + \alpha E + \beta)f(n) \qquad (-\infty < n < \infty) \tag{2.2}$$

with f as the indeterminate function, where

$$\text{(i)} \lim_{n\to\infty} \alpha(n) = A, \qquad \text{(ii)} \lim_{n\to\infty} \beta(n) = 1, \tag{2.3}$$

and

$$\sum_{n=0}^{\infty}(|\alpha(n+1) - \alpha(n)| + |\beta(n+1) - \beta(n)|) < \infty. \tag{2.4}$$

Assume here A is a complex number such that

$$A \notin [-2, 2]. \tag{2.5}$$

Then (2.2) can be factored as

$$(E - \varsigma_2)(E - \varsigma_1)f(n) \tag{2.6}$$

for large enough n, say $n \geq n_0$, where ς_1 and ς_2 are functions on integers. Moreover, writing t_{1n} and t_{2n} with $|t_{1n}| \geq |t_{2n}|$ for the roots of the equation

$$t^2 + \alpha(n)t + \beta(n) = 0, \tag{2.7}$$

we can choose ς_1 and ς_2 such that

$$\sum_{n=n_0}^{\infty} | t_{1n} - \varsigma_1(n) | < \infty . \tag{2.8}$$

It may be possible to gain some insignt into the meaning of the factorization in (2.6) by multiplying it out. We obtain

$$\begin{aligned}(E - \varsigma_2)(E - \varsigma_1)f(n) &= E^2 f(n) - \varsigma_2 E f(n) - E\varsigma_1 f(n) + \varsigma_2\varsigma_1 f(n) \\ &= f(n+2) - (\varsigma_2(n) + \varsigma_1(n+1))f(n+1) \\ &+ \varsigma_2(n)\varsigma_1(n)f(n).\end{aligned} \tag{2.9}$$

The equation

$$t^2 + At + 1 = 0 \tag{2.10}$$

is called the characteristic equation of the difference polynomial in (2.2). Condition (2.5) ensures that the roots of this polynomial have different absolute values, and this will be crucial for our method. Condition (2.3)(ii) is only technical, and it can be replaced with the more general condition

$$\lim_{n\to\infty} \beta(n) = B \neq 0, \tag{2.11}$$

but then (2.5) will have to be modified appropriately. The characteristic equation in this case is

$$t^2 + At + B = 0,$$

but this can be transformed back to the form in (2.10) by using the substitution $t = \sqrt{B}t'$. This substitution corresponds to the substitution

$$f(n) = B^{n/2} f'(n)$$

in (2.2), and so condition (2.11) can be reduced to condition (2.3)(ii).

As $A \neq \pm 2$ according to (2.5), the roots of equation (2.10) are distinct. Hence (2.4) implies

$$\sum_{n=0}^{\infty} (| t_{1,n+1} - t_{1n} | + | t_{2,n+1} - t_{2n} |) < \infty \tag{2.12}$$

in virtue of the case $m = 2$ of the following simple

LEMMA 2.2. *Assume that the roots $w_1, \ldots, w_m$ of the equation*

$$\sum_{j=0}^{m} u_j w^j = 0 \qquad (u_m = 1)$$

are pairwise distinct. Then the Jacobian

$$\det[\frac{\partial u_j}{\partial w_k}]_{0\le j\le m-1,\ 1\le k\le m}$$

is different from zero.

PROOF. Write

$$P(w) = \sum_{j=0}^{m} u_j w^j = \prod_{k=1}^{m} (w - w_k).$$

As the z_k 's are distinct, we have

$$\frac{\partial P(w)}{\partial w_k} = -\frac{P(w)}{w - w_k} = \sum_{j=0}^{m-1} w^j \frac{\partial u_j}{\partial w_k}.$$

Now the conclusion of the lemma follows from the observation that the vectors $[\partial u_j/\partial w_k]_{0\le j\le m-1}$ $(1 \le k \le m)$ are linearly independent. Namely, the equation

$$0 = \sum_{k=1}^{m} c_k \sum_{j=0}^{m-1} w^j \frac{\partial u_j}{\partial w_k} = -\sum_{k=1}^{m} c_k \frac{P(w)}{w - w_k}$$

implies $c_k = 0$ for each k; to see this, make $w \to w_k$. The proof is complete.

As we will apply Theorem 2.1 to obtain uniform asymptotics for orthogonal polynomials, we will need a more precise version of this theorem in that we will need an estimate for the remainder of the sum in (2.8).

THEOREM 2.3. *Assume the hypotheses of Theorem 2.1, and let* ρ *be a real with* $0 < \rho < 1$. *Let* n_0 *be such that*

$$|\,\beta(n)/t_{1n}\,| \le \rho \tag{2.13}$$

and

$$|\,t_{1n}\,| \ge 1 + \sum_{n=n_0}^{\infty} |\,t_{1,n+1} - t_{1n}\,| \tag{2.14}$$

hold for $n \ge n_0$. *Then* ς_1 *and* ς_2 *in (2.6) can be chosen such that*

$$\sum_{\nu=n}^{\infty} |\,t_{1\nu} - \varsigma_1(\nu)\,| \le \sum_{\nu=n_0}^{n-1} |\,t_{1,\nu+1} - t_{1\nu}\,| \,\rho^{n-1-\nu}/(1-\rho) \tag{2.15}$$

$$+ \sum_{\nu=n}^{\infty} |\,t_{1,\nu+1} - t_{1\nu}\,| /(1-\rho)$$

holds for every $n \geq n_0$.

Observe that an integer n_0 satisfying (2.13) and (2.14) does exist. Indeed, writing t_1 and t_2 with $|t_1| \geq |t_2|$ for the roots of the characteristic equation (2.10), we have $|t_1| > |t_2|$ in view of (2.5), and so we have $|t_1| > 1$, as $t_1t_2 = 1$. Now, since $\beta(n) \to 1$ and $t_{1n} \to t_1$, (2.13) will hold for large enough n. Moreover, writing $\eta = (t_1 - 1)/2$, we will have $|t_{1n}| \geq 1 + \eta$ for large enough n, and so (2.14) will hold for large enough n_0 in view of (2.12).

3. The Proof of the Factorization Theorem

The reason the Factorization Theorem is useful is that it reduces the solution of certain second orders difference equation to the successive solution of two first order difference equation. Under certain conditions, first order difference equations can be solved explicitly, as shown by the following simple

LEMMA 3.1. *Given the functions* f, g, *and* h *on integers, suppose that we have*

$$(E - g)f(n) = h(n) \qquad (-\infty < n < \infty). \tag{3.1}$$

Then

$$f(n) = \sum_{\ell=0}^{\infty} (E^{-1}g)^{\ell} E^{-1} h(n) \tag{3.2}$$

holds for each n *provided that*

$$\lim_{\ell \to \infty} (E^{-1}g)^{\ell} f(n) = 0 \tag{3.3}$$

holds for each n.

Observe that e.g.,

$$(E^{-1}g(n))^2 E^{-1} h(n) = g(n-1)g(n-2)h(n-3)$$

(cf. example (2.1) above). Condition (3.3) will be fulfilled in our applications, as we will have $f(n) = 0$ when n is a large enough negative number.

PROOF. Replacing $h(n)$ with the left-hand side of (3.1), the right-hand side of (3.2) becomes

$$\begin{aligned}\sum_{\ell=0}^{\infty}(E^{-1}g)^{\ell}E^{-1}(E - g)f(n) &= \sum_{\ell=0}^{\infty}((E^{-1}g)^{\ell} - (E^{-1}g)^{\ell+1})f(n) \\ &= \lim_{N\to\infty}(I - (E^{-1}g)^{N+1})f(n) = f(n),\end{aligned}$$

where I is the identity operator; the last qeuality follows from (3.3). This shows that (3.2) is indeed valid, completing the proof of the lemma.

Next we turn to the proof of the Factorization Theorem. As Theorem 2.1 follows from Theorem 2.3, it will be sufficient to present only the

PROOF OF THEOREM 2.3. As the expressions in (2.2) and on the right-hand side of (2.9) (which is the multiplied-out version of (2.6)) must agree for any choice of the function f, the respective coefficients must agree, i.e., we must have

$$-\alpha(n) = \varsigma_1(n+1) + \varsigma_2(n)$$

and

$$\beta(n) = \varsigma_2(n)\varsigma_1(n). \tag{3.4}$$

That is

$$\varsigma_1(n+1) = -\alpha(n) - \beta(n)/\varsigma_1(n). \tag{3.5}$$

Now, dividing equation (2.7) by t and replacing t with t_{1n} (which is a root of this equation), we obtain

$$t_{1n} = -\alpha(n) - \beta(n)/t_{1n}.$$

Subtracting (3.5) from this, we obtain

$$t_{1n} - \varsigma_1(n+1) = \frac{\beta(n)}{t_{1n}\varsigma_1(n)}(t_{1n} - \varsigma_1(n)),$$

that is

$$|\, t_{1,n+1} - \varsigma_1(n+1)\,| \le |\, t_{1,n+1} - t_{1n}\,| + \left|\frac{\beta(n)}{t_{1n}\varsigma_1(n)}\right| |\, t_{1n} - \varsigma_1(n)\,| \,. \tag{3.6}$$

Now choose

$$\varsigma_1(n_0) = t_{1n_0} \tag{3.7}$$

for the n_0 described in connection of (2.13) and (2.14), and define $\varsigma_1(n)$ for $n \ge n_0$ with the aid of (3.5). We will prove by induction on n that

$$|\,\varsigma_1(n)\,| \ge 1 \tag{3.8}$$

holds for $n \ge n_0$. This is true for $n = n_0$, since $\varsigma_1(n_0) = t_{1n_0} \ge 1$ holds in view of (2.14). Let $n \ge n_0$ and assume that (3.8) holds with ν such that $n_0 \le \nu \le n$ replacing n. Then

$$\left|\frac{\beta_1(\nu)}{t_{1\nu}\varsigma_1(\nu)}\right| \le \rho \qquad (n_0 \le \nu \le n)$$

holds in view of (2.13), and so (3.6) becomes

$$| t_{1,\nu+1} - \varsigma_1(\nu + 1) | \le | t_{1,\nu+1} - t_{1\nu} | + \rho | t_{1\nu} - \varsigma_1(\nu) | .$$

Using this for $\nu = n, n-1, \ldots, n_0$ repeatedly and noting that $t_{1n_0} - \varsigma_1(n_0) = 0$ by (3.7), we obtain

$$| t_{1,n+1} - \varsigma_1(n + 1) | \le \sum_{\nu=n_0}^{n} | t_{1,\nu+1} - t_{1\nu} | \rho^{n-\nu} . \tag{3.9}$$

As $\rho < 1$, this implies

$$| \varsigma_1(n + 1) | \ge | t_{1,n+1} | - \sum_{\nu=n_0}^{n} | t_{1,\nu+1} - t_{1\nu} | \ge 1, \tag{3.10}$$

where the last inequality holds in view of (2.14). Thus (3.8) holds with $n + 1$ replacing n. This completes the inductive argument, showing that (3.8) holds for all $n \ge n_0$. As a by-product, we also obtain that (3.9) holds for all $n \ge n_0$; actually it vacuously holds for $n = n_0 - 1$ as well in view of (3.7). In fact, (3.9) will be our key result, and (3.8) was needed only in order to establish it.

Conclusion (2.15) of the theorem to be proved now readily follows. In fact, using (3.9), for $n \ge n_0$ we obtain

$$\begin{aligned}\sum_{\nu=n}^{\infty} | t_{1\nu} - \varsigma_1(\nu) | &\le \sum_{\nu=n}^{\infty} \sum_{\ell=n_0}^{\nu-1} | t_{1,\ell+1} - t_{1\ell} | \rho^{\nu-1-\ell} \\ &= \sum_{\ell=n_0}^{n-1} | t_{1,\ell+1} - t_{1\ell} | \sum_{\nu=n}^{\infty} \rho^{\nu-1-\ell} + \sum_{\ell=n}^{\infty} | t_{1,\ell+1} - t_{1\ell} | \sum_{\nu=\ell+1}^{\infty} \rho^{\nu-1-\ell} .\end{aligned}$$

By evaluating the inner sums we obtain (2.15). The proof of Theorem 2.3 is complete.

PART II. AN APPLICATION TO ORTHOGONAL POLYNOMIALS

4. The Main Asymptotic Result

In what follows, by measure we will mean a positive finite measure α on the real line R whose support supp(α) (the smallest closed set $S \subset R$ with $\alpha(R \backslash S) = 0$) is an infinite set, and all the moments of which are finite, that is, for every integer $n \ge 0$ the integral

$$\int_{-\infty}^{\infty} x^n d\alpha(x)$$

exists (i.e., it is absolutely convergent). Associated with the measure α there is a unique sequence of orthonormal polynomials

$$p_n(x) = p_n(d\alpha, x) = \gamma_n x^n + \cdots \quad (\gamma_n = \gamma_n(d\alpha) > 0, \ n \ge 0) \tag{4.1}$$

(it is traditional to use the differential notation $d\alpha$ instead of α in these formulas) satisfying

$$\int_{-\infty}^{\infty} p_m(x)p_n(x)d\alpha(x) = 1 \quad or \quad 0 \tag{4.2}$$

according as $m = n$ or $m \neq n$ $(m, n \geq 0)$. These polynomials satisfy a recurrence relation

$$a_{n+2}p_{n+2}(z) + (b_{n+1} - z)p_{n+1}(z) + a_{n+1}p_n(z) = 0 \quad (n \geq -1), \tag{4.3}$$

where $p_{-1} = 0$, $p_0 = 0$, $a_0 = 0$, and $a_n > 0$ for $n \geq 0$ (cf. e.g., Freud [2, formula (I.2.4) on p. 17] or Szegö [7, formula (3.2.1) on p. 42]). If one wants to indicate the dependence of the coefficients a_n, b_n on the measure α, one may write $a_n(d\alpha)$ and $b_n(d\alpha)$ instead.

The connection between the behavior of the coefficients a_n, b_n and the properties of the measure α is frequently investigated. In the most studied cases, the limits

$$\lim_{n\to\infty} a_n = a\,(\neq 0) \quad and \quad \lim_{n\to\infty} b_n = b$$

exist (and are finite). By a linear change of variables, we may assume $a = \frac{1}{2}$ and $b = 0$ here. In this case, the support of α is [-1,1] plus countable many isolated atoms (singletons with positive measure); see Blumenthal's theorem in Chihara [1, Sections IV.3-4, pp. 113-124]. The set of measures for which $a = \frac{1}{2}$ and $b = 0$ is often called $M(0, 1)$, and it is studied in detail e.g., in Nevai [6] (from p. 10 on at several places).

In studying equation (4.3), one can make use of the corresponding algebraic equation:

$$a_{n+2}t^2 + (b_{n+1} - z)t + a_{n+1} = 0 \quad (n \geq 0) \tag{4.4}$$

(we will not consider the case $n = -1$, even though it is allowed in (4.3), because the fact that $a_0 = 0$ would cause complications). Define τ as a holomorphic function on $C\backslash[-1, 1]$, where C is the complex plane, by putting

$$\tau(z) = z + \sqrt{z^2 - 1}, \tag{4.5}$$

where that branch of τ is chosen for which

$$\lim_{z\to\infty} \tau(z) = \infty. \tag{4.6}$$

For $-1 \leq x \leq 1$ put e.g.,

$$\tau(x) = \lim_{y\to 0^+} \tau(x + iy).$$

Then the roots of equation (4.4) can be written as

$$t_{1n}(z) = \sqrt{\frac{a_{n+1}}{a_{n+2}}}\,\tau\left(\frac{z - b_{n+1}}{2\sqrt{a_{n+1}a_{n+2}}}\right) \quad (n \geq 0), \tag{4.7}$$

and

$$t_{2n}(z) = \frac{a_{n+1}}{a_{n+2}}(t_{1n}(z))^{-1} \quad (n \geq 0). \tag{4.8}$$

Using the Factorization Theorem of Section 2 (or, rather, its variant, Theorem 2.3), we will establish the following:

THEOREM 4.1. *Let α be a finite positive measure on the real line with finite moments such that* supp(α) *is an infinite set. Assume that, writing $a_n = a_n(d\alpha)$ and $b_n = b_n(d\alpha)$, we have*

$$(i)\ \lim_{n\to\infty} a_n = \frac{1}{2}, \quad (ii)\ \lim_{n\to\infty} b_n = 0 \tag{4.9}$$

and

$$\sum_{n=0}^{\infty}(|\, a_{n+1} - a_n \,| + |\, b_{n+1} - b_n \,|) < \infty. \tag{4.10}$$

Then, writing $p_n(z) = p_n(d\alpha, z)$, the limit

$$g(z) = \lim_{n\to\infty}\left(p_n(z)\prod_{\nu=0}^{n-1}(t_{1\nu}(z))^{-1}\right) \tag{4.11}$$

exists for every complex $z \notin [-1,1]$. Moreover, the convergence in (4.11) is uniform on every closed set $K \subset C\backslash[-1,1]$. The limit

$$L = \lim_{z\to\infty} g(z) \tag{4.12}$$

exists and $L \neq 0$.

When we say that a limit exists, we do, of course, require that it not be ∞. Given K as described, $t_{1\nu}^{-1}$ is holomorphic on K for large enough ν in view of (4.9). Hence the uniformness of the convergence in (4.11) implies that

$$g(z)\prod_{\nu=0}^{N} t_{1\nu}(z) \tag{4.13}$$

is holomorphic on K for large enough N. In fact, we do not quite need the uniformness of the convergence in (4.11) to reach this conclusion: clearly, it is sufficient to know that the convergence in (4.11) is uniform on compact subsets of K (this observation will be of some use in Section 6).

The above result is only partly new. It was setablished earlier as Theorem 1 in Máté-Nevai-Totik [5, formula (9) on p. 232] except that the uniformness of the convergence in (4.11) was established only for compact (and hence bounded) $K \subset C\backslash[-1,1]$, and the existence of the limit in (4.12) was not discussed. Condition (4.10) was first considered in Máté-Nevai [4] and (somewhat later) Dombrowski [2].

5. The Main Lemma

A major step toward establishing Theorem 4.1 is represented by

LEMMA 5.1. *Assume the hypotheses of Theorem 4.1, and let D be an open subset of the complex plain such that its closure $\overline{D}$ is disjoint from [-1,1]. Then for every large enough integer n_0 there are functions F_n and G_n satisfying*

$$p_n(z)\prod_{\nu=0}^{n-1}(t_{1\nu}(z))^{-1} = F_n(z)(p_{n_0}(z) + (p_{n_0+1}(z) - t_{1n_0}(z)p_{n_0}(z))G_n(z)) \tag{5.1}$$

for $n > n_0$ and $z \in D$ such that for certain function F, G, ψ_{1n}, and ψ_{2n} we have

$$F_n(z) = F(z)(1 + \psi_{1n}(z)) \tag{5.2}$$

and

$$G_n(z) = G(z) + \psi_{2n}(z) \tag{5.3}$$

for every $n > n_0$ and $z \in D$, and

$$\lim_{n\to\infty}\psi_{jn}(z) = 0 \quad (j = 1,2) \tag{5.4}$$

uniformly for $z \in D$. Moreover, the functions F and G are bounded on every compact subset of D.

It is clear from (5.2)-(5.4) that for any fixed $z \in D$ the limit of the right-and side of (5.1) exists as $n \to \infty$, and in fact this limit is

$$g(z) = F(z)(p_{n_0}(z) + (p_{n_0+1}(z) - t_{1n_0}(z)p_{n_0}(z))G(z)). \tag{5.5}$$

Moreover, if $K \subset D$ is a compact set, then this limit is uniform on K, since the functions F, G, p_{n_0}, p_{n_0+1}, and t_{1n_0} are bounded on K (for t_{1n_0} this is true in view of (4.7), since τ is bounded on compact sets – cf. (4.5)). Hence the pointwise existence and the uniformness on every compact set $K \subset C\backslash[-1,1]$ of the limit in (4.11) follows from the above lemma. To establish the uniformness of the convergence in (4.11) on unbounded K we need to study the behavior of the functions F, G, and G_n near infinity. This will be done in the next section.

PROOF. We are going to use Theorem 2.3 to factor the left-hand side of the recurrence equation

$$p_{n+2}(z) + \frac{b_{n+1} - z}{a_{n+2}} p_{n+1}(z) + \frac{a_{n+1}}{a_{n+2}} p_n(z) = 0; \tag{5.6}$$

this equation holds for $n \geq 0$ (actually for $n \geq -1$, but cf. the remark after (4.4)) according to (4.3). The roots of the corresponding algebraic equation (4.4) were given by (4.7) and (4.8).

Let D_1 be an open set such that $\overline{D} \subset D_1$ and $\overline{D}_1 \subset C\backslash[-1,1]$. As $1/z$ is bounded on D and the derivative of $\tau(z)$ (cf. (4.5)) is bounded on D_1, it is easy to conclude from (4.7), (4.9), and (4.10) that

$$\lim_{n\to\infty} \frac{1}{z} \sum_{\nu=n}^{\infty} | t_{1,\nu+1}(z) - t_{1\nu} | = 0 \tag{5.7}$$

uniformly for $z \in D$.

By (4.7) and (4.9) we have

$$\lim_{n\to\infty} \frac{t_{1n}(z)}{z} = \frac{\tau(z)}{z} \tag{5.8}$$

uniformly on D. As $| \tau(z) | > 1$ for $z \in C\backslash[-1,1]$, it follows that there are constants $\eta > 1$ and $C_1, C_2 > 0$ such that if n is sufficiently large, say $n \geq n_1$ for some n_1, then

$$| t_{1n}(z) | > \eta \tag{5.9}$$

and

$$C_1 | z | < | t_{1n}(z) | < C_2 | z | \tag{5.10}$$

hold for $z \in D$. The constants C_1 and C_2 (and $C_3, C_4, \ldots$ below) may of course depend on D and the measure α. (4.9)(i) and (5.9) imply that there are an integer $n_0 \geq n_1$ and a real ρ with $0 < \rho < 1$ such that

$$\left| \frac{a_{n+1}/a_{n+2}}{t_{1n}(z)} \right| \leq \rho \tag{5.11}$$

holds for $z \in D$ and $n \geq n_0$; e.g., one can take $\rho = 2/(1+\eta)$. Using (5.9) (for $| z |$ small) and (5.10) (for $| z |$ large) it is easy to conclude from (5.7) that

$$| t_{1n}(z) | \geq 1 + C_3 | z | + \sum_{\nu=n_0}^{\infty} | t_{1,\nu+1}(z) - t_{1\nu}(z) | \tag{5.12}$$

holds for $z \in D$ and $n \geq n_0$ with some positive constant C_3, provided n_0 is chosen large enough; hence the conditions analogous to (2.13) and (2.14) are satisfied by equation (4.4) replacing (2.7). Clearly, writing

$$\alpha(n) = \frac{b_{n+1} - z}{a_{n+2}} \quad \text{and} \quad \beta(n) = \frac{a_{n+2}}{a_{n+1}} \tag{5.13}$$

(these are the coefficients in equation (5.6)), conditions (2.3) and (2.4) are satisfied in view of (4.9) and (4.10). Finally, condition (2.5) corresponds to the relation $z \notin [-1, 1]$, which holds for $z \in D$.

Hence, according to Theorem 2.3, the left-hand side of (5.6) can be factored as

$$(E - \varsigma_{2z})(E - \varsigma_{1z})q_z(n) \quad (n \geq n_0), \tag{5.14}$$

where ς_{1z} and ς_{2z} are functions on integers (depending on z, as indicated by the second subscript) and

$$q_z(n) = p_n(z).$$

We wrote z as a subscript so as to retain the original use of the forward shift operator as acting on arguments.

If we choose n_0 above large enough, then $t_{1n}(z)$ will be holomorphic in D for $n \geq n_0$. Then, defining $\varsigma_{1z}(n)$ and $\varsigma_{2z}(n)$ as in the proof of Theorem 2.3, that is, by formulas (3.7), (3.5), and (3.4) (cf. (5.13) for $\alpha(n)$ and $\beta(n)$), these functions will also be holomorphic in D for $n \geq n_0$.

The analogue of (2.15) in Theorem 2.3 is satisfied, i.e.,

$$\sum_{\nu=n}^{\infty} | t_{1\nu}(z) - \varsigma_{1z}(\nu) | \leq \sum_{\nu=n_0}^{n-1} | t_{1,\nu+1}(z) - t_{1\nu}(z) | \rho^{n-1-\nu}/(1-\rho)$$
$$+ \sum_{\nu=n}^{\infty} | t_{1,\nu+1}(z) - t_{1\nu}(z) | /(1-\rho) \quad (z \in D, \; n \geq n_0). \tag{5.15}$$

In virtue of the uniformness of the convergence in (5.7), this implies that

$$\lim_{n\to\infty} \sum_{\nu=n}^{\infty} \left| \frac{t_{1\nu}(z)}{z} - \frac{\varsigma_{1z}(\nu)}{z} \right|$$
$$= \lim_{n\to\infty} \frac{1}{|z|} \sum_{\nu=n}^{\infty} | t_{1\nu}(z) - \varsigma_{1z}(\nu) | = 0 \tag{5.16}$$

uniformly for $z \in D$.

In view of inequality (5.10), this implies that

$$\lim_{n\to\infty} \sum_{\nu=n}^{\infty} | \log\frac{t_{1\nu}(z)}{z} - \log\frac{\varsigma_{1_z}(\nu)}{z} | = 0$$

uniformly for $z \in D$, i.e.,

$$\lim_{n\to\infty} \prod_{\nu=n}^{\infty} (\varsigma_{1_z}(\nu)/t_{1\nu}(z)) = 1 \tag{5.17}$$

uniformly for $z \in D$.

Now (5.14) factors the left-hand side of (5.6) for $n \geq n_0$; however, we need a factoring valid for all n, $-\infty < n < \infty$. To this end, put

$$\varsigma_{1_z}(n) = \varsigma_{2_z}(n) = 0 \tag{5.18}$$

for $z \in D$ and $-\infty < n < n_0$ (for $n \geq n_0$, $\varsigma_{1_z}(n)$ and $\varsigma_{2_z}(n)$ have already been defined), and write

$$r_z(n) = \begin{cases} p_n(z)(= q_z(n)) & \text{if } n \geq n_0, \\ 0 & \text{if } n < n_0. \end{cases} \tag{5.19}$$

Define $h_z(n)$ by the equation

$$(E - \varsigma_{2_z})(E - \varsigma_{1_z})r_z(n) = h_z(n) \tag{5.20}$$

for $z \in D$ and $-\infty < n < \infty$. Then

$$h_z(n) = 0 \quad \text{unless} \quad n = n_0 - 1 \text{ or } n = n_0 - 2, \tag{5.21}$$

$$h_z(n_0 - 2) = p_{n_0}(z), \tag{5.22}$$

and

$$h_z(n_0 - 1) = p_{n_0+1}(z) - \varsigma_{1_z}(n_0)p_{n_0}(z)$$

for $z \in D$. As

$$\varsigma_{1_z}(n_0) = t_{1n_0}(z) \tag{5.23}$$

according to (3.7), the last equation becomes

$$h_z(n_0 - 1) = p_{n_0+1}(z) - t_{1n_0}(z)p_{n_0}(z). \tag{5.24}$$

Using Lemma 3.1 twice, we can solve equation (5.20) for $r_z(n)$. The analogue of condition (3.3) is satisfied, since $r_z(n) = 0$ for $n < n_0$ according to (5.19). We obtain

$$r(n) = \sum_{k=0}^{\infty}\sum_{\ell=0}^{\infty}(E^{-1}\varsigma_1)^k E^{-1}(E^{-1}\varsigma_2)^\ell E^{-1}h(n). \tag{5.25}$$

To simplify our notation, everywhere in this formula we dropped the subscript z; that is, we wrote $r = r_z$, $\varsigma_1 = \varsigma_{1z}$, $\varsigma_2 = \varsigma_{2z}$, and $h = h_z$. Note that only finitely many terms are nonzero in the sums on the right-hand side in view of (5.21). In fact, according to (5.21), the only nonzero terms on the right-hand side are those for which $n - k - \ell - 2 = n_0 - 1$ or $n - k - \ell - 2 = n_0 - 2$. Moreover, by (5.18) we can see that the terms corresponding to the latter case are zero unless $\ell = 0$ (since otherwise this term contains $\varsigma_2(n_0 - 1)$ as a factor). Thus, assuming $n \geq n_0$, (5.25) becomes

$$r(n) = (E^{-1}\varsigma_1)^{n-n_0}E^{-2}h(n) + \sum_{k=0}^{n-n_0-1}(E^{-1}\varsigma_1)^k E^{-1}(E^{-1}\varsigma_2)^{n-n_0-1-k}E^{-1}h(n).$$

Eliminating the operator E as done in example (2.1), we obtain

$$\begin{aligned} r(n) &= h(n_0-2)\prod_{\lambda=1}^{n-n_0}\varsigma_1(n-\lambda) + h(n_0-1)\sum_{k=0}^{n-n_0-1}\left(\prod_{\mu=1}^{k}\varsigma_1(n-\mu)\right)\left(\prod_{\nu=k+2}^{n-n_0}\varsigma_2(n-\nu)\right) \\ &= \left(\prod_{\lambda=1}^{n-n_0}\varsigma_1(n-\lambda)\right)\left(h(n_0-2) + h(n_0-1)\sum_{k=0}^{n-n_0-1}\frac{1}{\varsigma_1(n-k-1)}\prod_{\nu=k+2}^{n-n_0}\frac{\varsigma_2(n-\nu)}{\varsigma_1(n-\nu)}\right). \end{aligned}$$

By introducing the new variables $j = n - \lambda$, $\ell = n - k - 1$, and $m = n - \nu$, this becomes

$$r(n) = \left(\prod_{j=n_0}^{n-1}\varsigma_1(j)\right)\left(h(n_0-2) + h(n_0-1)\sum_{\ell=n_0}^{n-1}\frac{1}{\varsigma_1(\ell)}\prod_{m=n_0}^{\ell-1}\frac{\varsigma_2(m)}{\varsigma_1(m)}\right). \tag{5.26}$$

Note that the denominators here are not zero in view of (3.8).

Write

$$F_n(z) = \left(\prod_{\nu=0}^{n_0-1}(t_{1\nu}(z))^{-1}\right)\prod_{j=n_0}^{n-1}\frac{\varsigma_{1z}(j)}{t_{1j}(z)} \tag{5.27}$$

and

$$G_n(z) = \sum_{\ell=n_0}^{n-1} \frac{1}{\varsigma_{1_z}(\ell)} \prod_{m=n_0}^{\ell-1} \frac{\varsigma_{2_z}(m)}{\varsigma_{1_z}(m)} \tag{5.28}$$

for $n \geq n_0$. Then (5.26) becomes

$$r_z(n) = \left(\prod_{\nu=0}^{n-1} t_{1\nu}(z)\right) F_n(z)(h_z(n_0 - 2) + h_z(n_0 - 1)G_n(z)).$$

Hence (5.1) follows by (5.19), (5.22), and (5.24).

Putting

$$F(z) = \lim_{n\to\infty} F_n(z) \quad \text{and} \quad G(z) = \lim_{n\to\infty} G_n(z), \tag{5.29}$$

we can see that

$$\lim_{n\to\infty} F_n(z)/F(z) = 1$$

uniformly for $z \in D$ according to (5.17). Thus (5.2) and (5.4) with $j = 1$ hold.

To show (5.3) and (5.4) with $j = 2$, i.e., that

$$\lim_{n\to\infty} G_n(z) = G(z) \tag{5.30}$$

uniformly for $z \in D$, note that if n is large enough, then

$$\varsigma_{1_z}(n) \geq (1 + \eta)/2 \tag{5.31}$$

for every $z \in D$ according to (5.9) (for $|z|$ small), (5.10) (for $|z|$ large), and (5.16); recall that $\eta > 1$. Moreover, as we have

$$\varsigma_{2_z}(n)\varsigma_{1_z}(n) = a_{n+1}/a_{n+2}$$

(cf. (3.4) and (5.13)) and the right-hand side here tends to 1 as $n \to \infty$ according to (4.9)(i), it follows from (5.31) that, if n is large enough, then

$$|\varsigma_{2_z}(n)| \leq 1 \tag{5.32}$$

for every $z \in D$. This and (5.31) imply that the convergence in (5.30) is uniform.

Finally, we have to show that, given a compact set $K \subset D$, F and G are bounded on K. In view of (5.2)-(5.4), it is sufficient to show for this that F_n and G_n are bounded on K for each $n > n_0$. For G_n this is so because G_n is holomorphic in D (cf. (3.8) and the paragraph preceding (5.15)). For F_n this is so because

$$H_n(z) = \prod_{j=n_0}^{n-1} \frac{\varsigma_{1j}(j)}{t_{1j}(z)} = F_n(z) \prod_{\nu=0}^{n_0-1} t_{1\nu}(z) \tag{5.33}$$

holomorphic in D (cf. the same paragraph), and $t_{1\nu}(z)$ is bounded away from zero on K for every $\nu \geq 0$ (since τ is bounded away from zero on the whole plane – cf. (4.5) and (4.7)). The proof Lemma 5.1 is complete.

6. Near Infinity

In order to complete the proof of Theorem 4.1, we are going to study the behavior of the functions F and G near infinity. We have

LEMMA 6.1. *Assume the hypotheses of Lemma 5.1 (and so, also those of Theorem 4.1), and suppose that the set D includes a deleted neighborhood of ∞. Let $K \subset D$ be closed. Then for the functions F, G, G_n, and the integer n_0 described in Lemma 5.1 we have*

$$C_4 \,|\, z \,|^{-n_0} < |\, F(z) \,| < C_5 \,|\, z \,|^{-n_0} \tag{6.1}$$

and

$$|\, G(z) - 1/t_{1n_0}(z) \,| \leq C_6 \,|\, z \,|^{-2} \tag{6.2}$$

for $z \in K$ with some positive constants C_4, C_5 and C_6, and

$$|\, G_n(z) - 1/t_{1n_0}(z) \,| \leq C_6 \,|\, z \,|^{-2} \tag{6.3}$$

for $z \in K$ and $n > n_0$.

PROOF. We are going to show (6.1) first. To this end, observe that there are positive constants C_7 and C_8 such that

$$C_7 \,|\, z \,| < |\, \varsigma_{1n}(z) \,| < C_8 \,|\, z \,| \tag{6.4}$$

holds for $n \geq n_0$ and $z \in D$. Indeed, the first inequality holds for $n > n_0$ with $C_7 = C_3$ in view of (3.10) and (5.12). For $n = n_0$, it holds in view of (5.10) and (5.23). As for the second inequality, it follows from (5.15) that

$$\sum_{\nu=n_0}^{\infty} |\, t_{1\nu}(z) - \varsigma_{1\nu}(z) \,| \leq \frac{1}{1-\rho} \sum_{\nu=n_0}^{\infty} |\, t_{1,\nu+1}(z) - t_{1\nu}(z) \,| \,.$$

The right-hand side here is less than

$$| t_{1n_0}(z) | /(1 - \rho)$$

(cf. (5.12)); thus the second inequality in (6.4) follows from (5.10).

We are now going to estimate $F(z)$ as given by (5.27) and (5.29). As

$$\lim_{z \to \infty} \tau(z)/z = 2$$

(cf. (4.5)), the limit

$$\lim_{z \to \infty} (z^{n_0} \prod_{\nu=0}^{n_0-1} (t_{1\nu}(z))^{-1})$$

exists and is different from 0 (cf. (4.7)). Now, according to (4.7), the expression after the limit is bounded, since $| \tau(z) | \geq 1$ for $z \in C$, and it is bounded away from zero on compact subsets of D since $\tau(z)$ is bounded on compact subsets of C and z is bounded away from zero on D. Therefore, we have

$$C_9 | z |^{-n_0} < | \prod_{\nu=0}^{n_0-1} (t_{1\nu}(z))^{-1} | < C_{10} | z |^{-n_0} \tag{6.5}$$

for $z \in D$ with some positive constants C_9 and C_{10}.

Moreover according to (5.17) there is an $N \geq n_0$ such that

$$\frac{1}{2} < | \prod_{j=N}^{\infty} \frac{\varsigma_{1z}(j)}{t_{1j}(z)} | < \frac{3}{2} \tag{6.6}$$

holds for $z \in D$. Finally we have

$$(\frac{C_7}{C_2})^{N-n_0} < | \prod_{j=n_0}^{N-1} \frac{\varsigma_{1z}(j)}{t_{1j}(z)} | < (\frac{C_8}{C_1})^{N-n_0} \tag{6.7}$$

for $z \in D$ according to (5.10) and (6.4). Now (6.1) follows from (5.27) and (5.29) with the aid of (6.5)-(6.7). Note that (6.1) actually holds for $z \in D$, and not only for $z \in K$.

As for (6.2), it is an obvious consequence of (6.3) and (5.29). To show (6.3), observe that in view of (5.23) we can write (5.28) as

$$G_n(z) = \frac{1}{t_{1n_0}(z)} + \sum_{\ell=n_0+1}^{n-1} \frac{1}{\varsigma_{1z}(\ell)} \prod_{m=n_0}^{\ell-1} \frac{\varsigma_{2z}(m)}{\varsigma_{1z}(m)}$$

for $n > n_0$ (and $z \in D$). From here (6.3) follows by virtue of (5.32) and (6.4) provided $| z |$ is large enough (so that $C_7 | z | > 1 + \epsilon$ for $z \in K$ in (6.4), where $\epsilon > 0$). For $z \in K$ not large (6.3) simply says that G_n is bounded uniformly in n on each compact subset of D; this is indeed so in view of the last sentence of Lemma 5.1 and the uniformness of the convergence in (5.30). The proof of Lemma 6.1 is complete.

We are now in the position to complete the

PROOF OF THEOREM 4.1. Let D be an open set with $K \subset D$ such that $\overline{D}$ is disjoint from $[-1,1]$ and D includes a deleted neighborhood of infinity. We have to establish (4.12) and the uniformness of the convergence in (4.11); the existence of the limit in (4.11) was pointed out right after Lemma 5.1. In what follows we assume $z \in K$.

As for (4.12), observe that

$$p_n(z) = \gamma_n z^n + O(|z|^{n-1})$$

holds for fixed n as $z \to \infty$, where $\gamma_n \neq 0$ (cf. (4.1)). Thus (5.5), (5.10), and (6.2) imply that

$$g(z) = F(z)(\gamma_{n_0+1} z^{n_0+1}/t_{1n_0}(z) + O(|z|^{n_0-1})) \tag{6.8}$$

as $z \to \infty$. Hence g is bounded away from 0 and ∞ in a deleted neighborhood of infinity, according to (5.10) and (6.1). Therefore, the existence of the limit L in (4.12) and $L \neq 0$ follow if we can show that g is holomorphic in a deleted neighborhood of ∞. Now g is indeed holomorphic in a deleted neighborhood of infinity, since the function in (4.13) is so and $t_{1\nu}(z)$ $(0 \leq \nu \leq N)$ has no zeros if $|z|$ is large enough. Here the remark made after (4.13) is significant, since we do not yet know the uniformness of the convergence in (4.11) on D, but we know it on compact subsets of D (cf. the discussion after Lemma 5.1).

Next we turn to the question of the uniformness of the convergence in (4.11). Writing $g_n(z)$ for the left-hand side of (5.1) and using (6.3) instead of (6.2) (and (5.1) instead of (5.5)), we obtain

$$g_n(z) = F_n(z)(\gamma_{n_0+1} z^{n_0+1}/t_{1n_0}(z) + O(|z|^{n_0-1}))$$

for $n > n_0$ as $z \to \infty$, where the bound implicit in the symbol $O(\cdot)$ is independent of n (it depends only on the constants in (5.10) and (6.2), and the coefficients of p_{n_0} and p_{n_0+1}). Thus, by (5.2) and (6.8) we have

$$|g_n(z) - g(z)| \leq F(z)(|\psi_{1n}(z)\gamma_{n_0+1} z^{n_0+1}/t_{1n_0}(z)| + (1 + |\psi_{n1}(z)|)C_{11}|z|^{n_0-1})$$

for $n > n_0$ and $|z| > R$, with some positive constants C_{11} and R. According to (5.10) and (6.1), the right-hand side here is less than

$$C_{12}|\psi_{1n}(z)| + C_{13}|z|^{-1} \qquad (|z| > R),$$

with some positive constants C_{13} and C_{14}. Given $\epsilon > 0$, this will be less than ϵ provided $|z|$ and n are large enough, say $|z| > R_1$ and $n > n_2$. That is

$$|g_n(z) - g(z)| < \epsilon \tag{6.9}$$

whenever $|z| > R_1$ and $n > n_2$. Since we know that the convergence

$$\lim_{n\to\infty} g_n(z) = g(z)$$

is uniform on each compact subset of D (cf. the discussion after Lemma 5.1), (6.9) now implies that this convergence is uniform on each closed subset of D. Thus the uniformness of the convergence on K in (4.11) follows. The proof of Theorem 4.1 is complete.

REFERENCES

1. T. S. Chihara, An Introduction to Orthogonal Polynomials, Gordon and Breach, New York-London-Paris, 1978.
2. J. M. Dombrowski, Tridiagonal matrix representations of cyclic self-adjoint operators, Pacific J. Math. 114 (1984), 325-334.
3. G. Freud, Orthogonal Polynomials, Akadémiai Kiadó, Budapest, and Pergamon Press, New York, 1971.
4. A. Máté and P. Nevai, Orthogonal polynomials and absolutely continuous measures. In: Approximation Theory IV (C.K. Chui, L.L. Schumaker, and J.D. Ward, eds.), Academic Press, New York, 1983; pp. 611-617.
5. A. Máté, P. Nevai, and V. Totik, Asymptotics for orthogonal polynomials defined by a recurrence relation, Constr. Approx. 1 (1985), 231-248.
6. P. Nevai, Orthogonal Polynomials, Mem. Amer. Math. Soc. 213 (1979), 1-185.
7. G. Szegö, Orthogonal polynomials, 4th ed., Amer. Math. Soc. Colloquium Publ. 23, Providence, Rhode Island, 1975.

GAUSS QUADRATURE FOR ANALYTIC FUNCTIONS

T.J. Rivlin
Mathematical Sciences Department
IBM Thomas J. Watson Research Center
Yorktown Heights, NY 10598
U.S.A.

Introduction.

There is a considerable literature concerning upper bounds for the error of Gauss quadrature of functions analytic in a domain which contains the interval of integration [-1, 1]. See, for example, [GAU-VAR], and, in particular, Gautschi's survey [GAU] and the references given there. Our selected survey will be focussed mainly on quadrature of functions which are analytic and bounded in either the interior of an ellipse with foci at -1 and 1, or in an open disc with center at the origin and radius not less than 1. We shall present lower bounds for the error of Gauss quadrature for such functions. This will be done by obtaining lower bounds for the error of optimal quadrature using Gaussian nodes for these classes of functions. Upper bounds for Gauss quadrature will also be derived and compared to the lower bounds. The case in which the functions are analytic and bounded in the unit disc, which has previously received less attention, will be presented in detail.

In the first section we shall introduce the notion of optimal quadrature in the context of the more general formulation of "optimal recovery" as given by Micchelli and Rivlin [MIC-RIV]. The second section deals with optimal and best quadrature of functions analytic and bounded in discs and ellipses containing the interval of integration. The third section presents results about Gauss quadrature and optimal quadrature using Gaussian information for functions analytic in discs. In the fourth and final section optimal quadrature using Gaussian nodes for bounded analytic functions in the unit disc is treated and definitive results for the ultraspherical nodes are presented.

1. Optimal Recovery.

By "optimal recovery" we mean estimating some required feature of a function, known to belong to some specified class of functions, from limited information about it as effectively as possible. In 1977 Micchelli and Rivlin [MIC-RIV] gave a general framework and theory for such problems, and many examples as well. Here is a simpler aspect of this general formulation relevant to our purposes here.

Let X be a linear space and Y and Z normed linear spaces. K is a subset of X, U a linear operator from X into Z (the *feature* operator) and I a linear operator from X into Y (the *information* operator). Suppose that for each $x \in K$ we know Ix and our objective is to determine the best possible estimate of Ux from this information. To this end let α be any function with domain IK and range in Z. We call such a function an *algorithm*. Our model is represented schematically in the following diagram.

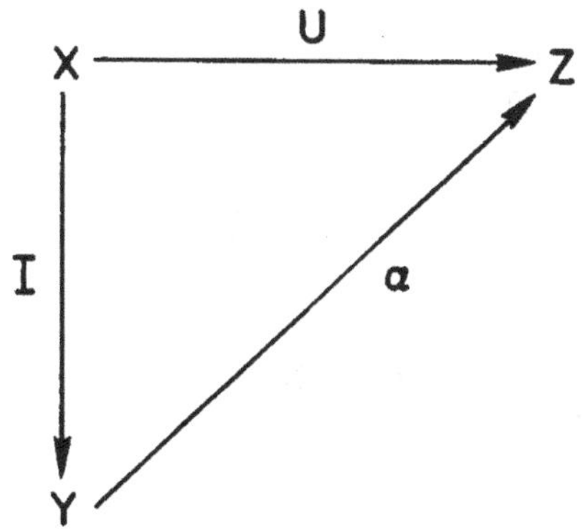

Each algorithm, α, produces an error

$$E_\alpha(K) = \sup\{ \| Ux - \alpha(Ix) \| : \ x \in K\}$$

and

(1) $$E(K) = \inf_\alpha E_\alpha(K)$$

is called the *intrinsic error* in the problem. If

$$E_{\alpha^*} = E,$$

then α^* is an *optimal algorithm* and is said to effect the optimal recovery of Ux.

The extremal problem posed by (1) seems rather intractable. However, an easy argument gives the following result: If K is a balanced convex subset of X then

(2) $$e(K) := \sup\{ \| Ux \| : x \in K, \ Ix = 0\} \leq E(K).$$

Moreover, it can be shown (cf. [MIC-RIV]) that if U is a linear functional (so that $Z = \mathbb{R}$ or $\mathbb{C}$), and some mild further restriction is placed on K, then equality holds in (2) and an optimal algorithm which is linear exists. This final remark is in force in the problems of numerical integration in which we are interested, so let us now turn to those particular cases.

Let $\mathscr{B}$ be a domain in the complex plane which contains the interval [-1, 1] in its interior. Put $X = H^\infty(\mathscr{B})$, the bounded analytic functions in $\mathscr{B}$ and $K(\mathscr{B}) = \{f \in X: \|f\| \le 1\}$, the norm being the sup norm on $\mathscr{B}$. Suppose $X_n = (x_1, \dots, x_n)$, $-1 < x_1 < \cdots < x_n < 1$, $n \ge 1$, is given and $If := (f(x_1), \dots, f(x_n))$, so that $Y = \mathbb{C}^n$. Let $w(t)$ be a non-negative weight function on [-1, 1] and

$$U(f, w) = \int_{-1}^{1} f(t)w(t)dt.$$

A linear algorithm is an approximation to U of the form

$$Q(f, X_n, a) = \sum_{i=1}^{n} a_i f(x_i)$$

$(a = (a_1, \dots, a_n))$, i.e. a quadrature formula. Now

$$E_a(K) = \sup\{ \mid U(f, w) - Q(f, X_n, a) \mid : f \in K\}$$

and so

(3) $$E(K) = \inf_a E_a(K),$$

the intrinsic error, is the error of optimal quadrature for the specified data. If the minimum in (3) is attained for $a = a^*$ we say that $Q(f, X_n, a^*)$ is an optimal quadrature formula. Since equality holds in (2) we have, upon writing $e(X_n, w)$ in place of $E(K)$, the duality theorem,

(4) $$E(K) =: e(X_n, w) = \sup\left\{ \mid \int_{-1}^{1} fw \mid : \|f\| \le 1,\ f(x_i) = 0,\ i = 1, \dots, n\right\}.$$

We intend to exploit (4) extensively in what follows.

It is natural at this point to consider minimizing the intrinsic error over all possible choices of X_n. This leads to

$$e(w) = \inf\{e(X_n, w): X_n\}.$$

If $e(w) = e(X_n^*, w)$ then we call an optimal quadrature formula for X_n^* a *best* quadrature formula (w and $\mathscr{B}$ being specified).

2. Optimal and Best Quadrature of Bounded Analytic Functions.

We shall now restrict the domain $\mathscr{B}$ to be either a disc or the interior of an ellipse. Let D_r be the open disc $|z| < r$, $r > 0$, and E_ρ $(\rho > 1)$ denote the interior of the ellipse with foci at -1 and 1 and sum of major and minor axes equal to 2ρ.

2.1. Let us choose $r \geq 1$ and put $\mathscr{B} = D_r$ ($\mathscr{B} = D$ when $r = 1$) in our previous formulation of optimal integration. Consider the finite Blaschke product

$$B_r(z) = \prod_{i=1}^{n} \frac{r(z - x_i)}{r^2 - x_i z}.$$

$\|B_r\| = 1$ and $|B_r(z)| = 1$ for $|z| = r$ (we put $B_r = B$ when $r = 1$), thus we might guess that B_r plays an important role in the extremal problem described in (4), since it "takes up" the constraints in that problem. Multiplication by B_r does not alter H^∞ norms so we may rewrite (4) in the present case as

$$e(X_n, w) = \sup\{ \left| \int_{-1}^{1} f(t)B_r(t)w(t)dt \right| : \|f\| \leq 1\}. \tag{5}$$

In the case that $\mathscr{B} = D$ and $w = 1$ some results about best quadrature are known. For example,

$$\frac{2}{3} e^{-6\sqrt{n}} \leq e(1) \leq 11\, e^{-\frac{\sqrt{n}}{2}}. \tag{6}$$

(A similar result holds for functions in H^p, $1 < p < \infty$. See [NEW] and the references given there. Even sharper bounds are in [AND].)

2.2. Suppose next that $\mathscr{B} = E_\rho$, $\rho > 1$. When w belongs to a broad class of weight functions, which includes the general non-negative weight functions considered in Gauss quadrature, Bakhvalov [BAK] shows that the error of a best quadrature formula satisfies

$$(7) \qquad e(w) \geq \frac{c_0(\rho, w)}{\rho^{2n}},$$

where c_0 is a constant depending only on ρ and w. He then recalls that if w is a Gaussian weight function, with the zeros of the corresponding orthogonal polynomial of degree n denoted by $X_n(w)$

$$(8) \qquad e(X_n(w), w) \leq \frac{c_1(\rho, w)}{\rho^{2n}},$$

where c_1 is a constant depending only on ρ and w. Observe that (7) and (8) imply that for functions analytic and bounded by 1 in E_ρ $(\rho > 1)$ Gauss quadrature exhibits the same dependence on n as a best quadrature formula. We can only refer the reader to [BAK] for the elaborate proof of (7), but give, next, the simple proof of (8).

Let w be non-negative and integrable on $[-1, 1]$ with finite moments

$$\int_{-1}^{1} t^k w(t)dt, \quad k = 0, 1, 2, \ldots .$$

Let $t_1, \ldots, t_n$ be the zeros of the orthogonal polynomial of degree n associated with w, and $\lambda_1, \ldots, \lambda_n$ the corresponding Christoffel numbers. We wish to show that if f is analytic and bounded by 1 in E_ρ, $\rho > 1$ and

$$R_n(f) = \int_{-1}^{1} f(t)w(t)dt - \sum_{i=1}^{n} \lambda_i f(t_i)$$

then

$$(9) \qquad | R_n(f) | \leq c_1(\rho, w)\, \rho^{-2n}.$$

Suppose that $1 < \rho' < \rho$, then

$$f(z) = \sum_{k=0}^{\infty} A_k T_k(z)$$

is absolutely convergent for z in $E_{\rho'}$ and

$$| A_k | \leq \frac{2}{(\rho')^k}, \quad k = 0, 1, 2 \ldots,$$

(cf. [RIV]). If we put

$$S_j(f) = \sum_{k=0}^{j} A_k T_k,$$

then for $-1 \leq t \leq 1$ we have

$$| f(t) - S_{2n-1}(f;t) | = | \sum_{k=2n}^{\infty} A_k T_k(t) | \leq \sum_{k=2n}^{\infty} | A_k | \leq \frac{2\rho'}{(\rho')^{2n}(\rho' - 1)}.$$

Keeping in mind that Gauss quadrature is exact for polynomials of degree less than $2n$, we have

$$R_n f = R_n(f - S_{2n-1}(f)) + R_n\, S_{2n-1}(f) = R_n(f - S_{2n-1}(f)).$$

But, from the definition of R_n we obtain

$$| R_n(f - S_{2n-1}(f)) | \leq \left(\int_{-1}^{1} w(t)dt + \sum_{k=1}^{n} | \lambda_k | \right) \| f - S_{2n-1}(f) \|_{\infty}.$$

Since

$$\sum_{k=1}^{n} | \lambda_k | = \sum_{k=1}^{n} \lambda_k = \int_{-1}^{1} w(t)dt,$$

we conclude, upon letting $\rho' \to \rho$, that

$$| R_n f | \leq c_1(\rho, w)\, \rho^{-2n}$$

where

$$c_1(\rho, w) = \left(4 \int_{-1}^{1} w(t)dt \right) \frac{\rho}{\rho - 1}.$$

3. Gauss Quadrature of Functions Analytic and Bounded in D_r.

We begin by observing that if $r > 1$

$$E_{\rho_1} \subset D_r \subset E_{\rho_2}$$

where $\rho_1 = r + (r^2 - 1)^{1/2}$ and $\rho_2 = r + (r^2 + 1)^{1/2}$. Hence if $f \in K(D_r)$, $r > 1$ then f is in $K(E_{\rho_1})$ and (9) yields an upper bound for Gauss quadrature namely,

$$\sup_{f \in K(D_r)} | R_n f | \leq c_3(r, w)\,(r + (r^2 - 1)^{1/2})^{-2n}, \tag{10}$$

where $c_3 = c_1(\rho_1, w)$. Similarly, (7) can be used to give a lower bound for $e(X_n(w), w)$ and hence for

$$\sup_{f \in K(D_r)} | R_n f | ,$$

namely,

$$e(X_n(w), w) \geq c_1(r, w)(r + (r^2 + 1)^{1/2})^{-2n} \tag{11}$$

where

$$c_1 = c_0(\rho_2, w).$$

Note the gap between (10) and (11).

We give next a slightly better lower bound than (11) - at least in the case that the nodes are zeros of ultraspherical polynomials - which, moreover, is valid for $r = 1$, in which case it is "optimal".

Suppose that $\mathscr{B} = D_r$, $r \geq 1$ and $X_n(\lambda)$ denotes the zeros $(x_1, \dots, x_n)$ (satisfying $-1 < x_1 < \cdots < x_n < 1$) of the ultraspherical polynomial $p_{n,\lambda}$ ($= P_n^{(\lambda)}$ in Szegö's notation and standardization [SZE]) $\lambda \geq 0$.

Let $w(x) = (1 - x)^\alpha (1 + x)^\beta$, $\alpha, \beta > -1$. We seek a bound from below for the intrinsic error, $e(X_n(\lambda), w(\alpha, \beta))$, which will then provide us with a lower bound for $\sup\{ | R_n f | : f \in K(D_r)\}$.

In view of (5) we have

$$e(X_n(\lambda), w(\alpha, \beta)) \geq \int_{-1}^{1} b^2(x) w(x) dx \tag{12}$$

where

$$b(x) = \prod_{i=1}^{n} \frac{r(x - x_i)}{r^2 - x_i x}.$$

Let us put $g(x) = b^2(x)$.

Lemma 1. If $n \geq 2$ and $x_n < x \leq 1$, then $g(x)$, $g'(x)$ and $g''(x)$ are positive.

Proof. That $g(x)$ is positive for $x_n < x \leq 1$ is obvious. Next consider

$$\frac{g'(x)}{g(x)} = 2\left(\sum_{i=1}^{n} \frac{1}{x - x_i} + \sum_{i=1}^{n} \frac{x_i}{r^2 - x_i x}\right). \tag{13}$$

It is easy to verify that

$$\frac{1}{x - x_i} > \frac{x_i}{r^2 - x_i x}, \quad i = 1, \dots, n,$$

hence

$$\frac{g'(x)}{g(x)} > 4 \sum_{i=1}^{n} \frac{x_i}{r^2 - x_i x}.$$

Since the x_i are symmetric about zero

$$\sum_{i=1}^{n} \frac{x_i}{r^2 - x_i x} = \sum_{i=1}^{\left[\frac{n}{2}\right]} \frac{2x_i^2 x}{r^4 - (x_i x)^2} > 0,$$

and we conclude that $g'(x) > 0$ since $g(x) > 0$.

Finally, differentiation of (13) yields

$$\frac{g''(x)}{g(x)} = 2\left(\sum_{i=1}^{n} \frac{x_i^2}{(r^2 - x_i x)^2} - \sum_{i=1}^{n} \frac{1}{(x - x_i)^2}\right) + \left(\frac{g'(x)}{g(x)}\right)^2. \tag{14}$$

But

$$\left(\frac{g'(x)}{g(x)}\right)^2 = 4\left(\left(\sum_{i=1}^{n}\frac{1}{x-x_i}\right)^2 + \left(\sum_{i=1}^{n}\frac{x_i}{r^2-x_ix}\right)^2 + 2\sum_{i=1}^{n}\frac{1}{x-x_i}\cdot\sum_{i=1}^{n}\frac{x_i}{r^2-x_ix}\right)$$

and we conclude that $g''(x)/g(x)$ is positive and, hence, so is $g''(x)$. Lemma 1 is established.

Thus $y = g(x)$ is positive, monotone increasing and convex in $x_n < x \leq 1$. The tangent line to this curve at $x = 1$ is $y = g(1) + g'(1)(x-1)$ and it crosses the x-axis at $x = x'$ where $x_n < x' < 1$. Moreover,

$$1 - x' = \frac{g(1)}{g'(1)}. \tag{15}$$

Furthermore, $g(x)$ lies above its tangent line at $x = 1$ for $x' \leq x \leq 1$. Thus

$$\begin{aligned}\int_{-1}^{1} g(x)w(x)dx &> \int_{x'}^{1} g(x)(1-x)^{\alpha}(1+x)^{\beta}dx \\ &> A_0(\beta)\int_{x'}^{1}(g(1)-g'(1)(1-x))(1-x)^{\alpha}dx,\end{aligned} \tag{16}$$

where $A_0(\beta) = \min(1, 2^{\beta})$. If we now integrate by parts in the last integral in (16) we obtain

$$\int_{-1}^{1} g(x)w(x)dx > A_0(\beta)\frac{g'(1)}{1+\alpha}\int_{x'}^{1}(1-x)^{1+\alpha}dx = \frac{A_0(\beta)}{(2+\alpha)(1+\alpha)}g(1)\left(\frac{g(1)}{g'(1)}\right)^{1+\alpha}. \tag{17}$$

We now need an upper bound for $g'(1)/g(1)$. But (13) gives, for $\lambda \geq 0$

$$\frac{g'(1)}{g(1)} < 4\sum_{i=1}^{n}\frac{1}{1-x_i} = 4\frac{p'_{n,\lambda}(1)}{p_{n,\lambda}(1)} = \frac{4n(n+2\lambda)}{2\lambda+1} \leq 4n^2, \tag{18}$$

where we use the second order differential equation for $p_{n,\lambda}$.

Finally, we provide a lower bound for

$$g(1) = r^{2n}\left(\prod_{i=1}^{n}\frac{1-x_i}{r^2-x_i}\right)^2.$$

Let $\bar{p}_{n,\lambda}$ be the orthogonal polynomial with respect to $(1-x^2)^{\lambda-½}$ renormalized so that

$$(19)\qquad \max_{-1\le x\le 1} \quad |\ \bar{p}_{n,\lambda}(x)\ | = 1.$$

For $\lambda \geq 0$ we also have $|\ \bar{p}_{n,\lambda}(1)\ | = 1$ (cf. [SZE]), so that

$$g(1) = r^{2n}\left(\frac{\bar{p}_{n,\lambda}(1)}{\bar{p}_{n,\lambda}(r^2)}\right)^2 = \frac{r^{2n}}{(\bar{p}_{n,\lambda}(r^2))^2}.$$

But, in view of (19) we know (cf. [RIV]) that

$$|\ \bar{p}_{n,\lambda}(r^2)\ | \leq T_n(r^2),$$

where $T_n(x)$ is the Chebyshev polynomial. Thus

$$(20)\qquad g(1) \geq \frac{r^{2n}}{(T_n(r^2))^2}$$

and in view of (18), (17) and (12) we obtain

$$(21)\qquad e(X_n(\lambda), w(\alpha, \beta)) > \frac{A_0(\beta)}{(2+\alpha)(1+\alpha)4^{1+\alpha}} \cdot \frac{1}{n^{2(1+\alpha)}} \frac{r^{2n}}{(T_n(r^2))^2}.$$

Observe that this result is independent of λ.

If we recall that

$$\begin{aligned} T_n(r^2) &= \frac{(r^2 + (r^4 - 1)^{½})^n + (r^2 - (r^4 - 1)^{½})^n}{2} \\ &\leq (r^2 + (r^4 - 1)^{½})^n, \end{aligned}$$

we then obtain

$$(22)\qquad e(X_n(\lambda), w(\alpha, \beta)) > \frac{A_0(\beta)}{(2+\alpha)(1+\alpha)4^{1+\alpha}} \frac{1}{n^{2(1+\alpha)}} (r + (r^2 - r^{-2})^{½})^{-2n}.$$

Note that when $r = 1$ and $\alpha = \beta = 0$ (21) yields

$$e(X_n(\lambda), 1) > \frac{1}{8n^2}.$$

Thus, the worst error using ultraspherical nodes, a fortiori using Gauss quadrature, is, in view of (6), far greater than the worst error of the best quadrature formula, for the weight function $w = 1$ and functions analytic and bounded by 1 in the unit disc (cf. [KOW-WER-WOZ]).

Remark. The lower bound approach we have just presented is based on using Lemma 1 and the consequent inequality (17). The assumption that the nodes are zeros of ultraspherical polynomials simplifies the proof of Lemma 1 and enables us to establish the estimate (18). We wish to point out now that an analog of Lemma 1 remains valid for an arbitrary choice of nodes, $X_n = (x_1, \dots, x_n)$, satisfying $-1 < x_1 < \cdots < x_n < 1$, and thus an inequality like (17) is always valid. The rub is that inequalities like (18) and (20) which are essential for meaningful asymptotic bounds require further specification of the nodes. We chose a strong restriction on them above in the interest of expository simplicity.

Let us sketch a proof of an analog of Lemma 1 for arbitrary X_n. Suppose $n \geq 2$. We say that X_n is *negative* if it contains more negative than non-negative points. We can now state

Lemma 2. Suppose $n \geq 2$. If X_n is negative then $g(x) > 0$, $g'(x) < 0$, $g''(x) > 0$ for $-1 \leq x < x_1$. If X_n is not negative then $g(x) > 0$, $g'(x) > 0$ and $g''(x) > 0$ for $x_n < x \leq 1$.

Proof. (Sketch).

(i) Suppose

$$F(x) = \frac{x-a}{R-ax} \cdot \frac{x-b}{R-bx}.$$

If $-1 < b \leq 0 \leq a < 1$, $b < a$ and $R > 1$ then differentiation of the partial fraction decomposition of $F(x)$ yields

$$F''(x) = \frac{R-ab}{a-b}\left(\frac{2a(R-a^2)}{(R-ax)^3} - \frac{2b(R-b^2)}{(R-bx)^3}\right) > 0, \quad -1 \leq x \leq 1.$$

(ii) If $-1 < b < a < 1$ then it is easy to see that $F(x) > 0$ for $-1 \leq x < b$ and $a < x \leq 1$, and $F'(x) < 0$ for $-1 \leq x < b$ while $F'(x) > 0$ for $a < x \leq 1$.

(iii) If $0 < b < a < 1$ then $F(x)$, $F'(x)$ and $F''(x) > 0$ in $a < x \leq 1$. If $-1 < b < a < 0$ then $F(x) > 0$, $F'(x) < 0$ and $F''(x) > 0$ in $-1 \leq x < b$.

(i), (ii) and (iii) give

I. If $-1 < b < a < 1$, $-1 \leq x \leq 1$ then

(a) $F(x) > 0$ in $x > a$ and $x < b$.

(b) $F'(x) < 0$ in $x < b$ and $F'(x) > 0$ in $x > a$.

(c) $F''(x) > 0$ in either $x > a$ or $x < b$ if $a, b > 0$ or $a, b < 0$ respectively.

Consider now

$$\frac{b(x)}{r^n} = \prod_{i=1}^{n} \frac{x - x_i}{R - x_i x}, \tag{23}$$

where we have written R for r^2. Let N denote the number of negative points in X_n and P the number of positive points. The number zero may be called "positive" or "negative" to suit our convenience. The idea now is to form as many pairs of points as possible each consisting of a positive and negative point. The remaining points are then all positive or all negative. Now write $r^{-n}b(x) = F_1(x)F_2(x)$ where $F_1(x)$ is the product of the factors in (23) corresponding to mixed pairs and $F_2(x)$ is the product of the remaining factors. A straightforward consequence of I is that if X_n is negative then if n is odd, $b(x) < 0$, $b'(x) > 0$ and $b''(x) < 0$ for $-1 \leq x < x_1$; while if n is even, $b(x) > 0$, $b'(x) < 0$ and $b''(x) > 0$ for $-1 \leq x < x_1$. If X_n is not negative then $b(x) > 0$, $b'(x) > 0$ and $b''(x) > 0$ for $x_n < x \leq 1$. The lemma now follows.

4. Optimal Quadrature of Functions in $H^\infty(D)$.

We now restrict our attention to functions which are analytic and bounded in the open unit disc, D, (cf. [DYN-MIC-RIV]). We consider only nodes $X_n(\lambda)$, $\lambda > 0$ and weights $(1 - x^2)^\alpha$, $\alpha \geq -(1/2)$. Thus (21) yields

$$e(X_n(\lambda),\ w(\alpha)) > \frac{M(\alpha)}{n^{2(1+\alpha)}}. \tag{24}$$

We turn, then, to the upper bound problem. Observe that in view of (5), with $r = 1$, we have

$$e(X_n(\lambda),\ w(\alpha)) \leq \int_{-1}^{1} \mid B(x) \mid (1-x^2)^\alpha\, dt. \tag{25}$$

Therefore, we seek an upper bound for $\mid B(x) \mid$. Note that

$$B(x) = \frac{p_{n,\lambda}(x)}{x^n p_{n,\lambda}\left(\frac{1}{x}\right)},$$

and consider Gegenbauer's formula (cf. [SEI-SZA])

$$p_{n,\lambda}(x) = C \int_0^\pi (x + (x^2-1)^{1/2}\cos\varphi)^n (\sin\varphi)^{2\lambda-1}\, d\varphi,$$

for all real x, where

$$C = p_{n,\lambda}(1)\, \frac{\Gamma(\lambda+1/2)}{\pi^{1/2}\Gamma(\lambda)}.$$

If $-1 \leq x \leq 1$ and $x = \cos\ \theta$ we obtain

$$\begin{aligned} x^n p_{n,\lambda}\left(\frac{1}{x}\right) &= C \int_0^\pi (1 + \sin\theta\cos\varphi)^n (\sin\varphi)^{2\lambda-1}\, d\varphi \\ &\geq C \int_0^{\pi/3} (1 + \sin\theta\cos\varphi)^n (\sin\varphi)^{2\lambda-1}\, d\varphi \\ &\geq c(\lambda)\, C\left(1 + \frac{\sin\theta}{2}\right)^n, \qquad -1 \leq x \leq 1, \end{aligned}$$

where

$$c(\lambda) = \int_0^{\pi/3} (\sin\varphi)^{2\lambda-1}\, d\varphi.$$

But we also have

$$\mid p_{n,\lambda}(x) \mid \leq p_{n,\lambda}(1), \qquad -1 \leq x \leq 1,$$

and so

$$| B(\cos\theta) | \leq \frac{\pi^{1/2}\,\Gamma(\lambda)}{c(\lambda)\Gamma(\lambda + 1/2)} (1 + (1/2)\sin\theta)^{-n}, \quad 0 \leq \theta \leq \pi.$$

Hence we have

$$\int_{-1}^{1} | B(x) | (1 - x^2)^{\alpha} dx \leq d_1(\lambda) \int_0^{\pi/2} (1 + (1/2)\ \sin\ \theta)^{-n} (\sin\ \theta)^{2\alpha+1} d\theta,$$

$$\leq d_2(\lambda, \alpha) \int_0^{\pi/2} \frac{\theta^{2\alpha+1}}{(1 + \frac{\theta}{\pi})^n} d\theta.$$

But

$$\int_0^{\pi/2} \frac{\theta^{2\alpha+1}}{(1 + \frac{\theta}{\pi})^n} d\theta = \frac{\pi^{2(1+\alpha)}}{n^{2(1+\alpha)}} \int_0^{n/2} \frac{t^{2\alpha+1}}{(1 + \frac{t}{n})^n} dt$$

$$\leq \frac{\pi^{2(1+\alpha)}}{n^{2(1+\alpha)}} \int_0^{n/2} t^{2\alpha+1} e^{-t/2} dt,$$

since $e^{(x/2)} \leq 1 + x, \quad 0 \leq x \leq 1$. Thus

$$\int_0^{\pi/2} \frac{\theta^{2\alpha+1}}{(1 + \frac{\theta}{\pi})^n} d\theta \leq \frac{(2\pi)^{2(1+\alpha)}}{n^{2(1+\alpha)}} \int_0^{\infty} t^{2\alpha+1} e^{-t} dt$$

$$\leq \frac{(2\pi)^{2(1+\alpha)}}{n^{2(1+\alpha)}} \Gamma(2(1 + \alpha)),$$

so that, finally,

(26) $$e(X_n(\lambda), w(\alpha)) \leq \frac{d_3(\lambda, \alpha)}{n^{2(1+\alpha)}}.$$

The case $\lambda = 0$, corresponding to zeros of the Chebyshev polynomial, is even simpler, for we then easily get

$$| B(x) | \leq \frac{2}{(1 + (1 - x^2)^{1/2})^n},$$

and so (26) remains valid for $\lambda = 0$ as well.

Upon comparing (26) with (24) we see that Gaussian nodes provide optimal information with respect to dependence on n, and the exponent of n depends only on the weight function. In particular, this is the case when $\alpha = \lambda - (1/2)$, and so we might well ask whether Gauss quadrature is then optimal, or even exhibits the same dependence on n for its error as does optimal quadrature for the same data. At this time we leave this as an open question for the reader.

References

[AND] Andersson, J-E., Optimal quadrature of H^p functions, Math. Z.**172**, 1980, pp. 55-62.

[BAK] Bakhvalov, N.S., On the optimal speed of integrating analytic functions, (English translation), U.S.S.R. Comput. Math. Math. Phys.7, 1967, pp. 63-75.

[DYN-MIC-RIV] Dyn, N., C.A. Micchelli and T.J. Rivlin, Blaschke products and optimal recovery in H^∞, to appear in CALCOLO.

[GAU] Gautschi, W., A survey of Gauss-Christoffel quadrature formulae, "E.B. Christoffel" (Eds. P.L. Butzer and F. Feher), Birkhauser Verlag, Basel, 1981, pp. 72-147.

[GAU-VAR] Gautschi, W., and R.S. Varga, Error bounds for Gaussian quadrature of analytic functions, SIAM J. Numer. Anal.**20**, 1983, pp. 1170-1186.

[KOW-WER-WOZ] Kowalski, M.A., A.G. Werschulz and H. Woźniakowski, Is Gauss quadrature optimal for analytic functions?, Numer. Math.**47**, 1985, pp. 89-98.

[MIC-RIV] Micchelli, C.A., and T.J. Rivlin, A survey of optimal recovery, Optimal Estimation in Approximation Theory, (Eds. C.A. Micchelli and T.J. Rivlin), Plenum Press, N.Y., 1977, pp. 1-54.

[NEW] Newman, D.J., Quadrature in H^p, Lectures III, IV, "Approximation with Rational Functions", CBMS Regional Conference Series in Math., No. 41, American Mathematical Society, Providence, R.I., 1979.

[RIV] Rivlin, T.J., "The Chebyshev Polynomials", Wiley, N.Y., 1974.

[SEI-SZA] Seidel, W., and O. Szász, On positive harmonic functions and ultraspherical polynomials, J. London Math. Soc.**26,** 1951, pp. 36-41.

[SZE] Szegö, G., "Orthogonal Polynomials", American Math. Soc., Providence, R.I., 1975.

ON THE GERONIMUS POLYNOMIAL SETS

W. A. Al-Salam and A. Verma

Department of Mathematics
University of Alberta
Edmonton, Canada T6G 2G1

1. INTRODUCTION

In [8] Geronimus considered polynomial sets (PS) of the form

$$P_n(x) = \sum_{k=o}^{n} a_{n-k} b_k \omega_k(x) \qquad n = 0,1,2,\ldots \tag{1.1}$$

where $\omega_o(x) = 1, \omega_k(x) = (x-x_1)(x-x_2)\ldots(x-x_k)$ for $k \geq 1$, $a_o = b_o = 1$ and $b_n \neq 0$ for all $n \geq 0$. He raised the following problem: For what sequences of real or complex numbers $\{x_n\}$, $\{a_n\}$, and $\{b_n\}$ is the p.s. (1.1) orthogonal?

A PS$\{P_n(x)\}$ is orthogonal (OPS) if and only if there are real sequences $\{\alpha_n\}$ and $\{\lambda_n\}$ with $\lambda_n > 0$ for all $n \geq 2$ so that

$$\frac{P_n(x)}{b_n} = \{x-\alpha_n\} \frac{P_{n-1}(x)}{b_{n-1}} - \lambda_n \frac{P_{n-2}(x)}{b_{n-2}} \qquad (n \geq 1) \tag{1.2}$$

$$P_o(x) = 1 , \; P_{-1}(x) = 0 .$$

Geronimus gave several examples of known p.s. which belong to this class. He stated that (1.2) are valid if and only if

$$\frac{a_{n-k}b_k}{b_n} = -\lambda_n \frac{a_{n-k-2}b_k}{b_{n-2}} + \frac{a_{n-k}b_{k-1}}{b_{n-1}} + \frac{a_{n-k-1}b_k}{b_{n-1}} (x_{k+1}-\alpha_n) \tag{1.3}$$

$k = 0,1,\ldots,n-1$; $n=1,2,\ldots$; $b_{-1} = a_{-1} = 0$.

Chihara [5,6] considered the case in which all $x_k = 0$ and found all OPS in this class (Brenke P.S.) . In [2] we considered the case $x_{2k+1} = x_1$, $x_{2k} = x_2$ where $x_1 \neq x_2$ and $a_1 = 0$. In that case we determined all OPS in that class. In [3] Askey and Wilson gave an example of a set in the case $x_k = \frac{1}{2}(aq^k + \frac{1}{aq^k})$.

In this note we consider the case $x_k = a+bq^{-k}$ for $k = 1,2,\ldots$ and find all OPS of the form (1.1) with this choice of x_k .

By simple calculation we can see that

$$\omega_k(x) = (-1)^k q^{-k(k-1)/2} b^k \left(\frac{x-a}{b};q\right)_k =$$

$$= (-1)^k q^{k(k-1)/2} a^k \left(\frac{x-b}{a}\,;\,q^{-1}\right)_k$$

where $(a;q)_o = 1 \quad (a;q)_k = (1-a)(1-aq)\ldots(1-aq^{k-1}) \qquad (k \geq 1)$.

Thus we see that whenever there is a solution of the problem there is a corresponding one with q being replaced by q^{-1}. We also see from the above that without loss of generality we can take $x_k = q^{-k+1}$.

Next we note that when $q = 0$ or ∞ our problem becomes that which Chihara solved in [5,6]. If on the other hand $q = -1$ then $x_{2k+1} = 1$ and $x_{2k} = -1$ and since in this case we may shift x by a constant, so that $a_1 = 0$, our problem becomes the one which we treated in [2]. Hence without loss of generality we may assume $0 < |q| < 1$. Furthermore we shall assume that we are dealing with real polynomials and that all the constants a_n, b_n, and q are real.

In the following we shall put $B_o = 0$, $B_n = b_{n-1}/b_n$ for $n \geq 1$. Without loss of generality we may assume that $a_o = b_o = 1$. We shall also write

$$(a;q)_\infty = \prod_{n=o}^{\infty} (1-aq^n)$$

and

$$_{p+1}\Phi_p \begin{bmatrix} a_1, a_2, \ldots, a_{p+1}; q, z \\ b_1, b_2, \ldots, b_p \end{bmatrix} =$$

$$= \sum_{k=o}^{\infty} \frac{(a_1;q)_k (a_2;q)_k \ldots (a_{p+1};q)_k}{(b_1;q)_k (b_2;q)_k \ldots (b_p;q)_k} \; \frac{z^k}{(q;q)_k}$$

2. NECESSARY AND SUFFICIENT CONDITIONS

Put $x_k = q^{-k+1}$ then it follows from (1.3) that for (1.1) to be an OPS it is necessary and sufficient that

(2.1) $$\alpha_k B_n + \beta_k B_{n-1} + a_2 a_{k-2} B_{n-2} - a_k B_{n-k} = C_k q^{-n} \qquad (n \geq k \geq 0),$$

or equivalently, if we put $\Gamma_n = B_n - B_{n-1}$,

(2.2) $$a_1(a_1 a_{k-2} - a_{k-1})\Gamma_n - a_2 a_{k-2}(\Gamma_n + \Gamma_{n-1}) + a_k(\Gamma_n + \ldots + \Gamma_{n-k+1}) = C_k q^{-n},$$

$(n \geq 3)$

where $\alpha_k = a_k+(a_1^2-a_2)a_{k-2}-a_1a_{k-1}$, $\beta_k = a_1a_{k-1}-a_1^2a_{k-2}$,

and $C_k = a_1q(1-q)a_{k-2}-q(1-q^{k-1})a_{k-1}$.

Put $k = 3$ then (2.1) becoms

$$\alpha_3\Gamma_n+(a_3-a_1a_2)\Gamma_{n-1}+a_3\Gamma_{n-2} = C_3q^{-n} \qquad (n \geq 3) \tag{2.3}$$

Our method of solving our problem is to use the necessary conditions (2.3) to find the candidates for the sequence $\{b_n\}$. With such sequences $\{b_n\}$ we use the necessary and sufficient conditions (2.1) to find the $\{a_n\}$ and then verify that the resulting PS satisfy (1.2).

To carry out our programm for the proof of the Main Theorem as outlined above we put

$$g(\lambda) = \alpha_3\lambda^2 + (a_3-a_1a_2)\lambda + a_3 = 0$$

to be the characteristic equation of the difference equation (2.3). Let λ_1,λ_2 be the two roots of this equation. Then the general solution of (2.3) is of the form

$$\Gamma_n = (k_1+k_2n)\lambda_1^n+k_3\lambda_2^n+(k_4+k_5n+k_6n^2)q^{-n} \qquad (n \geq 1) \tag{2.4}$$

where the constants $k_1,\dots,k_6$ are there depending on the character of the two roots λ_1 and λ_2. In fact we have the following six exclusive cases:

i. $C_3=0, \lambda_1 \neq \lambda_2$ so that $k_2=k_4=k_5=k_6=0$

ii. $C_3=0, \lambda_1=\lambda_2$ so that $k_3=k_4=k_5=k_6=0$

iii. $C_3\neq 0, \lambda_1=\lambda_2\neq q^{-1}$ so that $k_3=k_5=k_6=0$

iv. $C_3\neq 0, \lambda_1,\lambda_2,q^{-1}$ are pairwise distint so that $k_2=k_5=k_6=0$

v. $C_3\neq 0, \lambda_1=q^{-1}\neq\lambda_2$ so that $k_4=k_5=k_6=0$

vi. $C_3\neq 0, \lambda_1=\lambda_2=q^{-1}$, so that $k_1=k_2=k_3=0$

We shall need the following notation

$$\sigma_k(x) = \sum_{0}^{k-1} jx^{-j+1}, \qquad \tau_k(x) = \sum_{0}^{k-1} j^2x^{-j+1}$$

$$S_k(x) = a_1(a_1a_{k-2}-a_{k-1})-a_2a_{k-2}(1+\tfrac{1}{x}) + a_k\frac{1-x^k}{1-x}x^{1-k}$$

$$U_k(x) = (a_2a_{k-2}-a_k\tau_k(x))x^{-1}, \qquad T_k(x) = (a_2a_{k-2}-a_k\sigma_k(x))x^{-1}$$

3. THE MAIN THEOREM

We now proceed to prove the following:

Theorem. If $x_k = q^{-k+1}$ and $\{P_n(x)\}$ satisfy (1.1) and is an OPS if and only if it is one of the following:

a. Anyone of the OPS of Brenke type [5,6].

b. The OPS treated in [2], i.e. those that satisfy the three term recurrence relation

$$p_n(x) = (x-(-1)^n\alpha)p_{n-1}(x)-\lambda_n p_{n-2}(x)$$

where $\lambda_{2n} = Eq^n(1-\gamma q^{n-1})$, $\lambda_{2n+1} = E\gamma q^n(1-q^n)$,

$E > 0$, $\gamma > 0$, $0 < q < 1$.

c. A special Askey-Wilson OPS [3] defined by

$$P_n(x) = {}_3\Phi_2\left[\begin{matrix} q^{-n},0,x;p,q \\ q\alpha,\beta q \end{matrix}\right]$$

d. A q-Meixner P.S. [9] defined by

$$M_n(x,\beta,d) = {}_2\Phi_1\left[\begin{matrix} q^{-n},x;q,\frac{q^{n+1}}{d} \\ \beta q \end{matrix}\right],$$

e. The PS defined by

$$R_n(x;\beta) = \frac{1-q+q^{1+n}}{a^n(q;q)_n}\ {}_3\Phi_2\left[\begin{matrix} q^{-n},-(1-q)q^{-n},x;q,(1+\beta)q^{1+n} \\ q\beta,-(1-q)^{-n-1} \end{matrix}\right]$$

Proof: We first remark that (a) and (b) follow from our remark at the end of the Introduction. We next put (2.4) in (2.2) we get after some simplification

$$\{k_1S_k(\lambda_1)+k_2T_k(\lambda_1)\}\lambda_1^n + k_2S_k(\lambda_1)n\lambda_1^n+k_3S_k(\lambda_2)\lambda_2^n +$$

$$+k_6S_k(q^{-1})n^2q^{-n}+\{k_5S_k(q^{-1})+2k_6T_k(q^{-1})\}nq^{-n} \tag{3.1}$$

$$= \{C_k-k_4S_k(q^{-1})-k_5T_k(q^{-1})+k_6U_k(q^{-1})\}q^{-n}$$

Formula (3.1) implies that

$$k_1S_k(\lambda_1)+k_2T_k(\lambda_1) = 0 \quad , \quad k_2S_k(\lambda_1)=0 \quad , \quad k_3S_k(\lambda_2)=0 \tag{3.2}$$

$$k_6 S_k(q^{-1}) = 0 \quad , \quad k_5 S_k(q^{-1}) + 2k_6 T_k(q^{-1}) = 0$$

$$k_4 S_k(q^{-1}) + k_5 T_k(q^{-1}) - k_6 U_k(q^{-1}) = C_k$$

Lemma 1. $k_2 = 0$.

To prove this lemma we note that if not true then $k_2 \neq 0$ and $k_3 = 0$. Formulas (3.2) give $S_k(\lambda) = T_k(\lambda) = 0$ where λ is the common value of $\lambda_1 = \lambda_2$. Now $T_k(\lambda) = 0$ yields $a_2 a_{k-2} = a_k \sigma_k(\lambda)$. If we now calculate a_4 using $T_4(\lambda) = 0$ on one handand then using $S_4(\lambda) = 0$ we get that $\lambda(\lambda-1)^2 = 0$ which can easily seen to be imposible.

In the same way we can prove the following two lemmas,

Lemma 2. $k_6 = 0$.

Lemma 3. $k_5 = 0$.

Lemma 4. If $\Gamma_n = Aq^{-n}$ then the only possible solution are the OPS $\{U_n^{(a)}(x)\}$ and $\{V_n^{(a)}(x)\}$ of [1] .

The proof of this lemma follows from a characterization theorem of Ismail [10] in which he found the set of all OPS which are q-Sheffer. For in this case, if we first solve for B_n , we find that $b_n = [a^n q^{n(n-1)/2}]/(q;q)_n$. The use of the q-binomial theorem shows that $P_n(x)$ must be generated by $A(t)/(axt;q)_\infty$.

In view of the above lemmas the only cases which may lead to positive results are those for which $\lambda_1, \lambda_2, q^{-1}$, are pairwise distinct, i.e., cases (i) , (iv) and (v) .

In case (v) we must have $S_k(q^{-1}) = 0$, and $k_4 S_k(q^{-1}) = C_k$.

Lemma 5. If $k_4 = 0$ then $P_n(x)$ is the q-Meixner OPS .

To prove this lemma we see that when $k_4 = 0$ if follows that $C_k = 0$ for all k . This in turn implies that $a_n = a_1^n (1-q)^n/(q;q)_n$. Thus (2.1) takes the form

$$q^3(1-q^{k-1})(1-q^{k-2})\lambda^k - q(1-q^k)(1-q^{k-2})(1+q)\lambda^{k-1} +$$

$$+ (1-q^k)(1-q^{k-1})\lambda^{k-2} = (1-q)(1-q^2)$$

This implies that $\lambda = q^{-1}$, q^{-2} , so that $B_n = q^{-2n}E(1-q^n)(1-\beta q^n)$.
Thus

$$b_n = \frac{q^{n(n-1} b^n}{(q;q)_n(\beta q;q)_n}$$

where b,β are determined by the arbitrary constants b_1,b_2 . This clearly gives the assertion of lemma 5 and at the same time settles case (i) . The q-Meixner PS are known to be orthogonal and so there is no need to prove the converse.

Now the only remaining case is that when $C_3 \neq 0, \lambda_1, \lambda_2, q^{-1}$ are pairwise distinct.

Since $C_3 \neq 0$ we have $k_4 \neq 0$. We thus have two remaining cases, namely, when

$$\Gamma_n = k_1\lambda_1^n+k_3\lambda_2^n+k_4q^{-n} \quad \text{and} \quad \Gamma_n = k_1\lambda_1^n+k_4q^{-n} .$$

In the first case we have $S_k(\lambda_1) = 0$, $S_k(\lambda_2) = 0$ and $k_4S_k(q^{-1}) = C_3$. If we calculate a_4 in three different ways we get $(\lambda_1^2 -\lambda_2)(\lambda_1+\lambda_2) = 0$. If $\lambda_1 + \lambda_2 = 0$ we get inconsistant relations. Thus we must have $\lambda_2 = \lambda_1^2$. This leads to $\lambda_1 = q$, $\lambda_2 = q^2$. This in turn implies that

$$a_m = a_1 \frac{(1-q^{-1})}{1-q^{-m}} a_{m-1}$$

After some calculation we find that

$$b_n = \frac{(1-q)^n q^{n(n-1)/2} b^n}{(q;q)_n(q\alpha;q)_n(\beta q;q)_n}$$

The resulting PS is the special Askey-Wilson mentioned in the Theorem and which is known to be an OPS .

Finally when $k_3 = 0$ we only have two relations to satisfy . In this case we find that $\lambda_1 = q^{-2}$ and $\lambda_2 = (1-q+q^4)/q^3(1-q^3+q^4)$. We get

$$\text{(3.2)} \qquad b_n = \frac{q^{n(n+1)} b^n}{(q;q)_n(\beta q;q)_n} \quad \text{and} \quad a_n = \frac{1-q+q^{n+1}}{a^n(q;q)_n}$$

where $\beta+1 = a/b$.

This PS is believed to be new . We shall study it in the next section where we shall prove the orthogonality.

4. A SET OF ORTHOGONAL POLYNOMIALS

In view of the relations (3.2), (1.1) we see that

$$(4.1) \qquad R_n(x;\beta) = \frac{1-q+q^{1+n}}{a^n(q;q)_n}\, {}_3\Phi_2\left[\begin{matrix} q^{-n}, -(1-q)q^{-n}, x;q, (1+\beta)q^{1+n} \\ \beta q, -(1-q)q^{-n-1} \end{matrix}\right]$$

(4.2)

$$= \frac{1}{a^n(q;q)_n}\left\{ (1-q)\, {}_2\Phi_1\left[\begin{matrix} q^{-n}, x;q, (1+\beta)q^{n+2} \\ \beta q \end{matrix}\right] + q^{n+1}\, {}_2\Phi_1\left[\begin{matrix} q^{-n}, x;q, (1+\beta)q^{n+1} \\ \beta q \end{matrix}\right]\right\}$$

or equivalently and in terms of the q-Meixner PS

$$(4.3) \qquad R_n(x;\beta) = \frac{1}{a^n(q;q)_n}\, \{(1-q)M_n(x;\beta, \frac{1}{q(1+\beta)}|q) +$$

$$+ q^{n+1}M_n(x;\beta, \frac{1}{1+\beta}\ |q)\}$$

The leading coefficient of $R_n(x;\beta)$ is $q^{n(n+1)}(1+\beta)^n/(a^n(q;q)_n(\beta q;q)_n)$.

The monic set $\hat{R}_n(x,\beta) = \hat{R}_n(x)$ satisfies the three term recurrence relation

$$\hat{R}_n(x) = (x+\alpha_n)\hat{R}_{n-1}(x) - \lambda_n\hat{R}_{n-2}(x)$$

where

$$(4.3) \qquad \alpha_n = \frac{1}{1+\beta}\, \{q^{-2n}(1+q^3)-(1+\beta)(1+q^2)q^{-n}\}$$

$$(4.4) \qquad \lambda_n = \frac{q^{5-4n}}{(1+\beta)^2}\, (1-q^{n-1})(1-\beta q^{n-1})\,[1-(1+\beta)q^{n-1}]$$

This can be easily verified. We get orthogonality with respect a positive measure if

(i). $0 < q < 1$ and $\beta < q^{-1}-1$, or

(ii). $q > 1$ and $q^{-1}-1 < \beta < 0$, or

(iii). $q < -1$ and $0 < \beta q < 1$.

If $\{\tilde{R}_n(x;\beta)\}$ denotes the orthonormal set then we have

$$(4.5) \qquad x\tilde{R}_n(x;\beta) = k_n\tilde{R}_{n+1}(x;\beta) + \alpha_n\tilde{R}_n(x;\beta) + k_{n-1}\tilde{R}_{n-1}(x;\beta)$$

where

$$(4.6)\qquad k_n^2 = (1-q^{n+1})(1-\beta q^{n+1})(1-(1+\beta)q^{n+1})q^{-4n-3}(1+\beta)^{-2}$$

Orthogonality can be stated in terms of a linear functional $\mathcal{L}$ whose moments are

$$(4.7)\qquad \mu_n = \mathcal{L}(x^n) = \frac{1}{q}\sum_{k=o}^{n} \frac{(q^{-n};q)_k(\beta q;q)_k}{(q;q)_k(\beta+1)^k}\{1-(1-q)q^k\}$$

and

$$(4.8)\qquad \mathcal{L}[R_n(x)R_m(x)] = \frac{((1+\beta)q;q)_n q^n}{a^{2n}(\beta q;q)_n(q;q)_n}\delta_{nm}$$

To prove (4.8) we first introduce $\nabla C_k = C_k - C_{k+1}$. We can prove, by induction,

$$(4.9)\qquad \begin{aligned}(q^{k+1}\nabla)^j C_k &= q^{j(k+1)}\mathcal{L}[x^k(x;q)_j]\\ &= q^{j(k+1)}\sum_{r=o}^{j}(-1)^r q^{r(r-1)/2}\begin{bmatrix} j\\ r\end{bmatrix} C_{k+r}\end{aligned}$$

where $\begin{bmatrix} j\\ r\end{bmatrix}$ is the familiar Gaussian binomial coefficient.

Now to prove (4.8)

$$qa^n\mathcal{L}[x^kR_n(x)] = q\sum_{j=o}^{n}\frac{(1-q+q^{n-j+1})(1+\beta)^j q^{j(j+1-2k)/2}(-1)^j}{(q;q)_{n-j}(q;q)_j(\beta q;q)_j}(q^{k+1}\nabla)^j\mu_k$$

Using (4.9) and rearranging the terms we can get

$$qa^n\mathcal{L}[x^kR_n(x)] =$$

$$= \sum_{r=o}^{n}\frac{(q^{-k};q)_r(\beta q;q)_r}{(r;q)_r((\beta+1)^r}\{(1-q)(1-q^{n+r+1})\,{}_2\Phi_1\begin{bmatrix} q^{-n},\beta q^{1+r};q,q^{1+n-k}\\ \beta q\end{bmatrix}$$

$$-(1-q)^2q^r\,{}_2\Phi_1\begin{bmatrix} q^{-n},\beta q^{1+r};q,q^{2+n-k}\\ \beta q\end{bmatrix} + q^{n+1}\,{}_2\Phi_1\begin{bmatrix} q^{-n},\beta q^{1+r};q,q^{n-k}\\ \beta q\end{bmatrix}$$

Heine's transformation can be stated as

$${}_2\Phi_1\begin{bmatrix} q^{-n},\beta q^{1+r};q,z\\ \beta q\end{bmatrix} = \frac{1}{(\beta q;q)_r}\sum_{s=o}^{r}(-1)^s q^{s(s+1)/2}\begin{bmatrix} r\\ s\end{bmatrix}\beta^s(zq^{-n+s};q)_n$$

Applying this formula to each of the three $_2\phi_1$ on the right hand side of the previous formula we get that the right hand side is 0 when $k < n$. When $k = n$ we get the right hand side of (4.8). This complete the proof of (4.8).

This P S, in case $0 < q < 1$, belongs to a determined moment problem. The distribution function associated with it is discrete. To see the second assertion we note that (4.3) and (4.4) imply that $\frac{\lambda_{n+1}}{\alpha_n \alpha_{n+1}} \longrightarrow \frac{q^3}{(1+q^3)^2} < \frac{1}{8}$ and use [4,p. 119]. The first assertion follows from

$$\{\tilde{R}_n(1)\}^2 = \frac{q^{-n}(\beta q;q)_n (1-q+q^{n+1})^2}{(q;q)_n((1+\beta)q;q)_n}$$

and Theorem 2.3 in [7].

We are grateful to Professor M. Ismail who pointed out that one can find the spectral points of the distribution function in the following way. We first note that if denote

$$f(x) = \sum_{k=o}^{\infty} \frac{(x;q)_k(-1-\beta)^k}{(q;q)_k(\beta q;q)_k} q^{k(k+3)/2}$$

then it follows that

$$\tilde{R}_n(x) \sim \left(\frac{(\beta q;q)_\infty (1-q)^2}{((1+\beta)q;q)_\infty (q;q)_\infty}\right)^{1/2} \frac{f(x)}{q^{n/2}}$$

so that for all non-real z we have $\sum |\tilde{R}_n(z;\beta)|^2 = \infty$ except at the zeros (which are all real) of $f(x)$. At those zeros this sum is convergent whose reciprocal is the jump at that point.

REFERENCES

1. W.A. AL-SALAM and L. CARLITZ, Some orthogonal q-polynomials, Math. Nach., 30, 1965, pp. 47-61.

2. W.A. AL-SALAM and A. VERMA, On an orthogonal polynomial set, Proc. of the Konin. Nederl. Akademie van Wetensch., ser A, 85(3), 1982, pp. 335-340.

3. R. ASKEY and J. WILSON, Some basic hypergeometric orthogonal polynomials that generalize Jacobi polynomials, Memoir 319, Amer. Math. Society, 1985.

4. T.S. CHIHARA, An introduction to Orthogonal Polynomials, New York 1978.

5. T.S. CHIHARA, Orthogonal polynomials with Brenke type generating function, Duke Math. J., 35, 1968, pp. 505-518.

6. T.S. CHIHARA, The orthogonality of a class of Brenke polynomials. Duke Math. J., 38, 1971, pp. 599-603.

7. G. FREUD, Orthogonal Polynomials, Budapest, 1971.

8. J. GERONIMUS, The orthogonality of some systems of polynomials, Duke Math. J., 14, 1947, 503-510.

9. W. HAHN, Über Orthogonalpolynome, die q-Differenzengleichungen genügen, Math. Nachrichten , 2, 1949, pp. 4-34.

10. M. ISMAIL, Orthogonal polynomials in a certain class of polynomials Bul. Inst. Polit. Din Iaşi, Ser. I, 20, 1974, pp. 45-50.

THE BOUNDS FOR THE ERROR TERM OF AN ASYMPTOTIC APPROXIMATION OF JACOBI POLYNOMIALS

Paola BARATELLA
and
Luigi GATTESCHI
Universitá di Torino
Italy

Abstract. We consider a new asymptotic approximation of Jacobi polynomials $P_n^{(\alpha,\beta)}(\cos\theta)$ and we obtain a realistic and explicit bound for the corresponding error term. The approximation is of Hilb's type and is uniformly valid for $0 < \theta \leq \pi-\varepsilon$, $\varepsilon > 0$. Bounds for the error term in the asymptotic approximation of the zeros of $P_n^{(\alpha,\beta)}(\cos\theta)$ are also given.

1. Introduction. In May 1984, we started this work with the main object of obtaining simple uniform representations of Jacobi polynomials $P_n^{(\alpha,\beta)}(x)$, as $n\to\infty$, in terms of Bessel functions with realistic and explicit upper bounds for the error terms. We were not aware of a uniform asymptotic expansion with bounds given by Frenzen and Wong [4]. Indeed this result, together an application to the zeros of the Jacobi polynomials, was communicated by Wong [5] at a Conference of Special Functions held in Turin in October 1984. In the meantime we obtained by using Olver's differential equations theory [11, Chapter 12], as Elliot [3] had done for the asymptotics of $P_n^{(\alpha,\beta)}(z)$ when z is a complex, a representation of the form

$$(1.1)\qquad \left(\sin\frac{\theta}{2}\right)^{\alpha+\frac{1}{2}}\left(\cos\frac{\theta}{2}\right)^{\beta+\frac{1}{2}} P_n^{(\alpha,\beta)}(\cos\theta) = 2^{-\frac{1}{2}} N^{-\alpha}\,\frac{\Gamma(n+\alpha+1)}{n!}\,.$$

$$\cdot\left[\theta^{\frac{1}{2}} J_\alpha(N\theta)\sum_{s=0}^{m}\frac{A_s(\theta)}{N^{2s}} + \theta^{\frac{3}{2}} J_{\alpha+1}(N\theta)\sum_{s=0}^{m-1}\frac{B_s(\theta)}{N^{2s+1}} + \varepsilon_m(N,\theta)\right],$$

where $\alpha > -1$, β arbitrary and real,

$$(1.2)\qquad N = n + \frac{\alpha+\beta+1}{2}$$

This work was supported by the Consiglio Nazionale delle Ricerche of Italy and by the Ministero della Pubblica Istruzione of Italy.

and $J_\alpha(t)$ denotes the Bessel function of first kind of order α. The functions $A_s(\theta)$ and $B_s(\theta)$ are analytic for $0\le\theta<\pi$ and defined recursively, starting from $A_0(\theta) = 1$, by

$$\theta B_s(\theta) = -\frac{1}{2}A_s'(\theta) - \frac{1+2\alpha}{2}\int_0^\theta \frac{A_s'(t)}{t}\,dt + \frac{1}{2}\int_0^\theta f(t)A_s(t)dt,$$

$$A_{s+1}(\theta) = \frac{1}{2}\theta B_s'(\theta) - \alpha B_s(\theta) - \frac{1}{2}\int_0^\theta tf(t)B_s(t)dt,$$

with

$$f(t) = \frac{(1/4)-\alpha^2}{t^2} - \frac{(1/4)-\alpha^2}{4\sin^2(t/2)} - \frac{(1/4)-\beta^2}{4\cos^2(t/2)},$$

for $s=0,1,\ldots,$ where the constant of integration is chosen so that $A_{s+1}(0) = 0$ for $s=0,1,\ldots$. As $n\to\infty$ the remainder term $\varepsilon_m(N,\theta)$ can be estimated. Precisely we have

$$(1.3)\qquad \varepsilon_m(N,\theta) = \begin{cases} \theta^{\alpha+5/2}\,O(N^{-2m+\alpha}), & \text{if } 0 < \theta \le c/N, \\ \theta\,.\,O(N^{-2m-3/2}), & \text{if } c/N \le \theta \le \pi-\delta, \end{cases}$$

being c and δ fixed positive constants.

When $m=0$, the expansion (1.1) with the remainder (1.3) reduces to the well-known Szegö formula [12, p.214] of Hilb's type for Jacobi polynomials.

The first coefficients in (1.1) are given by

$$(1.4)\qquad \begin{aligned} &A_0(\theta) = 1, \qquad \theta B_0(\theta) = \frac{1}{4}g(\theta), \\ &A_1(\theta) = \frac{1}{8}g'(\theta) - \frac{1+2\alpha}{8}\frac{g(\theta)}{\theta} - \frac{1}{32}g^2(\theta), \end{aligned}$$

where

$$g(\theta) = (\tfrac{1}{4}-\alpha^2)(\cot\tfrac{\theta}{2} - \tfrac{2}{\theta}) - (\tfrac{1}{4}-\beta^2)\tan\tfrac{\theta}{2}.$$

The explicit determination of the other coefficients A_s and B_s, as well as of an upper bound for $\varepsilon_m(N,\theta)$, seems to be rather difficult.

As a consequence of (1.1), by using a general formula due to Olver [11, p.453], we obtained the following representation for the zeros $\theta_{n,k}(\alpha,\beta)$, $k=1,2,\ldots,$ of $P_n^{(\alpha,\beta)}(\cos\theta)$.

<u>Theorem 1.1</u> Let $-1/2 \le \alpha \le 1/2$ and $-1/2 \le \beta \le 1/2$. Then as $n\to\infty$

$$(1.5)\qquad \theta_{n,k}(\alpha,\beta) = \frac{j_{\alpha,k}}{N} - \frac{1}{4N^2}\Big[(\tfrac{1}{4}-\alpha^2)(\tfrac{2}{t}-\cot\tfrac{t}{2}) + (\tfrac{1}{4}-\beta^2)\tan\tfrac{t}{2}\Big] + t\,O(n^{-4}),$$

where $j_{\alpha,k}$ is the k-th positive zero of the Bessel function $J_\alpha(x)$,

$t = j_{\alpha,k}/N$ and $N = n+(\alpha+\beta+1)/2$. The O-term is uniformly bounded for all values of $k = 1,2,\dots,[\gamma n]$, where γ is a fixed number in $(0,1)$.

Under less restrictive conditions for α and β and with a different estimation of the O-term, formula (1.5) had been obtained earlier by Frenzen and Wong [4, p. 982], as a consequence of the following interesting two-term expansion.

Theorem 1.2. For $\alpha > -1/2$, $\alpha-\beta > -4$ and $\alpha+\beta \geq -1$, we have

$$(\sin \tfrac{\theta}{2})^{\alpha}(\cos \tfrac{\beta}{2})^{\beta} P_n^{(\alpha,\beta)}(\cos\theta) =$$

$$= \frac{\Gamma(n+\alpha+1)}{n!}\left(\frac{\theta}{\sin\theta}\right)^{1/2}\left[\frac{J_\alpha(N\theta)}{N^\alpha} + \theta B_o(\theta)\frac{J_{\alpha+1}(N\theta)}{N^{\alpha+1}} + \sigma_2\right], \tag{1.6}$$

where $B_o(\theta)$ has the same meaning as in (1.4) and

$$|\sigma_2| \leq E_2\theta^{2+\alpha}N^{-2} \quad , \qquad 0 \leq \theta \leq \pi/2 \ , \tag{1.7}$$

being E_2 a constant.

In the special case $\alpha = \beta = 0$, one easily obtains $E_2 = 0.1253\dots$ Unfortunately, for general α and β, the evaluation of E_2 is more involved by using the formulas given in the paper of Frenzen and Wong. The purpose of this paper is to show that another approach, essentially based on the Liouville-Stekloff method [12, Chapter 8], allows us to obtain a representation of $P_n^{(\alpha,\beta)}(\cos\theta)$, similar to (1.6) but with a different error bound which can be expressed very simply in terms of α and β. As a consequence we will give also a proof of (1.5), with an upper bound for the error term.

2. A comparison equation for Jacobi polynomials.

In the asymptotic study of Jacobi polynomials $P_n^{(\alpha,\beta)}(x)$, as $n \to \infty$, it is convenient to use the differential equation

$$\frac{d^2u}{d\theta^2} + \left[N^2 + \frac{1/4-\alpha^2}{4\sin^2\frac{\theta}{2}} + \frac{1/4-\beta^2}{4\cos^2\frac{\theta}{2}}\right]u = 0 \ , \qquad N = n + \frac{\alpha+\beta+1}{2} \ , \tag{2.1}$$

which is satisfied by

$$u_n(\theta) = (\sin \tfrac{\theta}{2})^{\alpha+1/2}(\cos \tfrac{\theta}{2})^{\beta+1/2} P_n^{(\alpha,\beta)}(\cos\theta) \ .$$

Equation (2.1) is usually compared, for instance when we are interested in considering values of θ in the neighbourhood of $\theta = 0$, with Bessel differencial equation

(2.2) $$\frac{d^2z}{d\theta^2} + \left[N^2 + \frac{1/4-\alpha^2}{\theta^2}\right] z = 0 .$$

Here, instead of equation (2.2), we will consider the differential equation

(2.3) $$\frac{d^2v}{d\theta^2} + k(\theta)v = 0 ,$$

with

(2.4) $$k(\theta) = \frac{1}{2}\frac{f'''(\theta)}{f'(\theta)} - \frac{3}{4}\left[\frac{f''(\theta)}{f'(\theta)}\right]^2 + \frac{1-4\alpha^2}{4}\left[\frac{f'(\theta)}{f(\theta)}\right]^2 + [f'(\theta)]^2$$

which is satisfied by

$$v_1(\theta) = \left[\frac{f(\theta)}{f'(\theta)}\right]^{1/2} J_\alpha[f(\theta)] \quad \text{and} \quad v_2(\theta) = \left[\frac{f(\theta)}{f'(\theta)}\right]^{1/2} J_{-\alpha}[f(\theta)] ,$$

($J_{-\alpha}[f(\theta)]$ is to be replaced by $Y_\alpha[f(\theta)]$, if α is a integer or zero) and, as we will see, is "very close" to (2.1) for a proper $f(\theta)$. For the choice of $f(\theta)$ we observe that, when θ is fixed and $N \to \infty$,

$$J_\alpha(N\theta)+\theta B_o(\theta)N^{-1}J_{\alpha+1}(N\theta) = [1+\alpha B_o(\theta)N^{-2}]J_\alpha(N\theta)-\theta B_o(\theta)N^{-1}J'_\alpha(N\theta) =$$

$$= [1+\alpha B_o(\theta)N^{-2}]\ [J_\alpha(N\theta)-\theta B_o(\theta)N^{-1}J'_\alpha(N\theta)+O(N^{-3})].$$

By Taylor's theorem, for some τ,

$$J_\alpha(N\theta)-\theta B_o(\theta)N^{-1}J'_\alpha(N\theta) = J_\alpha[N\theta-\theta B_o(\theta)N^{-1}] + O(N^{-2})J''_\alpha(N\tau),$$

and according to (1.6), a well-founded choice of $f(\theta)$ is

(2.5) $$f(\theta) = N\theta + \frac{1}{16N}\left[(1-4\alpha^2)\left(\frac{2}{\theta} - \cot\frac{\theta}{2}\right) + (1-4\beta^2)\tan\frac{\theta}{2}\right].$$

In the following we shall set

$$A = 1 - 4\alpha^2 , \qquad B = 1 - 4\beta^2 ,$$

$$a(\theta) = \frac{2}{\theta} - \cot\frac{\theta}{2} , \qquad b(\theta) = \tan\frac{\theta}{2} ,$$

Thus (2.5) will be written in the form

(2.6) $$f(\theta) = N\theta + \frac{1}{16N}[A\,a(\theta) + B\,b(\theta)] .$$

We will see that with this choice of $f(\theta)$ the coefficient $k(\theta)$ in (2.3) differs from the coefficient of $u(\theta)$ in (2.1) by a quantity which is $O(N^{-2})$.

Throughout this paper we shall assume $-1/2 \leq \alpha \leq 1/2$, $-1/2 \leq \beta \leq 1/2$. Without loss of generality we may also restrict our attention to $0 < \theta \leq \pi/2$, since $P_n^{(\alpha,\beta)}(x) = (-1)^n P_n^{(\beta,\alpha)}(-x)$. First we observe that

$$[f'(\theta)]^2 = N^2 + \frac{A}{16\sin^2\frac{\theta}{2}} + \frac{B}{16\cos^2\frac{\theta}{2}} - \frac{A}{4\theta^2} + \frac{[Aa'(\theta)+Bb'(\theta)]^2}{256N^2}$$

and write (2.1) in the form

$$\frac{d^2u}{d\theta^2} + k(\theta)\,u = F(\theta)\,u \,. \tag{2.7}$$

Next we notice that $a(\theta)$, $b(\theta)$ and all their derivatives are positive increasing functions of θ in $0 < \theta \leq \pi/2$ and the following expansions [1, p. 75] hold

$$a(\theta) = \frac{\theta}{6} + \frac{\theta^3}{8.45} + \ldots + \frac{(-1)^{n-1}2^{2n}B_{2n}}{(2n)!}\left(\frac{\theta}{2}\right)^{2n-1} + \ldots \quad , \qquad |\theta| < 2\pi ,$$

$$b(\theta) = \frac{\theta}{2} + \frac{\theta^3}{24} + \ldots + \frac{(-1)^{n-1}2^{2n}(2^{2n}-1)B_{2n}}{(2n)!}\left(\frac{\theta}{2}\right)^{2n-1} + \ldots \quad , |\theta| < \pi$$

where B_m is the m-th Bernoulli number.

The above properties of the functions $a(\theta)$ and $b(\theta)$ allow us to see that the function $F(\theta)$ is $O(N^{-2})$. A more precise study which requires accurate calculations shows that the functions $F(\theta)$ is positive and that

$$\frac{1}{16N^2}(\lambda_1 A+\lambda_2 B) \leq F(\theta) \leq \frac{1}{16N^2}(\mu_1 A+\mu_2 B) \ , \quad 0<\theta\leq\pi/2 \ , \ n \geq 5 \ , \tag{2.8}$$

$$\lambda_1=0.0082467, \quad \lambda_2=0.12270, \quad \mu_1=0.10536, \quad \mu_2 =1.0744 \,.$$

Throughout this paper we assume $n \geq 5$ which is not restrictive because we are dealing with asymptotic representations. Here we give only some essential steps of the proof of (2.8).

Set

$$F(\theta) = F_1(\theta) + F_2(\theta) \ ,$$

where

$$F_1(\theta) = \frac{1}{2}\ \frac{Aa'''(\theta)+Bb'''(\theta)}{16N^2+Aa'(\theta)+Bb'(\theta)} - \frac{3}{4}\left[\frac{Aa''(\theta)+Bb''(\theta)}{16N^2+Aa'(\theta)+Bb'(\theta)}\right]^2$$

$$F_2(\theta) = \frac{A}{2\theta^3}\ \frac{Aa'(\theta)\ +Bb'(\theta)\theta-Aa(\theta)-Bb(\theta)}{16N^2+Aa(\theta)/\theta+Bb(\theta)/\theta} \ \cdot$$

$$\cdot\left[1+ \frac{1}{2}\ \frac{Aa'(\theta)\ +Bb'(\theta)\theta-Aa(\theta)-Bb(\theta)}{16N^2\theta+Aa(\theta)+Bb(\theta)}\right] + \frac{[Aa'(\theta)+Bb'(\theta)]^2}{256N^2} \ .$$

Then

$$\frac{1}{16N^2}\,(\gamma_1 A+\gamma_2 B) \leq F_1(\theta) \leq \frac{1}{16N^2}\,(\delta_1 A+\delta_2 B) \ , \qquad n \geq 5 \ ,$$

$$\gamma_1 = 0.0082467, \quad \gamma_2 = 0.12270, \quad \delta_1 = 0.014466, \quad \delta_2 = 1,$$

$$\frac{1}{16N^2}(\varepsilon_1 A^2 + \varepsilon_2 AB + \varepsilon_3 B^2) \leq F_2(\theta) \leq \frac{1}{16N^2}(\eta_1 A^2 + \eta_2 AB + \eta_3 B^2), \quad n \geq 5,$$

$$\varepsilon_1 = 0.0045082, \quad \varepsilon_2 = 0.051999, \quad \varepsilon_3 = 0.015625,$$

$$\eta_1 = 0.0053812, \quad \eta_2 = 0.097350, \quad \eta_3 = 0.062500.$$

By summation one gets

$$\frac{1}{16N^2}(\gamma_1 A + \gamma_2 B + \varepsilon_1 A^2 + \varepsilon_2 AB + \varepsilon_3 B^2) \leq F(\theta) \leq \frac{1}{16N^2}(\delta_1 A + \delta_2 B + \eta_1 A^2 + \eta_2 AB + \eta_3 B^2)$$

and finally (2.8) follows.

In conclusion we have proved the following result.

Theorem 2.1. Let $-1/2 \leq \alpha \leq 1/2$, $-1/2 \leq \beta \leq 1/2$, $N = n + (\alpha+\beta+1)/2$ and

$$f(\theta) = N\theta + \frac{1}{16N}\left[(1-4\alpha^2)\left(\frac{2}{\theta} - \cot\frac{\theta}{2}\right) + (1-4\beta^2)\tan\frac{\theta}{2}\right].$$

Then the Jacobi differential equation

$$\frac{d^2u}{d\theta^2} + \left[N^2 + \frac{1/4-\alpha^2}{4\sin^2\frac{\theta}{2}} + \frac{1/4-\beta^2}{4\cos^2\frac{\theta}{2}}\right] u = 0$$

can be written in the form

$$\frac{d^2u}{d\theta^2} + k(\theta)u = F(\theta)u$$

where, if $0 < \theta \leq \pi/2$ and $n \geq 5$,

$$\frac{1}{16N^2}[\lambda_1(1-4\alpha^2) + \lambda_2(1-4\beta^2)] \leq F(\theta) \leq \frac{1}{16N^2}[\mu_1(1-4\alpha^2) + \mu_2(1-4\beta^2)],$$

with λ_1, λ_2, μ_1 and μ_2 given by (2.8).

The function $k(\theta)$ is given by (2.4) and the equation $v''+k(\theta)v=0$ is satisfied by the fundamental system of solutions

$$(2.9) \qquad v_1(\theta) = \left[\frac{f(\theta)}{f'(\theta)}\right]^{-1/2} J_\alpha[f(\theta)], \quad v_2(\theta) = \left[\frac{f(\theta)}{f'(\theta)}\right]^{1/2} J_{-\alpha}[f(\theta)]$$

($J_{-\alpha}[f(\theta)]$ is to be replaced by $Y_0[f(\theta)]$ if α is equal to zero).

3. An integral equation for $P_n^{(\alpha,\beta)}(\cos\theta)$. Theorem 2.1 can be used to obtain, by means of Liouville-Stekloff method, an integral equation for $P_n^{(\alpha,\beta)}(\cos\theta)$.

According to this method and taking into account that the differential equation (2.3) admits the fundamental system (2.9), the Jacobi polynomial $P_n^{(\alpha,\beta)}(\cos\theta)$ satisfies the following integral equation of Volterra type

$$(\sin\tfrac{\theta}{2})^{\alpha+1/2}(\cos\tfrac{\theta}{2})^{\beta+1/2}P_n^{(\alpha,\beta)}(\cos\theta) =$$

$$\text{(3.1)}\quad = C_1\left[\frac{f(\theta)}{f'(\theta)}\right]^{1/2} J_\alpha[f(\theta)] + C_2\left[\frac{f(\theta)}{f'(\theta)}\right]^{1/2} J_{-\alpha}[f(\theta)] +$$
$$+\left[\frac{f(\theta)}{f'(\theta)}\right]^{1/2}\int_{\theta_o}^{\theta}\left[\frac{f(t)}{f'(t)}\right]^{1/2}\frac{J_\alpha[f(t)]J_{-\alpha}[f(\theta)]-J_\alpha[f(\theta)]J_{-\alpha}[f(t)]}{W(t)}$$
$$.F(t)\,(\sin\tfrac{t}{2})^{\alpha+1/2}(\cos\tfrac{t}{2})^{\beta+1/2}P_n^{(\alpha,\beta)}(\cos t)dt\ .$$

Here C_1 and C_2 are certain constants that we shall determine later and $W(t)$ indicates, as usually, the Wronskian of the solutions $\left[\frac{f(\theta)}{f'(\theta)}\right]^{1/2}J_\alpha[f(\theta)]$ and $\left[\frac{f(\theta)}{f'(\theta)}\right]^{1/2}J_{-\alpha}[f(\theta)]$, that is

$$\text{(3.2)}\qquad W(t)=\begin{cases}-\dfrac{2\sin\alpha\pi}{\pi}\ , & \alpha\neq 0\\[2mm] \dfrac{2}{\pi}\ , & \alpha = 0\end{cases}$$

(Obviously, if $\alpha = 0$, $J_{-\alpha}$ has to be replaced by Y_o in (3.1)).

Let N be fixed, then (2.5) gives $f(\theta) = O(\theta)$ and $\frac{f(\theta)}{f'(\theta)} = \theta+O(\theta^3)$. Moreover it is well known that, as $z\to 0$, $J_\alpha(z) = O(z^\alpha)$, $\alpha\neq 0$, and $Y_o(z) = O(\lg z)$. Hence the integral at the right hand side of (3.1) is convergent for $\theta_o = 0$, so that we may assume $\theta_o = 0$. As $\theta\to 0$ it becomes:

$$O(1)\int_o^\theta t^{1/2}(t^\alpha\theta^{-\alpha}+\theta^\alpha t^{-\alpha})t^{\alpha+1/2}dt = O(\theta^{\alpha+2})\ ,\qquad \alpha\neq 0$$

or

$$O(1)\int_o^\theta t^{1/2}(\lg\theta-\lg t)t^{1/2}dt = O(\theta^2\lg\theta)\ ,\qquad \alpha = 0\ .$$

To find the constants C_1 and C_2 let us divide (3.1) by $\left[\frac{f(\theta)}{f'(\theta)}\right]^{\alpha+1/2}$ and let $\theta\to 0$. If $\alpha\neq 0$, we have

$$2^{-\alpha-1/2}P_n^{(\alpha,\beta)}(1)+O(\theta^2) = \frac{2^{-\alpha}C_1N^\alpha}{\Gamma(\alpha+1)}\left(1+\frac{C}{16N^2}\right)^\alpha + O(\theta^2) +$$
$$+ C_2\left[\frac{f'(\theta)}{f(\theta)}\right]^\alpha J_{-\alpha}[f(\theta)]+O(\theta^2)\ ,$$

where $C = A/6+B/2$.

It follows

$$(3.3)\qquad C_1 = 2^{-1/2}\,\frac{\Gamma(n+\alpha+1)}{n!}\,N^{-\alpha}\left(1+\frac{C}{16N^2}\right)^{-\alpha}, \qquad C_2 = 0 .$$

If $\alpha = 0$, the previous identity becomes

$$2^{-1/2}+O(\theta^2) = C_1\left[1+O(\theta^2)\right]+C_2 O(\lg\left[f(\theta)\right])+O(\theta^2 \lg\theta) ,$$

which furnishes the values

$$(3.4)\qquad C_1 = 2^{-1/2} , \qquad C_2 = 0 .$$

Finally we observe that, by using the well-known formula

$$Y_\alpha(x) = \left[J_\alpha(x)\cos\alpha\pi - J_{-\alpha}(x)\right]/\sin\alpha\pi ,$$

we have

$$J_\alpha[f(t)]J_{-\alpha}[f(\theta)]-J_\alpha[f(\theta)]J_{-\alpha}[f(t)] =$$
$$= \{-J_\alpha[f(t)]Y_\alpha[f(\theta)] + J_\alpha[f(\theta)]Y_\alpha[f(t)])\}\sin\alpha\pi .$$

So, from (3.1), we obtain that the following integral equation

$$(3.5)\qquad \begin{aligned} \left[\frac{f(\theta)}{f'(\theta)}\right]^{-1/2} u_n(\theta) &= C_1 J_\alpha[f(\theta)] - \frac{\pi}{2}\int_0^\theta \left[\frac{f(t)}{f'(t)}\right]^{1/2} \Delta(t,\theta)F(t)u_n(t)\,dt, \\ C_1 &= 2^{-1/2}\,N^{-\alpha}\,\frac{\Gamma(n+\alpha+1)}{n!}\left[1+\frac{1}{16N^2}\left(\frac{A}{6}+\frac{B}{2}\right)\right]^{-\alpha}, \end{aligned}$$

with

$$(3.6)\qquad \Delta(t,\theta) = J_\alpha[f(\theta)]Y_\alpha[f(t)]-J_\alpha[f(t)]Y_\alpha[f(\theta)] ,$$

is satisfied by

$$(3.7)\qquad u_n(\theta) = \left(\sin\frac{\theta}{2}\right)^{\alpha+1/2}\left(\cos\frac{\theta}{2}\right)^{\beta+1/2} P_n^{(\alpha,\beta)}(\cos\theta).$$

4. <u>Asymptotic formula for</u> $P_n^{(\alpha,\beta)}(\cos\theta)$. In this section we shall give an estimate for the integral in (3.5), i.e.

$$I = \frac{\pi}{2}\int_0^\theta \left[\frac{f(t)}{f'(t)}\right]^{1/2} \Delta(t,\theta)F(t)\left(\sin\frac{t}{2}\right)^{\alpha+1/2}\left(\cos\frac{t}{2}\right)^{\beta+1/2} P_n^{(\alpha,\beta)}(\cos t)\,dt$$

where $f(t)$ and $\Delta(t,\theta)$ are defined by (2.5) and (3.6) respectively.

Let θ^* be the root of the transcendental equation $f(\theta) = \pi/2$. Due to the monotonicity of the function $f(\theta)$ in $0 < \theta \leq \pi/2$ such root exists and is unique. Since

$$(4.1)\qquad N\theta \leq f(\theta) \leq N\theta + \frac{\theta}{2N\pi^2}$$

it is easy to verify that, if $n \geq 5$,

(4.5) $$0.99797 \frac{\pi}{2N} < \theta^* < \frac{\pi}{2N} .$$

We shall derive two different upper bounds for $|I|$, depending whether θ belongs to the interval $0 < \theta \leq \theta^*$ or to the interval $\theta^* \leq \theta \leq \pi/2$.

i) Let us consider the first case $0 < \theta \leq \theta^*$.

Lemma 4.1. Let $-1/2 \leq \nu \leq 1/2$ and $0 < z < \pi/2$. Then the first zero of the cylinder function $C_\nu(x) = J_\nu(z)Y_\nu(x) - J_\nu(x)Y_\nu(z)$ is $x = z$.

For the proof see Gatteschi [6, p. 147] observing that the result holds also if $\nu = -1/2$.

Since $f(t) < \pi/2$, $t \in [0,\theta]$, $\theta < \theta^*$, and since for any $t \in (0,\pi/2]$ both functions $f(t)$ and $f'(t)$ are positive and increasing, it follows from the above lemma that the function $\Delta(t,\theta)$ does not change sign in $[0,\theta]$. Hence, if we denote by M and upper bound for the absolute value of

$$F(t) \left[\frac{\sin t/2}{f(t)/f'(t)}\right]^{-\alpha+1/2} P_n^{(\alpha,\beta)}(\cos t) \left[\frac{1}{f'(t)}\right]^{\alpha+2}$$

in $0 \leq t \leq \theta$, it is

(4.3) $$|I| \leq M \frac{\pi}{2} \left| \int_0^\theta f^{\alpha+1}(t) f'(t) \Delta(t,\theta) dt \right| .$$

According to the formulas

$$\frac{d}{dx}\left[x^\alpha J_\alpha(x)\right] = x^\alpha J_{\alpha-1}(x) , \qquad \frac{d}{dx}\left[x^\alpha Y_\alpha(x)\right] = x^\alpha Y_{\alpha-1}(x)$$

and making use of

(4.4) $$\lim_{x\to+0} x^\nu J_\nu(x) = 0 , \qquad \lim_{x\to+0} x^\nu Y_\nu(x) = -\frac{2}{\sin \nu\pi \; \Gamma(1-\nu)} , \qquad \pi > 0 ,$$

where $\sin \nu\pi \; \Gamma(1-\nu)$ has to be replaced by its limiting values as $\nu \to m$ we easily get (see [9])

$$\int_0^\theta f^{\alpha+1}(t)\Delta(t,\theta) df(t) = -\frac{2}{\pi f(\theta)} f^{\alpha+1}(\theta) + 2^{\alpha+1} \frac{\Gamma(\alpha+1)}{\pi} J_\alpha[f(\theta)].$$

Hence, since using the series representation of the Bessel function $J_\alpha(x)$ two terms cancel,

$$(4.5)\quad \int_0^\theta f^{\alpha+1}(t)\Delta(t,\theta)df(t) = 2^{\alpha+1}\,\frac{\Gamma(1+\alpha)}{\pi}\left[\frac{f(\theta)}{2}\right]^\alpha \sum_1^\infty \frac{(-1)^k}{k!\Gamma(\alpha+k+1)}\left[\frac{f(\theta)}{2}\right]^{2k}.$$

The terms of this series are decreasing and alternating in sign and $f(\theta) \leq \pi/2$ for $\theta \leq \theta^*$. Thus

$$0 < -\int_0^\infty f^{\alpha+1}(t)\Delta(t,\theta)df(t) \leq \frac{2}{\pi}\Gamma(1+\alpha)f^\alpha(\theta)\left(\frac{\pi}{4}\right)^2 \frac{1}{\Gamma(\alpha+2)} = \frac{\pi}{8(1+\alpha)}f^\alpha(\theta),$$

that is, at least for $n \geq 5$,

$$(4.6)\quad \left|\int_0^\theta f^{\alpha+1}(t)\Delta(t,\theta)df(t)\right| \leq \frac{\pi}{8(1+\alpha)}\, N^\alpha\theta^\alpha(1.0017)^\alpha .$$

For the constant M, if we suppose again $n \geq 5$ and observe that, when $0 \leq t \leq \theta^*$, $f'(t) \leq 1.00171N$ and $|P_n^{(\alpha,\beta)}(\cos t)| \leq$ $\leq P_n^{(\alpha,\beta)}(1) = \binom{n+\alpha}{n}$, we have

$$(4.7)\quad M \leq \frac{\mu_1(1-4\alpha^2)+\mu_2(1-4\beta^2)}{16N^2}\,(0.50086)^{\alpha+1/2}\binom{n+\alpha}{n}\frac{1}{N^{\alpha+2}}$$

with μ_1 and μ_2 given by (2.9).

By substitution of (4.6) and (4.7) into (4.3) we obtain the following estimate for $|I|$ as $0 < \theta \leq \theta^*$

$$|I| \leq \frac{\theta^\alpha}{N^4}\binom{n+\alpha}{n}\left[0.0081172(1-4\alpha^2)+0.082774(1-4\beta^2)\right].$$

ii) Let us now consider the case $\theta^* \leq \theta \leq \frac{\pi}{2}$. It is convenient to divide the integration interval into the subintervals $[0,\theta^*]$ and $[\theta^*,\theta]$ and to deal separately with the two resulting integrals, that we shall denote by I_1 and I_2, respectively.

For I_1 the analogous of inequality (4.3) holds

$$(4.8)\quad |I_1| \leq M\,\frac{\pi}{2}\left|\int_0^{\theta^*} f^{\alpha+1}(t)f'(t)\Delta(t,\theta)dt\right| .$$

Again consider identities (4.4) and obtain

$$\int_0^{\theta^*}\Delta(t,\theta)f^{\alpha+1}(t)df(t) =$$

$$= \{J_\alpha[f(\theta)]Y_{\alpha+1}[f(\theta^*)]-J_{\alpha+1}[f(\theta^*)]Y_\alpha[f(\theta)]\}f^{\alpha+1}(\theta^*) +$$

$$+ \frac{2^{\alpha+1}}{\pi}\,\Gamma(\alpha+1)J_\alpha[f(\theta)].$$

With the aid of inequality [11,(7.31.5)]

$$(4.9)\quad |J_\alpha(x)| \leq \left(\frac{2}{\pi x}\right)^{1/2}$$

and of the analogous for $|Y_\alpha(x)|$, which can be easily derived, we have

$$\left|\int_0^{\theta^*} \Delta(t,\theta) f^{\alpha+1}(t)\,df(t)\right| \leq$$

$$\leq \left[\frac{2}{\pi f(\theta)}\right]^{1/2} \{[|Y_{\alpha+1}[f(\theta^*)]|+|J_{\alpha+1}[f(\theta^*)]|]f^{\alpha+1}(\theta^*)+\frac{2^{\alpha+1}\Gamma(\alpha+1)}{\pi} \leq$$

$$\leq \left[\frac{2}{\pi f(\theta)}\right]^{1/2} \{(\tfrac{\pi}{2})^{\alpha+1} 2^{1/2}[J^2_{\alpha+1}(\tfrac{\pi}{2})+Y^2_{\alpha+1}(\tfrac{\pi}{2})]^{1/2} + \frac{2^{\alpha+1}\Gamma(\alpha+1)}{\pi}\}$$

from which, since [13, p. 449]

$$J^2_{\alpha+1}(x)+Y^2_{\alpha+1}(x) \leq \frac{2}{\pi x}\left[1+\frac{4(\alpha+1)^2-1}{8x^2}\right]$$

it follows that

$$(4.10)\quad \left|\int_0^{\theta^*} f^{\alpha+1}(t)\delta(t,\theta)\,df(t)\right| \leq \left[\frac{2}{\pi f(\theta)}\right]^{1/2} \frac{2}{\pi}\{(\tfrac{\pi}{2})^{\alpha+1}[2(1+\tfrac{4}{\pi^2})]^{1/2}+$$

$$+ 2^{\alpha}\Gamma(\alpha+1)\} \leq (N\theta)^{-1/2}\, 3.6335\, \frac{2}{\pi}\,.$$

Making use of (4.7) (which is still valid in this case) and (4.10), inequality (4.8) becomes

$$|I_1| \leq \frac{\mu_1 A+\mu_2 B}{16N^2}\binom{n+\alpha}{n}(0.50086)^{\alpha+1/2}N^{-\alpha-2}N^{-1/2}\theta^{-1/2}\, 3.6335$$

i.e.

$$(4.11)\quad |I_1| \leq N^{-\alpha-9/2}\theta^{-1/2}\binom{n+\alpha}{n}[0.023927(1-4\alpha^2)+0.24399(1-4\beta^2)].$$

Next consider the integral I_2 and observe that in this case $\Delta(t,\theta)$ is not necessarily one-signed throughout $[\theta^*,\theta]$. From (2.8) and (4.9) it follows that

$$|I_2| \leq \frac{\pi}{2}\,\frac{4}{\pi}\left[\frac{1}{f(\theta)}\right]^{1/2}\frac{\mu_1 A+\mu_2 B}{16N^2}\,.$$

$$(4.12)\quad \cdot\int_{\theta^*}^{\theta}\left|\frac{1}{f(t)}\right|^{1/2}\left|\frac{f(t)}{f'(t)}\right|^{1/2}\left|(\sin\tfrac{t}{2})^{\alpha+1/2}(\cos\tfrac{t}{2})^{\beta+1/2}P_n^{(\alpha,\beta)}(\cos t)\right|dt$$

We now recall that a result obtained by Baratella [2] implies

$$\left|(\sin\tfrac{\theta}{2})^{\alpha+1/2}(\cos\tfrac{\theta}{2})^{\beta+1/2}P_n^{(\alpha,\beta)}(\cos\theta)\right| \leq \binom{n+\alpha}{n}N^{-\alpha-1/2}2.821,$$

for every $\theta \in (0,\frac{\pi}{2}]$; on the other hand we know that $f(\theta) > N\theta$, $0 < \theta \leq \frac{\pi}{2}$, (see (4.1)), and it is easy to verify that $\frac{f(t)}{f'(t)} < t$, $0 < t \leq \frac{\pi}{2}$. Hence we have

$$|I_2| \leq 2\left[\frac{1}{f(\theta)}\right]^{1/2} \frac{\mu_1 A+\mu_2 B}{16N^2} \binom{n+\alpha}{n} N^{-\alpha-1/2}\, 2.821 \int_{\theta*}^{\theta} \frac{t^{1/2}}{(Nt)^{1/2}}\, dt \quad ,$$

that is

$$(4.13) \qquad |I_2| \leq N^{-\alpha-7/2}\theta^{1/2}\binom{n+\alpha}{n} \left[0.0372(1-4\alpha^2)+0.379(1-4\beta^2)\right].$$

Summing up (4.11) and (4.13), we have, for $\theta* \leq \theta \leq \frac{\pi}{2}$,

$$|I| \leq \theta^{-1/2}N^{-\alpha-9/2}\binom{n+\alpha}{n} \left[0.0240(1-4\alpha^2)+0.244(1-4\beta^2)\right] +$$

$$+ \theta^{1/2} N^{-\alpha-7/2}\binom{n+\alpha}{n}\left[0.0372(1-4\alpha^2)+0.379(1-4\beta^2)\right] ,$$

which, recalling (4.2), can be simplified as follows

$$(4.14) \qquad |I| \leq N^{-\alpha-7/2}\theta^{1/2}\binom{n+\alpha}{n} \left[0.0526(1-4\alpha^2)+0.535(1-4\beta^2)\right] ,$$

$$\theta* \leq \theta \leq \frac{\pi}{2} .$$

Thus the final result can be stated by the following theorem:

Theorem 4.1. Let $-1/2 \leq \alpha,\beta \leq 1/2$ and let $\theta*$ be the root of the transcendental equation $f(\theta) = \pi/2$. Then the following asymptotic representation holds

$$(4.15) \qquad \begin{aligned} &\left[\frac{f(\theta)}{f'(\theta)}\right]^{-1/2} \left(\sin \frac{\theta}{2}\right)^{\alpha+1/2} \left(\cos \frac{\theta}{2}\right)^{\beta+1/2} P_n^{(\alpha,\beta)}(\cos\theta) = \\ &= 2^{-1/2}N^{-\alpha} \frac{\Gamma(n+\alpha+1)}{n!} \left[1+ \frac{1}{32N^2}\left(\frac{A}{3}+B\right)\right]^{-\alpha} J_\alpha[f(\theta)] - I , \end{aligned}$$

where

$$|I| \leq \begin{cases} \theta^{\alpha}N^{-4}\binom{n+\alpha}{n}(0.00812A+0.828B) , & 0 < \theta \leq \theta* , \\ \theta^{1/2}N^{-\alpha-7/2}\binom{n+\alpha}{n}(0.0526A+0.535B) , & \theta* \leq \theta \leq \frac{\pi}{2} , \end{cases}$$

with $A = 1-4\alpha^2$, $B = 1-4\beta^2$.

In the ultraspherical case, $\alpha = \beta$, we have the following corollary:

Corollary 4.1. Let $-1/2 \leq \alpha,\beta \leq 1/2$ and let $\theta*$ be the root of the transcendental equation $f(\theta) = \pi/2$. Then the following asymptotic representation holds:

$$\left[\frac{f(\theta)}{f'(\theta)}\right]^{-1/2} (\sin \theta)^{\alpha+1/2} P_n^{(\alpha,\alpha)}(\cos\theta) =$$

$$= 2^{\alpha}N^{-\alpha} \frac{\Gamma(n+\alpha+1)}{n!} \left[1+ \frac{1}{24N^2}(1-4\alpha^2)\right]^{-\alpha} J_\alpha[f(\theta)] - I ,$$

where

$$|I| \leq \begin{cases} \theta^{\alpha} N^{-4} \binom{n+\alpha}{n} (1-4\alpha^2)\, 0.091 , & 0 < \theta \leq \theta^* , \\ \theta^{1/2} N^{-\alpha-7/2} \binom{n+\alpha}{n} (1-4\alpha^2)\, 0.59 , & \theta^* \leq \theta \leq \frac{\pi}{2} . \end{cases}$$

5. <u>The representation of the zeros</u>. In this section we shall obtain an asymptotic representation of the zeros $\theta_{n,k} \equiv \theta_{n,k}(\alpha,\beta)$, in increasing order, $0 < \theta_{n,1} < \theta_{n,2} < \ldots < \theta_{n,[n/2]} \leq \pi/2$, of $P_n^{(\alpha,\beta)}(\cos\theta)$, $-1/2 \leq \alpha \leq 1/2$ and $-1/2 \leq \beta \leq 1/2$. The bounds for the error term in this representation will also be given.

We first observe that, from (4.15), the zeros of $P_n^{(\alpha,\beta)}(\cos\theta)$ coincide with the zeros of the function

$$(5.1) \qquad K_n(\alpha,\beta;\theta) = J_\alpha[f(\theta)] + E_n(\alpha,\beta)\theta^{1/2}N^{-\alpha-7/2} ,$$

where

$$(5.2) \qquad |E_n(\alpha,\beta)| \leq 2^{1/2} \frac{1}{\Gamma(\alpha+1)} N^\alpha \left[1 + \frac{1}{32N^2}\left(\frac{A}{3}+B\right)\right]^\alpha (c_1 A + c_2 B) ,$$

$$A = 1-4\alpha^2 , \qquad B = 1-4\beta^2 ,$$

and $f(\theta)$ has the previous meaning, i.e.

$$(5.3) \qquad f(\theta) = N\theta + \frac{1}{16N}\left[A\left(\frac{2}{\theta} - \cot\frac{\theta}{2}\right) + B\tan\frac{\theta}{2}\right] .$$

According to Theorem 4.1, being the zeros $\theta_{n,k}$ greater than the abscissa θ^* , the constants c_1 and c_2 in (5.2) are given by

$$(5.4) \qquad c_1 = 0.0526 , \qquad c_2 = 0.535 .$$

We establish now a number of preliminary results.

<u>Lemma 5.1</u>. If $-1/2 \leq \alpha = 1/2$ and $-1/2 \leq \beta = 1/2$, then

$$\frac{j_{\alpha,k}}{N}\left(1-\frac{1}{8N^2}\right) < \theta_{n,k} \leq \frac{j_{\alpha,k}}{N} ,$$

where $j_{\alpha,k}$ is the k-th positive zero of $J_\alpha(x)$. Here the equality sign holds when $\alpha^2 = \beta^2 = 1/4$.

The upper bound is readily derived from the inequality (see [7])

$$j_{\alpha,k}\left[N^2+\frac{1}{4}-\frac{\alpha^2+\beta^2}{2}-\frac{1-4\alpha^2}{2}\right]^{-1/2} < \alpha_{n,k} \leq j_{\alpha,k}\left[N^2+\frac{1-\alpha^2-3\beta^2}{12}\right]^{-1/2}$$

For the lower bound we have

$$\theta_{n,k} > \frac{j_{\alpha,k}}{\sqrt{N^2+1/4}} > \frac{j_{\alpha,k}}{N}\left(1 - \frac{1}{8N^2}\right).$$

Lemma 5.2. Let $-1/2 \leq \alpha \leq 1/2$, $-1/2 \leq \beta \leq 1/2$ and let $f(\theta)$ be the function defined by (5.3). If $\tau_{n,k} \equiv \tau_{n,k}(\alpha,\beta)$ denotes the root lying in $(0,\pi/2)$ of the equation $f(\theta) = j_{\alpha,k}$ then

$$\text{(a)} \quad \tau_{n,k} \leq \theta_{n,k}, \qquad \text{(b)} \quad \tau_{n,k} > \frac{j_{\alpha,k}}{N}\left(1 - \frac{1}{8N^2}\right);$$

that is, $\tau_{n,k}$ and $\theta_{n,k}$ belong to the same interval

$$\frac{j_{\alpha,k}}{N}\left(1 - \frac{1}{8N^2}\right) < \theta \leq \frac{j_{\alpha,k}}{N}. \quad \#$$

The inequality (a) has been proved under more general hypotheses by Gatteschi [9]. For the proof of (b) we observe that $(2\theta^{-1}-\cot\frac{\theta}{2})/(\theta/6)$ and $(\tan\frac{\theta}{2})/(\theta/2)$ are monotonically increasing functions of θ in $0<\theta<\pi$. Then, by using the inequality (a) and the lemma 5.1, we obtain

$$\frac{2}{\tau_{n,k}} - \cot\frac{\tau_{n,k}}{2} < \frac{j_{\alpha,k}}{6N}\,\frac{4\pi^{-1}-1}{\pi/12} = 2\,\frac{j_{\alpha,k}(4-\pi)}{N\pi^2},$$

$$\tan\frac{\tau_{n,k}}{2} < \frac{j_{\alpha,k}}{2N}\,\frac{4}{\pi} = 2\,\frac{j_{\alpha,k}}{N\pi}.$$

Hence

$$\tau_{n,k} = \frac{j_{\alpha,k}}{N} - \frac{1}{16N^2}\left[A\left(\frac{2}{\tau_{n,k}} - \cot\frac{\tau_{n,k}}{2}\right) + B\tan\frac{\tau_{n,k}}{2}\right] >$$

$$> \frac{j_{\alpha,k}}{N} - \frac{1}{8N^3}\,j_{\alpha,k}\left(\frac{4-\pi}{\pi^2} + \frac{1}{\pi}\right) > \frac{j_{\alpha,k}}{N}\left(1-\frac{1}{8N^2}\right).$$

Lemma 5.3. Let $-1/2 \leq \alpha \leq 1/2$. Then we have

$$(5.5) \quad J'_\alpha(x) = -\left(\frac{2}{\pi x}\right)^{1/2}\left\{\left[1-\eta_1\frac{(4\alpha^2+15)(4\alpha^2-1)}{2^5(2x)^2}\right]\sin\left(x-\frac{\alpha\pi}{2}-\frac{\pi}{4}\right) + \right.$$

$$\left. + \eta_2\,\frac{4\alpha^2+3}{8x}\cos\left(x-\frac{\alpha\pi}{2}-\frac{\pi}{4}\right)\right\},$$

with $0 < \eta_1 < 1$ and $0 < \eta_2 < 1$. #

This follows from the formula $J'_\alpha(x) = \frac{1}{2}[J_{\alpha-1}(x) - J_{\alpha+1}(x)]$, and the well-known expansion of Hankel type complete with error bound [13,p.206].

Lemma 5.1 and Lemma 5.2 assure us that if we set $\theta_{n,k} = \tau_{n,k} + \varepsilon$, then $0 < \varepsilon < j_{\alpha,k}/8N^3$.

We observe now that (5.1) with $\theta = \theta_{n,k} = \tau_{n,k} + \varepsilon$ gives

$$(5.6) \qquad \varepsilon\, J'_\alpha[f(\xi)]\Big\{N + \frac{1}{16N}\Big[A\Big(\frac{1}{2\sin^2\frac{\theta}{2}} - \frac{2}{\theta^2}\Big) + \frac{B}{2\cos^2\frac{\theta}{2}}\Big]_{\theta=\xi}\Big\} +$$

$$+ E_n(\alpha,\beta)\theta_{n,k}^{1/2} N^{-\alpha-(7/2)} = 0,$$

where ξ is between $\tau_{n,k}$ and $\theta_{n,k}$.

Since $f(\theta)$ and $f'(\theta)$ are monotonically increasing functions, we have

$$j_{\alpha,k} = f(\tau_{n,k}) < f(\xi) < f(\tau_{n,k} + \frac{j_{\alpha,k}}{8N^3}) \leq$$

$$\leq f(\tau_{n,k}) + \frac{j_{\alpha,k}}{8N^2} + \frac{j_{\alpha,k}}{8N^3}\frac{1}{16N}\Big[A\Big(\frac{1}{2\sin^2\frac{\pi}{4}} - \frac{8}{\pi^2}\Big) + \frac{B}{2\cos^2\frac{\pi}{4}}\Big] \leq$$

$$\leq f(\tau_{n,k}) + \frac{j_{\alpha,k}}{8N^2} + \frac{j_{\alpha,k}}{64N^4}\Big(1 - \frac{4}{\pi^2}\Big).$$

Therefore, with $n \geq 5$,

$$f(\xi) < f(\tau_{n,k}) + \frac{j_{\alpha,k}}{8N^2}\Big[1 + \frac{1}{8.25}\Big(1 - \frac{4}{\pi^2}\Big)\Big],$$

i.e.

$$(5.7) \qquad j_{\alpha,k} < f(\xi) < j_{\alpha,k}\Big(1 + \frac{c_3}{8N^2}\Big), \qquad c_3 = 1 + \frac{1}{200}\Big(1 - \frac{4}{\pi^2}\Big).$$

This last result, together with the inequalities [13, p.490]

$$(5.8) \qquad k\pi - \frac{\pi}{4} + \frac{1}{2}\alpha\pi \leq j_{\alpha,k} \leq k\pi - \frac{\pi}{8} + \frac{1}{4}\alpha\pi, \quad k=1,2,\ldots, \quad -1/2 \leq \alpha \leq 1/2,$$

allows us to obtain a lower bound for $|J'_\alpha[f(\xi)]|$, when $\tau_{n,k} < \xi < \theta_{n,k}$.

Indeed, (5.8) and (5.7) show that

$$k\pi - \frac{\pi}{2} < f(\xi) - \frac{1}{2}\alpha\pi - \frac{\pi}{4} < k\pi - \frac{3}{8}\pi - \frac{1}{4}\alpha\pi + \Big(k\pi - \frac{\pi}{8} + \frac{1}{4}\alpha\pi\Big)\frac{c_3}{8N^2},$$

that is, for $k \leq n/2 \leq N/2$,

$$k\pi - \frac{\pi}{2} < f(\xi) - \frac{1}{2}\alpha\pi - \frac{\pi}{4} < k\pi - \frac{\pi}{2} + \frac{\pi}{8} - \frac{1}{4}\alpha\pi + \frac{c_3\pi}{16N};$$

hence, as it is easily verified, the two trigonometric functions on the right hand side of (5.5) have opposite signs and more precisely, assuming again $n \geq 5$, we find

$$|J'_\alpha[f(\xi)]| > \left[\frac{2}{\pi f(\xi)}\right]^{1/2}\left[\sin(\frac{\pi}{4} - \frac{d_3}{80}\pi) - \frac{4\alpha^2+3}{8f(\xi)}\cos(\frac{\pi}{4} - \frac{c_3}{80}\pi)\right].$$

Applying (5.7) and (5.8) yields

$$[f(\xi)]^{-1/2} > j_{\alpha,k}^{-1/2}\left[1 + \frac{c_3}{8N^2}\right]^{-1/2} \geq j_{\alpha,k}^{-1/2}\left[1 + \frac{c_3}{200}\right]^{-1/2},$$

$$\frac{4\alpha^2+3}{8f(\xi)} < \frac{4\alpha^2+3}{8j_{\alpha,k}} \leq \frac{1}{\pi} .$$

Therefore,

$$(5.9) \qquad |J'[f(\xi)]| > \left[\frac{2}{\pi j_{\alpha,k}}\right]^{1/2} c_4, \qquad c_4 = 0.44383$$

when $n \geq 5$ and $-1/2 \leq \alpha \leq 1/2$.

By using (5.2) and (5.9), and observing that

$$[1 + \frac{1}{32N^2}(\frac{A}{3} + B)]^\alpha \leq \left[1 + \frac{1}{24N^2}\right]^{1/2},$$

we obtain from (5.6) the following preliminary result.

Theorem 5.1. Let $-1/2 \leq \alpha \leq 1/2$, $-1/2 \leq \beta \leq 1/2$ and let $\tau_{n,k} \equiv \tau_{n,k}(\alpha,\beta)$, $k=1,2,\ldots$, denote the roots in increasing order of the equation

$$(5.10) \qquad \begin{array}{l} N\theta + \frac{1}{16N}[A(\frac{2}{\theta} - \cot\frac{\theta}{2}) + B\tan\frac{\theta}{2}] = j_{\alpha,k} \\ 0<\theta\leq\pi/2, \quad N=n+(\alpha+\beta+1)/2, \quad A=1-4\alpha^2, \quad B=1-4\beta^2. \end{array}$$

Then for the zeros $\theta_{n,k}(\alpha,\beta)$ of $P_n^{(\alpha,\beta)}(\cos\theta)$ we have

$$\theta_{n,k}(\alpha,\beta) = \tau_{n,k}(\alpha,\beta) + \varepsilon_n^*(\alpha,\beta)N^{-5}$$

where

$$(5.11) \qquad 0 < \varepsilon_n^*(\alpha,\beta) < \frac{1}{\Gamma(\alpha+1)} j_{\alpha,k}(r_1A+r_2B).$$

Here r_1 and r_2 are certain constants and when $n\geq 5$ we may assume

$$r_1 < 0.211, \qquad r_2 < 2.14. \qquad \#$$

To represent $\theta_{n,k}(\alpha,\beta)$ explicitely in terms of the zeros $j_{\alpha,k}$ of $J_\alpha(x)$, we write the equation (5.10) in the form

$$\theta = g(\theta) = t_{n,k} - \frac{1}{16N^2}\left[A\left(\frac{2}{\theta} - \cot\frac{\theta}{2}\right) + B\tan\frac{\theta}{2}\right]$$

where $t_{n,k} \equiv t_{n,k}(\alpha,\beta) = j_{\alpha,k}/N$.

Then, for some $\bar\theta$ between $\tau_{n,k}$ and $t_{n,k}$,

$$\tau_{n,k}-g(t_{n,k}) = g(\tau_{n,k})-g(t_{n,k}) = (\tau_{n,k}-t_{n,k})g'(\bar\theta) =$$

$$= (t_{n,k}-\tau_{n,k})\,\frac{1}{16N^2}\left[\frac{A}{2}\left(\frac{1}{\sin^2\bar\theta/2} - \frac{4}{\bar\theta^2}\right) + \frac{B}{2\cos^2\bar\theta/2}\right].$$

Replacing $\bar\theta$ by $\pi/2$ and observing that, from Lemma 5.2 , $0 < t_{n,k}-\tau_{n,k} < j_{\alpha,k}/8N^3$, we obtain

$$0 < \tau_{n,k}-g(t_{n,k}) < \frac{j_{\alpha,k}}{128N^5}\left[A\left(1-\frac{8}{\pi^2}\right) + B\right].$$

Thus, we can state the main result of this section.

Theorem 5.2. Under the conditions and with the notations mentioned in Theorem 5.1, we have

$$\theta_{n,k}(\alpha,\beta)=t_{n,k}-\frac{1}{16N^2}\left[A\left(\frac{2}{t_{n,k}} - \cot\frac{t_{n,k}}{2}\right) + B\tan\frac{t_{n,k}}{2}\right]+\varepsilon_n(\alpha,\beta)N^{-5}, \tag{5.12}$$

where

$$0 \leq \varepsilon_n(\alpha,\beta) \leq j_{\alpha,k}\left[A\,0.240 + B2.43\right] \tag{5.13}$$

and $t_{n,k} = j_{\alpha,k}/N$. The equality sign in (5.13) holds if and only if $\alpha^2 = \beta^2 = 1/4$. In the ultraspherical case $\alpha = \beta$, we obtain:

Corollary 5.1. Let $-1/2 < \alpha < 1/2$ and let $\theta_{n,k}(\alpha)$ be the k-th zero of the ultraspherical polynomial $P_n^{(\alpha,\alpha)}(\cos\theta)$. We have

$$\theta_{n,k}(\alpha) = \frac{j_{\alpha,k}}{N} - \frac{1-4\alpha^2}{8N^2}\left(\frac{N}{j_{\alpha,k}} - \cot\frac{j_{\alpha,k}}{N}\right) + \varepsilon_n(\alpha)N^{-5}, \tag{5.14}$$

$$k=1,2,\ldots,[n/2], \qquad N = n+\alpha+1/2 ,$$

with

$$0 \leq \varepsilon_n(\alpha) \leq (1-4\alpha^2)j_{\alpha,k}2.67, \qquad n \geq 5 .$$

Theorem 5.2 gives very sharp numerical results, in particular for the first few zeros of $P_n^{(\alpha,\beta)}(\cos\theta)$, that is for those zeros which generally are difficult to evaluate. For example, by using (5.14), the first zero $\theta_{n,1}$ of the Legendre polynomial $P_n(\cos\theta)$ can be obtained with an error which is positive and less than $6.43(n+1/2)^{-5}$.

Remark. If we are interested only in a qualitative estimate of the error term $\varepsilon_n(\alpha,\beta)N^{-5}$ in (5.12) it is easy to see that the asymptotic representation, as $n \to \infty$,

$$\theta_{n,k}(\alpha,\beta) = t_{n,k} - \frac{1}{16N^2}\left[A\left(\frac{2}{t_{n,k}} - \cot\frac{t_{n,k}}{2}\right) + B\tan\frac{t_{n,k}}{2}\right] + \rho_n(\alpha,\beta;k),$$

with $\rho_n(\alpha,\beta;k) = j_{\alpha,k}O(N^{-5})$, $-1/2 \leq \alpha \leq 1/2$ and $-1/2 \leq \beta \leq 1/2$, holds for all the zeros belonging to the interval $0 < \theta \leq \theta - a$, with fixed a.

This gives another proof of (1.5) obtained by using a general formula due to Olver.

Acknowledgment. We would like to thank Professor R. Wong for careful reading of the manuscript and several helpful remarks.

References

[1] M. Abramowith and I.A. Stegun (eds.), Handbook of Mathematical Functions, Applied Mathematics Series, 55, National Bureau of Standards, Washington, DC , 1964.

[2] P. Baratella, Bounds for the error term in Hilb formula for Jacobi polynomials, Atti Acc. Scienze Torino, Cl.Sci.Fis.Mat.Natur., 120 (1986).

[3] D. Elliot, Uniform asymptotic expansion of the Jacobi polynomials and an associated function, Math. Comp. 25 (1971), 309-314.

[4] C.L.Frenzen and R.Wong, A uniform asymptotic expansion of the Jacobi polynomials with error bounds, Can.J.Math.37(1985),979-1007.

[5] C.L. Frenzen and R. Wong, The Lebesgue constants for Jacobi series, Rend. Semin. Mat. Univ. Politc. Torino, Fa . spec. "Special Functions: Theory and Computation", (1985), 117-148.

[6] L. Gatteschi, Il termine complementare nella formula di Hilb-Szegö ed una nuova valutazione asintotica degli zeri dei polinomi ultrasferici, Ann. Mat. Pura ed Appl., 36 (1954), 143-158.

[7] L. Gatteschi, Una nuova disuguaglianza per gli zeri dei polinomi di Jacobi, Atti Acc. Sci. Torino, Cl. Sci. Fis. Mat. Natur., 103 (1968-69), 259-265.

[8] L. Gatteschi, On the zeros of Jacobi polynomials and Bessel functions, Rend. Semin. Mat. Univ. Politec. Torino, Fasc. spec. "Special Functions: Theory and Computation", (1985), 149-177.

[9] L. Gatteschi, New inequalities for the zeros of Jacobi polynomials, to appear on SIAM J. Math. Anal.

[10] N.W.Mc Lachlan, Bessel functions for engineers, Oxford Univ. Press, 1934.

[11] F.W.Olver, Asymptotics and Special Funcions, Academic Press, inc., New York, 1974.

[12] G. Szegö, Orthogonal Polynomials, Colloquium Publications, Vol. 23, 4-th ed., Amer. Math. Soc., Providence, RI, 1975.

[13] G.N. Watson, A treatise on the theory of the Bessel Functions, 2nd ed., Univ. Press, Cambridge, 1966.

THE DISTRIBUTION OF ZEROS OF THE POLYNOMIAL EIGENFUNCTIONS OF ORDINARY DIFFERENTIAL OPERATORS OF ARBITRARY ORDER.

E. BUENDIA, J.S. DEHESA and F.J. GALVEZ
Departamento de Física Nuclear, Universidad de Granada
E-18071 Granada. Spain.

Abstract

The distribution of zeros of the polynomial eigenfunctions of ordinary differential operators of arbitrary order with polynomial coefficients is calculated via its moments directly in terms of the parameters which characterize the operators. Some results of K.M. Case and the authors are extended. In particular, the restriction for the degree of the polynomial coefficient of the ith-derivative to be not greater than i is relaxed. Applications to the Heine polynomials, the Generalized Hermite polynomials and the Bessel type orthogonal polynomials (which no trivial properties of zeros were known of) are shown.

I. INTRODUCTION

In this paper we shall be concerned with the polynomial solutions of ordinary differential equations (ODE) of arbitrary order $n \geqslant 2$ which have polynomial coefficients. These objects appear in many physical applications, especially for low values of n, and have a close conection with the general Weyl-Titchmarsh theory of higher order differential equations. The study of these polynomials was considered already in the early thirties by H.L. Krall [1,2] but it has become increasingly important in the last few years due to the works of A.M. Krall [3,4,5], L.L. Littlejohn [6,7] and others [8,9,10,11,12].

K.M. Case [13] initiated the analysis of the distribution of zeros of these polynomials with the following two assumptions: (a) the zeros are simple and (b) the coefficients of the ith-derivative is a polynomial of degree not greater than i. He generated sum rules

y_r for the powers of the zeros in a recurrent way and applied them to the orthogonal polynomial satisfying an ODE of second order, i.e. Hermite, Laguerre and Jacobi. Although the found recursion formulas for the sum rules y_r are valid for any value of n and in principle solve the problem of the distribution of zeros of the polynomial in terms of the coefficients of their ODE, in practice they are simple to use only in the case n=2. The reason is that they involve non-typical J-sum rules which are related to the y_r's in a very complicated way. We [14,15] have studied in detail the relation between these two kinds of sum rules, so making usable Case's method for ODE's of arbitrary order. Besides we applied it [14,15] to the orthogonal polynomials satisfying ODEs of fourth and sixth order.

In a second paper Case [16] calculated the sum rules y_r of the zeros of the Lame polynomials which satisfy a ODE of second order in which the assumption (b) is not fulfilled, so indicating the way to extend his method. Here we will generalise this idea to any polynomial satisfying an ODE of arbitrary order in which the assumption (b) is relaxed. Then we will consider polynomials $P_N(x)$ satisfyng an ODE of the form

$$\sum_{i=0}^{n} g_i(x)\, P_N^{(i)}(x) = 0 \qquad (1)$$

where $P_N^{(i)}(x)$ denotes the ith-derivative of $P_N(x)$, and $g_i(x)$ is the polynomial of degree c_i defined by

$$g_i(x) = \sum_{j=0}^{c_i} a_j^{(i)} x^j \qquad (2)$$

We will assume that all zeros of $P_N(x)$ are simple. We will be interested in the normalized-to-unity discrete density function $\rho_N(x)$ of the zeros of the polynomial $P_N(x)$ which will be characterised by means of its moments around the origin, that is

$$\mu'_r = \frac{1}{N}\, y_r = \frac{1}{N} \sum_{i=1}^{N} x_i^r \qquad (3)$$

First of all we will find general basic relations for these moments which allow us to calculate in a recurrent way the quantities y_r with $r \geq q+1$, $q = \max\{c_i - i;\ i=0,1,\dots,n\}$ in terms of the $a_j^{(i)}$-coefficients provided one knows the first q values $y_o=N$, y_1, $y_2,\dots,y_q$. Then one has to have an alternative way to calculate moments; all this is described in Section II. Finally the new method shown in the previous section is applied to the study of the distribution of zeros of the Heine polynomials [17] (Section III), the Generalized Hermite polynomials [17,18] (Section IV) and the recently introduced Bessel type polynomials [19] (Section V). Up to now, no trivial properties of these polynomials were known.

II. GENERALIZED CASE METHOD

Let us write the polynomial $P_N(x)$ in the form

$$P_N(x) = (\text{const.}) \prod_{l=1}^{N} (x - x_l) \tag{4}$$

Inserting this expression into Eq.(1) and (2), dividing by $P_N^{(1)}(x)$ and evaluating the result at a zero x_{l_1} of $P_N(x)$ we obtain

$$\sum_{i=2}^{n} i \sum_{j=0}^{c_i} a_j^{(i)} x_{l_1}^j \sum_{\neq}{}' \frac{1}{(x_{l_1}-x_{l_2})(x_{l_1}-x_{l_3})\dots(x_{l_1}-x_{l_i})} = - \sum_{j=0}^{c_1} a_j^{(1)} x_{l_1}^j$$

where $\sum_{\neq}'$ denotes a sum over all l_j other than l_1, subject to the condition that all l's are different.

Multiplication by x_{l_1} and summation over l_1 lead to the result

$$\sum_{i=2}^{n} i \sum_{j=0}^{c_i} a_j^{(i)} J_{r+j}^{(i)} = - \sum_{j=0}^{c_1} a_j^{(1)} y_{r+j} \quad , r=0,1,\dots \tag{5}$$

where we have used Eq.(3) and the notation

$$J_s^{(i)} = \sum_{\neq} \frac{x_{l_1}^s}{(x_{l_1}-x_{l_2})(x_{l_1}-x_{l_3})\dots(x_{l_1}-x_{l_i})} \tag{6}$$

Here $\sum_{\neq}$ means to sum over all l's subject to none of them being equal. Taking into account the fact that [13] $J_s^{(i)}=0$ for $0 \le s \le i-2$, one can transform Eq.(5) into

$$\sum_{i=2}^{n} i \sum_{m=-1}^{r+c_i-i-1} a_{i+m+1-r}^{(i)} J_{i+m}^{(i)} = - \sum_{j=0}^{c_1} a_j^{(1)} y_{r+j-1} \qquad (7)$$

for $r=1,2,\ldots$. The $J_r^{(i)}$-quantities may be expressed in terms of the sum rules y_t with $t \le r-i+1$. Explicit relations for $i=2$ [13] and $i=3,4$ and 5 [14,15] have been found. Also formulae between the $J_{i+m}^{(i)}$ and y-sum rules for $m=-1,0,+1$ [13] and $m=2$ and 3 [15] are known. A general recurrent way to obtain the J-sum rules in terms of the y-quantities have recently been found [20].

Furthermore, notice that the left-hand side of Eqs.(5) or (7) involves quantities y_s with $0 \le s \le r+q$, and

$$q = \max \left\{ c_i - i;\ i = 2,3,\ldots,n \right\} \qquad (8)$$

and the right-hand side has quantities y_s with $r-1 \le s \le r+c_1-1$. Therefore one can calculate the moments y_t, $q+1 \le t \le r+q$ successively by means of the recursion relations (5) or (7) provided that the moments of lower orders $y_o=N$, $y_1,\ldots,y_q$ are previously known. Briefly, the basic relations (5) or (7) of this method permit us to evaluate the quantities y_s with $s \ge q+1$ recursively in terms of y_o, $y_1,\ldots,y_q$. Then, it is clear that one needs an alternative way to calculate the moments of low order y_t, $t \le q$. Let us describe it. One can write Eq.(4) as

$$P_N(x) = (\text{const}) \sum_{k=0}^{N} (-1)^k \alpha_k\, x^{N-k} \qquad (9)$$

normalised so that $\alpha_o = 1$. The α-coefficients are related to the zeros of $P_N(x)$ by

$$\alpha_k = \frac{1}{k!} \sum_{\neq} x_{l_1} x_{l_2} \cdots\cdots x_{l_k} = \frac{(-1)^k}{k!} Y_k(-y_1,-y_2,-2y_3,\ldots,-(n-1)!y_n) \qquad (10)$$

where the Y_k-symbols denote the well-known Bell polynomials of the number theory [21]. The first few terms are

$$\begin{aligned}
\alpha_1 &= y_1 \\
\alpha_2 &= (y_1^2 - y_2)/2 \\
\alpha_3 &= (y_1^3 - 3y_1y_2 + 2y_3)/6 \\
\alpha_4 &= (y_1^4 - 6y_1^2y_2 + 8y_1y_3 + 3y_2^2 - 6y_4)/24 \\
\alpha_5 &= (y_1^5 - 10y_1^3y_2 + 20y_1^2y_3 + 15y_1y_2^2 - 30y_1y_4 - 20y_2y_3 + 24y_5)/120
\end{aligned} \tag{11}$$

Deriving i times Eq.(9) one has

$$P_N^{(i)}(x) = (\text{const}) \sum_{k=0}^{N-i} (-1)^k \alpha_k \frac{(N-k)!}{(N-k-i)!} x^{N-i-k}$$

and replacing this into the differential equation (1)-(2) one gets the following identity

$$\sum_{i=0}^{n} \sum_{j=0}^{c_i} \sum_{k=0}^{N-i} (-1)^k a_j^{(i)} \alpha_k \frac{(N-k)!}{(N-k-i)!} x^{N+j-k-i} = 0$$

Therefore, it is fulfilled that

$$\sum_{i=0}^{n} \frac{N!}{(N-i)!} a_{i+q}^{(i)} = 0 \tag{12a}$$

$$\sum_{m=0}^{s} (-1)^{s-m} \alpha_{s-m} \sum_{i=0}^{n} \frac{(N-s+m)!}{(N-s+m-i)!} a_{i+q-m}^{(i)} = 0 \; ; \; s > 1 \tag{12b}$$

Eq.(12a) is a necessary condition for a differential equation of n-order to have polynomial solutions since it involves only coefficients which characterize such an equation. Eq.(12b) gives

$$\alpha_s = - \frac{\sum_{m=1}^{s} (-1)^m \alpha_{s-m} \sum_{i=0}^{n} \frac{(N-s+m)!}{(N-s+m-i)!} a_{i+q-m}^{(i)}}{\sum_{i=0}^{n} \frac{(N-s)!}{(N-s-i)!} a_{i+q}^{(i)}} \tag{13}$$

Eqs. (10) and (13) give an alternative way to calculate the quantities y_r in terms of the coefficients $a_j^{(i)}$ of the differential equation(1)-(2). Clearly it is highly non-linear and cumbersome. This is the reason to use it to evaluate only the moments of low order.

In summary the method described in this section to study the distribution of zeros of an ordinary differential equation of arbitrary order (1)-(2) via its moments about the origin has two parts: (i) the first q moments are calculated by means of Eqs.(10) and (13), and (ii) the rest of N-q moments are evaluated recursively by means of Eq.(5) or (7). Notice that for q=0 this method reduces to that of [13] used and extended in [14,15].Also, the application of this method to the differential equation fulfilled by the Lamé polynomials gives the results of [16] as one can easily show. In the next sections we illustrate the applicability of the method to three different non clasical systems of orthogonal polynomials which are relevant in different fields.

III. THE HEINE POLYNOMIALS

These polynomials were introduced by Heine [22] in his treatise on spherical harmonics and are considered in a more precised way by Chihara [17]. Recently [23] they have been used in connection with the theory of non-linear phenomena. They form an orthogonal system in the sense

$$\int_0^{\alpha} P_m(x)\, P_n(x)\, w(x)\, dx = 0 \;, \; m \neq n$$

where $w(x) = \left[x(\alpha - x)(\beta - x)\right]^{-1/2}$, $0 < \alpha < \beta$. Also, they fulfil the differential equation

$$2w^2(x)(x-\gamma)f''(x)+\left\{(x-\gamma)\left[w^2(x)\right]'-2w^2(x)\right\}f'(x)+\left[a+bx-N(2N-1)x^2\right]f(x)=0 \quad (14)$$

which is of the form (1)-(2) with the coefficients

$$a_4^{(2)}=2 \;, \; a_3^{(2)}=-2(\alpha+\beta+\gamma), \; a_2^{(2)}=2(\alpha\beta+\alpha\gamma+\beta\gamma), \; a_1^{(2)}=-2\alpha\beta\gamma \;, \; a_0^{(2)}=0$$

$$a_3^{(1)}= 1 \;, \; a_2^{(1)} = -3\gamma \;, \; a_1^{(1)} = 2(\alpha\gamma+\beta\gamma)-\alpha\beta \;, \; a_0^{(1)} = -\alpha\beta\gamma$$

$$a_2^{(0)}= -N(2N-1) \;, \; a_1^{(0)} = b \;, \; a_0^{(0)} = a$$

According to Eq.(8), we have q=2 in this case. Then, one has to calculate y_1 and y_2 by means of Eqs. (10) or (11) and (13). The latter equation gives

$$\alpha_1 = \frac{1}{4N-3}\left[2(\alpha+\beta+\gamma)N^2 - (2\alpha+2\beta-\gamma)N - b\right]$$

$$\alpha_2 = -\frac{1}{8N-10}\left[\left\{-2(\alpha+\beta+\gamma)N^2 + 3(2\alpha+2\beta-\gamma)N + (b-4\alpha-4\beta-\gamma)\right\}\alpha_1 - \left\{2(\alpha\beta+\alpha\gamma+\beta\gamma)N^2 - 3\alpha\beta N + a\right\}\right]$$

This together with the first two expressions of Eq.(11) lead to

$$y_1 = \frac{1}{4N-3}\left[2(\alpha+\beta+\gamma)N^2 - (2\alpha+2\beta-\gamma)N - b\right]$$
$$y_2 = y_1^2 - \frac{1}{4N-5}\left[2(\alpha+\beta+\gamma)N^2 - 3(2\alpha+2\beta+\gamma)N + 4\alpha+4\beta+\gamma-b\right]y_1 - \frac{1}{4N-5}\left[2(\alpha\beta+\alpha\gamma+\beta\gamma)N^2 - 3\alpha\beta N + a\right] \quad (15)$$

On the other hand, the basic recursion relation (5) gives

$$4\left[J_{r+4}^{(2)} - (\alpha+\beta+\gamma)J_{r+3}^{(2)} + (\alpha\beta+\alpha\gamma+\beta\gamma)J_{r+2}^{(2)} - \alpha\beta\gamma\, J_{r+1}^{(2)}\right]$$
$$= \alpha\beta\gamma\, y_r - (2\alpha\gamma + 2\beta\gamma - \alpha\beta)y_{r+1} + 3\gamma\, y_{r+2} - y_{r+3}$$

for r = 0,1,2,.... Since $J_0^{(2)} = 0$, $J_1^{(2)} = N(N-1)/2$, $J_2^{(2)} = (N-1)y_1$ and

$$J_s^{(2)} = \frac{1}{2}\left[(2N-s)y_{s-1} + \sum_{j=1}^{s-2} y_{s-1-j}\, y_j\right], \quad s \geq 3 \quad (16)$$

this relation immediately leads to the following expressions

$$y_3 = \frac{1}{4N-7}\left\{\left[4(\alpha+\beta+\gamma)N - 3(2\alpha+2\beta+\gamma)\right]y_2 - 4y_1y_2 + 2(\alpha+\beta+\gamma)y_1^2 + 5\alpha\beta + 2\alpha\gamma + 2\beta\gamma - 4(\alpha\beta+\alpha\gamma+\beta\gamma)N\, y_1 + 2\alpha\beta\gamma N^2 - \alpha\beta\gamma N\right\} \quad (17a)$$

$$y_4 = \frac{1}{4N-9}\left\{-4y_3y_1 - 2y_2^2 + \left[4(\alpha+\beta+\gamma)N - 8\alpha - 8\beta - 5\gamma\right]y_3 + 4(\alpha+\beta+\gamma)y_2y_1 + \left[7\alpha\beta + 4\alpha\gamma + 4\beta\gamma - 4(\alpha\beta+\alpha\gamma+\beta\gamma)N\right]y_2 - 2(\alpha\beta+\alpha\gamma+\beta\gamma)y_1^2 + (4\alpha\beta\gamma N - 3\alpha\beta\gamma)y_1\right. \quad (17b)$$

and

$$y_{r+3} = \frac{1}{4N-7-2r} \Bigg\{ \alpha\beta\gamma\, y_r - (2\alpha\gamma + 2\beta\gamma - \alpha\beta) y_{r+1} + 3\gamma y_{r+2}$$

$$-2\left[\sum_{i=1}^{r+2} y_{r+3-i}\, y_i - (\alpha+\beta+\gamma)\left\langle (2N-3-r)y_{r+2} + \sum_{i=1}^{r+1} y_{r+2-i}\, y_i \right\rangle\right.$$

$$+(\alpha\beta + \alpha\gamma + \beta\gamma)\left\langle (2N-2-r)y_{r+1} + \sum_{i=1}^{r} y_{r+1-i}\, y_i \right\rangle$$

$$\left. - \alpha\beta\gamma \left\langle (2N-1-r)y_r + \sum_{i=1}^{r-1} y_{r-i}\, y_i \right\rangle \right] \Bigg\} \tag{17c}$$

valid for $r \geq 2$. By means of these three expressions we can easily evaluate all the quantities y_s, $s > 2$ in terms of the already known values (15) of y_1 and y_2. From Eqs.(15) and (17) one can evaluate in a trivial manner the moments about the origin $\mu'_r = y_r/N$ of the density of zeros of the Heine polynomials of order N which uniquely determine the spectrum.

The asymptotic values (i.e. for $N \to \infty$) of these moments are of the form $\mu'_r = (\text{const.}) + O(N^{-1})$. In particular we found that

$$\mu'_1 = \frac{1}{2}(\alpha+\beta+\gamma) + O(N^{-1})$$

$$\mu'_2 = \frac{1}{2}\left[\frac{3}{2}(\alpha^2+\beta^2+\frac{1}{2}\gamma^2) + 2\alpha\beta + \frac{5}{4}(\alpha\gamma+\beta\gamma)\right] + O(N^{-1})$$

for $N \to \infty$.

IV. THE GENERALIZED HERMITE POLYNOMIALS

These polynomials were introduced by Szego [18] and studied extensively by Chihara [17] and others [24]. They are denoted by $H_N^{(\mu)}(x)$, $\mu = 0$ being a parameter. For $\mu = 0$ they are the well-known classical orthogonal polynomials.

The polynomials $H_N^{(\mu)}(x)$ form an orthogonal system and fulfil the differential equation

$$xf''(x) + 2(\mu - x^2)f'(x) + (2Nx - \theta_N x^{-1})f(x) = 0$$

where $\theta_{2N} = 0$ and $\theta_{2N+1} = 2$. The distribution of zeros of these polynomials is the same as that of the polynomial solution of the equation

$$x^2 f''(x) + 2(\mu x - x^3) f'(x) + (2Nx^2 - \theta_N) f(x) = 0$$

This expression is of the form (1)-(2) with the coefficients

$$a_2^{(2)} = 1, \quad a_1^{(2)} = 0, \quad a_0^{(2)} = 0$$

$$a_3^{(1)} = -2, \quad a_2^{(1)} = 0, \quad a_1^{(1)} = 2\mu, \quad a_0^{(1)} = 0$$

$$a_2^{(0)} = 2N, \quad a_1^{(0)} = 0, \quad a_0^{(0)} = -\theta_N$$

Here, again we have q=2 according to Eq.(8). Therefore we have to calculate y_1 and y_2 before using the recursion relation (5) or (7). Eqs.(10) and (13) give $y_1 = 0$ and

$$y_2 = \frac{1}{2}\left[N(N-1) + 2\mu N - \theta_N\right] \tag{18a}$$

The other quantities y_s, $s \geqslant 3$ can be easily calculated by means of the basic relation (5), which gives $y_3 = 0$ and

$$y_{r+3} = J_{r+2}^{(2)} + \mu\, y_{r+1}, \quad r \geqslant 1$$

Then, taking into account (16) one obtains

$$\begin{aligned} y_{2s+1} &= 0, \quad s \geqslant 1 \\ y_{2s+2} &= (N + \mu - s - 1/2) y_{2s} + \frac{1}{2} \sum_{i=1}^{s-1} y_{2(s-i)} y_{2i}, \quad s \geqslant 1 \end{aligned} \tag{18b}$$

The moments about the origin $\mu_r' = y_r/N$ inmediately follow from these expressions. They determine the density of zeros of the polynomial $H_N^{(\mu)}(x)$. In particular all odd moments μ_{2s+1}' are zero what indicates that the zeros are symmetrically distributed around the origin. This should not be surprising since the generalized Hermite polynomials are known to be odd or even [17]. We should remark that the y-quantities of $H_N^{(\mu)}(x)$ given by Eqs.(18a,b) reduce to the corresponding sum rules of the classical Hermite polynomial [13] when $\mu=0$.

Asymptotically one can easily check that the even moments are given by

$$\mu'_{2r} = \frac{N^r}{2^r(r+1)} \binom{2r}{r} \quad \text{for } N \to \infty \quad \text{and } r=0,1,2,\ldots$$

Therefore the asymptotic density of zeros of the polynomial $H_N(x)$ is the semicircular function

$$\rho_\infty(x\sqrt{N}) = \frac{1}{\pi}\left(2 - x^2\right)^{1/2}, \quad -\sqrt{2} \leq x \leq +\sqrt{2}$$

V. THE BESSEL TYPE POLYNOMIALS

Let us consider the polynomials $R_n(a;z)$ defined by

$$R_n(a;z) = \frac{(-1)^n}{(a+n)_n} y_n(-2z;a+1); \quad n = 0,1,2,\ldots$$

where $a \neq -1,-2,\ldots$ and $y_n(x;b)$ denotes the well-known generalized Bessel polynomials [17,25], i.e.

$$y_n(x;b) = \sum_{k=0}^{n} 2^{-k} \binom{n}{k} (n+b-1)_k \, x^k$$

Here we used the Pochhammer's symbol $(a)_0 = 1$ and $(a)_n = a(a+1)\ldots(a+n-1)$. Then

$$R_n(a,z) = \frac{(-1)^n}{(a+n)_n} \sum_{k=0}^{n} (-n)_k (a+n)_k \frac{z^k}{k!}$$

Recently Hendriksen [19] has introduced a new set of polynomials $Q_n(z;a,\rho)$, $\rho \in \mathbb{C}$, defined by

$$Q_n(z;a,\rho) = R_n(a-1;z) + (-1)^{n-1} \rho \frac{(a)_{n-1}}{(n-1)!} z \, R_{n-1}(a+1;z)$$

with $n = 1, 2, \ldots$. It is called the Bessel type orthogonal polynomial system because as the other "classical type" orthogonal polynomial systems [4,12] they are obtained from the classical ones by adding a linear combination of at most two Dirac delta functions to the real weight function of the corresponding classical system. Indeed, the polynomials $Q_n(z;a,\rho)$ are orthogonal with respect to the distributional weight function $w(z) + \rho\,\delta(z)$ where $\delta(z)$ is the Dirac delta function and $w(z)$ is a real weight function for the generalized Bessel polynomials $R_n(a-1;z)$.

These polynomials fulfil [19] the following second order differential equation

$$(t_n z^3 + r_n z^2)\, Q_n''(z) + \left[a t_n z^2 + (r_n(a+1) - t_n) z - r_n\right] Q_n'(z)$$
$$- \left[n(a+n-1) t_n z + n(a+n-1)(r_n+1)\right] Q_n(z) = 0$$

where

$$r_n = (-1)^{n-1} \frac{n!}{\rho\,(a)_{n-1}}$$

$$t_n = -\left[r_n(a - r_n - 1) + n(a+n-1)\right]$$

This differential equation is of the form (1)-(2) with the coefficients

$$a_3^{(2)} = t_n, \quad a_2^{(2)} = r_n, \quad a_1^{(2)} = 0, \quad a_0^{(2)} = 0$$

$$a_2^{(1)} = a t_n, \quad a_1^{(1)} = r_n(a+1) - t_n, \quad a_0^{(1)} = -r_n$$

$$a_1^{(0)} = -n(a+n-1) t_n, \quad a_0^{(0)} = -n(a+n-1)(r_n+1)$$

It turns out that $q=1$. Since $y_0=N$ for a polynomial $Q_N(x)$, it only remains to calculate y_1 by means of Eqs.(10) and (13) and then to apply the generalized Case relation (5). From Eqs.(10) and (13) one easily has

$$y_1 = \frac{N(t_N - r_N)}{(2N+a-2)t_N} + \frac{N(a+N-1)}{(2N+a-2)t_N}$$

The rest of moments are given by Eq.(5) which in this case reduces as

$$y_2 = \frac{1}{(2N+a-3)}\left[y_1 - y_1^2 - r_N(N+a-1)/t_N\right]$$

and

$$y_{r+2} = \frac{1}{(2N-3-r+a)t_N}\left\{-t_N \sum_{i=1}^{r+1} y_{r+2-i}y_i - r_N \sum_{i=1}^{r} y_{r+1-i}y_i + r_N y_r + (t_N - r_N(2N+a-r-1))y_{r+1}\right\}$$

valid for r = 1,2,.... It is important to notice that for $\rho=0$, these sum rules reduce to the corresponding quantities of the classical Bessel polynomials [25] $y_n(-2z;a)$ as expected.

VI. CONCLUSION

We have shown a method to study the global properties of the spectrum of zeros of a polynomial with simple zeros which satisfies an ordinary differential equation of arbitrary order with polynomial coefficients. In particular the density of zeros is calculated via its moments about the origin directly in terms of the coefficients of the differential equation.

The applicability of the method is illustrated for various sets of polynomials: the Heine polynomials, the Generalized Hermite polynomials and the Bessel type polynomials. The moments of the density of zeros of these polynomials are evaluated in a recursive manner.

ACKNOWLEDGEMENTS

This work has been supported by the CAICYT (Comisión Asesora de Investigación Científica y Técnica), Spain.

REFERENCES

1 H.L. KRALL, Certain Differential Equations for Tchebycheff Polynomials, Duke Math. J. 4(1938),705-718.

2 H.L. KRALL, On Orthogonal Polynomials Satisfying a Certain Fourth Order Differential Equation", Pennsylvania State College Studies, No.6, Pennsylvania St. College, State College. (1940).

3 A.M. KRALL, Tchebycheff Sets of Polynomials which Satisfy an Ordinary Differential Equation, SIAM Rev. 22(1980),436-441

4 A.M. KRALL, Orthogonal Polynomials Satisfying Fourth Order Differential Equations, Proc. Roy. Soc. Edinburgh, A87 (1981),271-288

5 A.M. KRALL, A Review of Orthogonal Polynomials Satisfying Boundary Value Problems, Contribution to this volume.

6 L.L. LITTLEJOHN, On the Classification of Differential Equations Having Orthogonal Polynomial Solutions. I, Annali di Matem. 138 (1984),35-53. II, to appear in Annali di Matem. (1986).

7 L.L. LITTLEJOHN, Orthogonal Polynomials Solution to Ordinary Partial Differential Equations, Contribution to this volume.

8 W. HAHN, On Differential Equation For Orthogonal Polynomials, Funkcialaj Ekvacioj 21(1978),1-9

9 F.V. ATKINSON and W.N. EVERITT, Orthogonal Polynomials Satisfying Second Order Differential Equations, Christoffel Festschrift (1981) Birkhauser Verlag, Basel), p. 173-181

10 R. RASALA, The Rodrigues Formula and Polynomial Differential Operators, J. Math. Anal. Appl. 84(1981),443-482.

11 R. SMITH, An Abundance of Orthogonal Polynomials,IMA J. Appl. Math. 28(1982)161-167.

12 T.H. KOORNWINDER, Orthogonal Polynomials with Weight Function $(1-x)^{\alpha}(1+x)^{\beta} + M\delta(x+1) + N\delta(x-1)$.Canad. Math. Bull. 27 (1984),205-214

13 K.M. CASE, Sum Rules for Zeros of Polynomials. I,J. Math. Phys. 21(1980),702-708

14 J.S. DEHESA, E. BUENDIA and M.A. SANCHEZ BUENDIA, On the Polynomial Solutions of Ordinary Differential Equations of the Fourth Order, J. Math. Phys. 26(1985),1547-1552

15 E. BUENDIA, J.S. DEHESA and M.A. SANCHEZ BUENDIA, On the Zeros of Eigenfunctions of Polynomial Differential Operators,J. Math. Phys. 26(1985),2729-2736.

16 K.M. CASE, Sum Rules for Zeros of Polynomials. II, J. Math. Phys. 21(1980),709-714

17 T.S. CHIHARA,An Introduction to Orthogonal Polynomials, Gordon and Breach, New York, 1977

18 G. SZEGO, Ein Beitrag zur Theorie der Polynome von Laguerre und Jacobi, Math. Zeit. 1(1918),341-356.

19 E. HENDRIKSEN, A Bessel Type Orthogonal Polynomial System, Indagations Math. 46(1984),407-414

20 E. BUENDIA and J.S. DEHESA, A New Class of Sum Rules for Zeros of Polynomials, Preprint of Granada University, 1986.

21 J. RIORDAN, An Introduction to Combinatorial Analysis. Wiley, N. Y., 1958.

22 N.E. HEINE, Handbuch der Kugelfunctionen, I. Reiman, Berlin, 1878.

23 K. AOMOTO, Lax Equation and the Spectral Density of Jacobi Matrices for Orthogonal Polynomials, Nagoya preprint

24 A.M. KRALL, On the Generalized Hermite Polynomials, Indiana Math. J. 30(1981)73-78.

25 F.J. GALVEZ and J.S. DEHESA, Some Open Problems of Generalized Bessel Polynomials, J. Phys. A: Math. Gen.17(1984),2759-2766.

ORTHOGONAL POLYNOMIALS AND JUMP MODIFICATIONS

M.A. Cachafeiro
Departamento de Matemática Aplicada
E.T.S. Ingenieros Industriales
36280 Vigo, España

F. Marcellán
Departamento de Matemática Aplicada
E.T.S. Ingenieros Industriales
28006 Madrid, España

Abstract: Let $\{\hat{P}_n(z,d\alpha)\}$ be a system of orthonormal polynomials on the unit circle U with respect to a measure α and let $\{\hat{P}_n(z,d\beta)\}$ be a system of orthonormal polynomials on U with respect to β ($\beta=\alpha+\varepsilon\delta(t)$, where $t\in U$, $\varepsilon>0$, and $\delta(t)$ the unit measure supported at $z=t$). If $(e_n(\alpha))^{-\frac{1}{2}}$ and $(e_n(\beta))^{-\frac{1}{2}}$ are their leading coefficients, in this paper we present some properties of the difference $e_n(\beta)-e_n(\alpha)$ and we show that both sequences $e_n(\alpha)$ and $e_n(\beta)$ have the same limit by using the recurrence relation.

1. We establish some properties of the difference $e_n(\beta)-e_n(\alpha)$ and the ratio $e_n(\alpha)/e_n(\beta)$.

Proposition 1:

$$K_n(z,y;d\beta)=K_n(z,y;d\alpha)-\frac{\varepsilon K_n(t,y;d\alpha)}{1+\varepsilon K_n(t,t;d\alpha)}K_n(z,t;d\alpha) \tag{1}$$

Proof: $K_n(z,y;d\beta)=\sum_{j=0}^{n}\overline{\alpha_j(y)}\hat{P}_j(z,d\alpha)$ where $\overline{\alpha_j(y)}=\overline{\hat{P}_j(y,d\alpha)-\varepsilon K_n(t,y;d\beta)\cdot}$ $\overline{\cdot\hat{P}_j(t,d\alpha)}$

Taking $z=t$ we obtain

$$K_n(t,y;d\beta)=\frac{K_n(t,y;d\alpha)}{1+\varepsilon K_n(t,t;d\alpha)} \tag{2}$$

and then (1) holds.

Corollary 1:

$$\frac{e_n(\alpha)}{e_n(\beta)} = \frac{1+\varepsilon K_{n-1}(t,t;d\alpha)}{1+\varepsilon K_n(t,t;d\alpha)} \tag{3}$$

Proof: By identifying in (1) the coefficients of $\bar{y}^n$ we get

$$\hat{P}_n(z,d\beta) = \left(\frac{e_n(\beta)}{e_n(\alpha)}\right)^{\frac{1}{2}}\left[\hat{P}_n(z,d\alpha) - \frac{\varepsilon\hat{P}_n(t,d\alpha)K_n(z,t;d\alpha)}{1+\varepsilon K_n(t,t;d\alpha)}\right] \tag{4}$$

The comparison of leading coefficients in (4) shows that $\left(\frac{e_n(\alpha)}{e_n(\beta)}\right)^{\frac{1}{2}} = \left(\frac{e_n(\beta)}{e_n(\alpha)}\right)^{\frac{1}{2}}\left[1- \frac{\varepsilon|\hat{P}_n(t,d\alpha)|^2}{1+\varepsilon K_n(t,t;d\alpha)}\right]$ which proves (3). (It follows from (3) that $e_n(\alpha)<e_n(\beta)$).

Corollary 2:

$$e_n(\beta)-e_n(\alpha) = \varepsilon\tilde{P}_n(t,d\beta)\overline{\tilde{P}_n(t,d\alpha)} \tag{5}$$

Proof: The comparison of the coefficients of $\bar{y}^n$ in (2) shows that

$$\frac{\hat{P}_n(t,d\beta)}{\hat{P}_n(t,d\alpha)} = \frac{(e_n(\beta))^{\frac{1}{2}}}{(e_n(\alpha))^{\frac{1}{2}}(1+\varepsilon K_n(t,t;d\alpha))} \tag{6}$$

After substituting relation (6) in (4) we obtain $\hat{P}_n(z,d\beta) = \left(\frac{e_n(\beta)}{e_n(\alpha)}\right)^{\frac{1}{2}}\hat{P}_n(z,d\alpha)-\varepsilon\hat{P}_n(t,d\beta)K_n(z,t;d\alpha)$ and by comparison of the coefficients of z^n on both sides we obtain (5).

(It follows from (5) that $\tilde{P}_n(t,d\beta)$ and $\tilde{P}_n(t,d\alpha)$ have the same argument).

Corollary 3:

$$\frac{e_n(\beta)-e_n(\alpha)}{4^n} \leqslant \varepsilon$$

Proof: Taking into consideration (5) and the following relations: $|\tilde{P}_n(t,d\alpha)|\leqslant 2|\tilde{P}_{n-1}(t,d\alpha)|$ and $|\tilde{P}_n(t,d\beta)|\leqslant 2|\tilde{P}_{n-1}(t,d\beta)|$, we obtain $e_n(\beta)-e_n(\alpha)<4(e_{n-1}(\beta)-e_{n-1}(\alpha))$, and the result follows.

Corollary 4:

$$e_n(\beta)-e_n(\alpha) = \frac{\varepsilon e_n(\beta)|\hat{P}_n(t,d\alpha)|^2}{1+\varepsilon K_n(t,t;d\alpha)} = \frac{\varepsilon e_n(\alpha)|\hat{P}_n(t,d\alpha)|^2}{1+\varepsilon K_{n-1}(t,t;d\alpha)}$$

Proof: Writing (6) for monic polynomials we get

$$\tilde{P}_n(t,d\beta) = \frac{e_n(\beta)\tilde{P}_n(t,d\alpha)}{e_n(\alpha)(1+\varepsilon K_n(t,t;d\alpha))} \tag{7}$$

Substituting (7) in (5) we obtain the first equality, and the second follows from (3).

Corollary 5:

$$\frac{e_n(\beta)}{e_n(\alpha)} = \frac{\tilde{P}_n(t,d\beta)\tilde{P}_{n+1}(t,d\alpha)}{\tilde{P}_{n+1}(t,d\beta)\tilde{P}_n(t,d\alpha)}$$

Proof: From (7)

$$\frac{\tilde{P}_n(t,d\beta)\tilde{P}_{n+1}(t,d\alpha)}{\tilde{P}_n(t,d\alpha)\tilde{P}_{n+1}(t,d\beta)} = \frac{e_n(\beta)e_{n+1}(\alpha)(1+\varepsilon K_{n+1}(t,t;d\alpha))}{e_n(\alpha)e_{n+1}(\beta)(1+\varepsilon K_n(t,t;d\alpha))}$$ and using (3) we obtain the equality.

2. We shall prove in several steps that the sequences $\{e_n(\alpha)\}$ and $\{e_n(\beta)\}$ have the same limit.

Lemma 1:

$$\frac{|\tilde{P}_n(t,d\alpha)|^2}{K_n(t,t;d\alpha)} \leqslant e_n(\alpha)$$

Proof:

$$\frac{|\tilde{P}_n(t,d\alpha)|^2}{K_n(t,t;d\alpha)} = \frac{|\tilde{P}_n^*(t,d\alpha)|^2}{K_n(t,t;d\alpha)} = \frac{|K_n(t,0;d\alpha)|^2}{|K_n(0,0;d\alpha)|^2 K_n(t,t;d\alpha)} =$$

$$= \frac{e_n(\alpha)|K_n(t,0;d\alpha)|^2}{K_n(0,0;d\alpha)K_n(t,t;d\alpha)} \leqslant e_n(\alpha)$$

Lemma 2:

$$0 < \frac{A|\tilde{P}_n(t,d\alpha)|^2}{K_n(t,t;d\alpha)} \leqslant e_n(\beta)-e_n(\alpha) \leqslant \frac{2|\tilde{P}_{n-1}(t,d\alpha)|^2}{K_{n-1}(t,t;d\alpha)}$$ with $A =$

$$= \frac{\varepsilon|\hat{P}_0(t,d\alpha)|^2}{1+\varepsilon|\hat{P}_0(t,d\alpha)|^2}$$

Proof:

$$e_n(\beta)-e_n(\alpha) = \frac{\varepsilon|\tilde{P}_n(t,d\alpha)|^2}{1+\varepsilon K_{n-1}(t,t;d\alpha)} \leq \frac{2\varepsilon|\tilde{P}_{n-1}(t,d\alpha)|^2}{1+\varepsilon K_{n-1}(t,t;d\alpha)} \leq \frac{2|\tilde{P}_{n-1}(t,d\alpha)|^2}{K_{n-1}(t,t;d\alpha)}$$

On the other hand:

$$e_n(\beta)-e_n(\alpha) \geq \frac{\varepsilon|\tilde{P}_n(t,d\alpha)|^2}{1+\varepsilon K_n(t,t;d\alpha)} = \frac{|\tilde{P}_n(t,d\alpha)|^2}{K_n(t,t;d\alpha)} \frac{1}{[1+\frac{1}{\varepsilon K_n(t,t;d\alpha)}]} \geq$$

$$\geq \frac{|\tilde{P}_n(t,d\alpha)|^2}{K_n(t,t;d\alpha)} A$$

Lemma 3:

$$\text{If } \lim_{n\to\infty} e_n(\alpha)>0 \text{ then } \lim_{n\to\infty} \left|\frac{\hat{P}_n(t,d\alpha)}{\hat{P}_{n-1}(t,d\alpha)}\right| = 1$$

Proof: Assume that $\lim_{n\to\infty} e_n(\alpha) = e(\alpha)>0$. Since $e_n(\alpha) = \frac{1}{K_n(0,0;d\alpha)}$, we have $\lim_{n\to\infty} K_n(0,0;d\alpha) = \frac{1}{e(\alpha)} > 0$ and $\lim_{n\to\infty} |\tilde{P}_n(0,d\alpha)|^2 = \lim_{n\to\infty} K_{n-1}(0,0;d\alpha)\cdot$ $\cdot(e_{n-1}(\alpha)-e_n(\alpha)) = 0$; hence $\lim_{n\to\infty} \frac{\tilde{P}_n(z,d\alpha)}{z\tilde{P}_{n-1}(z,d\alpha)} = 1$ uniformly for $|z|\geq 1$ [see 4]. In particular $\lim_{n\to\infty} \frac{\tilde{P}_n(t,d\alpha)}{t\tilde{P}_{n-1}(t,d\alpha)} = 1$, which implies

$$\lim_{n\to\infty} \left|\frac{\tilde{P}_n(t,d\alpha)}{\tilde{P}_{n-1}(t,d\alpha)}\right| = 1.$$

As $\left|\frac{\hat{P}_n(t,d\alpha)}{\hat{P}_{n-1}(t,d\alpha)}\right| = \left|\left(\frac{e_{n-1}(\alpha)}{e_n(\alpha)}\right)^{\frac{1}{2}} \frac{\tilde{P}_n(t,d\alpha)}{\tilde{P}_{n-1}(t,d\alpha)}\right|$, then $\lim_{n\to\infty} \left|\frac{\hat{P}_n(t,d\alpha)}{\hat{P}_{n-1}(t,d\alpha)}\right| = 1.$

Proposition 2:

$$\lim_{n\to\infty} e_n(\alpha) = \lim_{n\to\infty} e_n(\beta)$$

Proof: If $\lim_{n\to\infty} e_n(\alpha) = 0$ (case D), it follows from lemma 1 that $\lim_{n\to\infty} \frac{|\tilde{P}_n(t,d\alpha)|^2}{K_n(t,t;d\alpha)} = 0$. Applying lemma 2 we obtain $\lim_{n\to\infty}(e_n(\beta)-e_n(\alpha)) = 0$, which completes the proof.

Assume that $\lim_{n\to\infty} e_n(\alpha) = e(\alpha) > 0$ (case C) and take into consideration that:

$$\frac{e_n(\beta)}{e_n(\alpha)} = 1+\frac{\varepsilon|\hat{P}_n(t,d\alpha)|^2}{1+\varepsilon K_{n-1}(t,t;d\alpha)} = 1+\left|\frac{\hat{P}_n(t,d\alpha)}{\hat{P}_{n-1}(t,d\alpha)}\right|^2 \frac{\varepsilon|\hat{P}_{n-1}(t,d\alpha)|^2}{1+\varepsilon K_{n-1}(t,t;d\alpha)} =$$

$$= 1+\left|\frac{\hat{P}_n(t,d\alpha)}{\hat{P}_{n-1}(t,d\alpha)}\right|^2 \left(1-\frac{1+\varepsilon K_{n-2}(t,t;d\alpha)}{1+\varepsilon K_{n-1}(t,t;d\alpha)}\right) = 1+\left|\frac{\hat{P}_n(t,d\alpha)}{\hat{P}_{n-1}(t,d\alpha)}\right|^2 \left(1-\frac{e_{n-1}(\alpha)}{e_{n-1}(\beta)}\right)$$

From Lemma 3, it is clear that $\lim_{n\to\infty} e_n(\beta) = e(\beta) > 0$; hence

$$\frac{e(\beta)}{e(\alpha)} = 2+\frac{e(\alpha)}{e(\beta)} \Longrightarrow e(\alpha) = e(\beta)$$

Corollary 6:

$$\lim_{n\to\infty} e_n(\alpha) \frac{|\hat{P}_n(t,d\alpha)|^2}{K_n(t,t;d\alpha)} = 0$$

Proof: It follows inmediately from Lemma 2 and Proposition 2.

Remark In case C it follows from Corollary 6: $\lim_{n\to\infty} \frac{|\hat{P}_n(t,d\alpha)|^2}{K_n(t,t;d\alpha)} = 0$,

but in case D we can only conclude:

$$\frac{|\hat{P}_n(t,d\alpha)|^2}{K_n(t,t;d\alpha)} = o(K_n(0,0;d\alpha))$$

References:

[1] CACHAFEIRO, M.A.; MARCELLAN, F. "Polinomios ortogonales y medidas singulares sobre curvas". Comunicación a las XI Jornadas Hispano-Lusas de Matemáticas. Badajoz, 1.986.

[2] FREUD, G. "Orthogonal Polynomials". Pergamon Press. New York, 1.971.

[3] GERONIMUS, L. "Orthogonal Polynomials". Consultants Bureau. New York, 1.961.

[4] MATE, A.; NEVAI, P. "Remarks on E. A. Rahmanov's Paper: "On the asymptotics of the ratio of Orthogonal Polynomials"". Journal of Approximation Theory. 36 (1.982) pp. 64-72.

[5] RAHMANOV, E.A. "On the asymptotics of the ratio of Orthogonal Polynomials II". Math. U.S.S.R. Sb. 46 (1.983) pp. 105-117.

ELECTROSTATIC INTERPRETATION OF ZEROS

E. Hendriksen and H. van Rossum
University of Amsterdam. The Netherlands

1. INTRODUCTION

Stieltjes was the first to give an electrostatic interpretation of the zeros of the Jacobi polynomials [4]. In Szegö's account of this work, ([5], p. 140-142) also the cases of Laguerre- and Hermite polynomials are treated. The work of Stieltjes, who also gave an interesting treatment of the polynomial solutions of the Lamé equation (see [3], p. 73-88), was confined to the case of the real line. In the present paper we consider point charge distributions in the complex plane. The electric potential is assumed to be logarithmic.

Our definitions for the classical orthogonal polynomials differ slightly from those given in Szegö's book [5]. For convenience we will list here the polynomials involved, their differential equations and the confluences connecting the different types of polynomials.

1. Jacobi polynomials

$$J_n(a,c;x) = (-1)^n \frac{(a+1)_n}{(c+1)_n} {}_2F_1(-n,c+n;a+1;x) \qquad (c > a > -1).$$

(1.1) $$x(1-x)y'' + [a+1-(c+1)x]y' + n(c+n)y = 0 \quad , \quad y = J_n(a,c;x)$$

2. Laguerre polynomials

$$L_n^{(a)}(x) = (-1)^n (a+1)_n \, {}_1F_1(-n,a+1;x) \qquad (a > -1) .$$

(1.2) $$xy'' + (a+1-x)y' + ny = 0 \quad , \quad y = L_n^{(a)}(x) .$$

3. Bessel polynomials

$$R_n(c,x) = \frac{(-1)^n}{(c+n)_n} {}_2F_0(-n,c+n;x) \qquad (-c \notin \{0,1,\ldots\}).$$

(1.3) $$x^2y'' + [(c+1)x-1]y' - n(c+n)y = 0 \quad , \quad y = R_n(c;x) .$$

4. Generalized ultrasferical polynomials $P_n(a,c;x)$.

$$P_{2n}(a,c;x) = J_n(a,c;x^2) \; ; \quad P_{2n+1}(a,c;x) = xJ_n(a+1,c+1;x^2) \; .$$

(1.4a) $$x(1-x^2)y'' + [2a+1-(2c+1)x^2]y' + 2n(2c+2n)xy = 0 \, , \quad y = P_{2n}(a,c;x).$$

(1.4b) $$x^2(1-x^2)y'' + [2a+1-(2c+1)x^2]xy' + [(2n+1)(2c+2n+1)x^2 - (2a+1)]y = 0 \quad , \quad y = P_{2n+1}(a,c;x) \; .$$

If $a = -\frac{1}{2}$ we have the polynomials usually called ultraspherical - or Gegenbauer polynomials.

5. <u>Generalized Hermite polynomials</u> $H_n^{(a)}(x)$ $(a > -1)$

$$H_{2n}^{(a)}(x) = L_n^{(a)}(x^2) \quad , \quad H_{2n+1}^{(a)}(x) = xL_n^{(a+1)}(x^2) \; .$$

(1.5) $$x^2y''+(2a+1-2x^2)xy'+(2nx^2-\rho)y = 0 \quad , \quad y = H_n^{(a)}(x) \quad ,$$

$$\rho = \begin{cases} 0, n & \text{even} \\ 2a+1, n & \text{odd} \end{cases}$$

If $a = -\frac{1}{2}$ we have the ordinary Hermite polynomials.

6. <u>Lacunary Bessel polynomials</u> $S_n(c;x)$ $(-c \notin \{0,1,\ldots\})$.

$$S_{2n}(c;x) = \frac{(-1)^n}{(c+n)_n} \; {}_2F_0(-n,c+n;x^2) \quad ;$$

$$S_{2n+1}(x) = \frac{(-1)^n x}{(c+n+1)_n} \; {}_2F_0(-n,c+n+1;x^2) \; .$$

(1.6) $$x^4y''+x[(2c+1)x^2-2]y' + [\rho-n(n+2c)x^2]y = 0 \quad ,$$

$$y = S_n(c;x) \; , \qquad \rho = \begin{cases} 0, n & \text{even} \\ 2, n & \text{odd} \end{cases}$$

<u>Confluence relations</u>

$$\lim_{c\to\infty} c^n P_{2n}(a,c;c^{-1/2}x) = \lim_{c\to\infty} c^n J_n(a,c;c^{-1}x^2) = L_n^{(a)}(x^2) = H_{2n}^{(a)}(x) .$$

$$\lim_{c\to\infty} c^{n+\frac{1}{2}} P_{2n+1}(a,c;c^{-1/2}x) = \lim_{c\to\infty} c^n xJ_n(a+1,c+1;c^{-1}x^2) =$$

$$= xL_n^{(a+1)}(x^2) = H_{2n+1}^{(a)}(x) \; .$$

$$\lim_{a\to\infty} a^{-n}P_{2n}(a,c;a^{1/2}x) = \lim_{a\to\infty} a^{-n}J_n(a,c;ax^2) =$$

$$= R_n(c;x^2) = S_{2n}(c;x).$$

$$\lim_{a\to\infty} a^{-n-1/2}P_{2n+1}(a,c;a^{1/2}x) = \lim_{a\to\infty} a^{-n}xJ_n(a+1,c+1;ax^2) =$$

$$= xR_n(c+1;x^2) = S_{2n+1}(c;x) \quad .$$

R. Smith [1,2] considered lacunary orthogonal polynomials P satisfying differential equations of two types:

(1.7) $$(1-x^m)xP''-\gamma k-1+\delta x^m)P'+n(n+\delta-1)x^{m-1}P = 0 \quad ,$$

k and m positive integers $k < m$, $\delta > 1-k$.

(1.8) $$xP'' - (k-1+mx^m)P'+mnx^{m-1}P = 0 \ ,$$

conditions on k, m and δ as above. If $m = 1$ it is not necessary for k to be a positive integer and the solutions to the two classes of equations are Jacobi and Laguerre polynomials (see Smith [2] , loc. cit.). For all polynomials mentioned above, we give electrostatic interpretations of their zeros and of the confluences connecting the different classes of polynomials.

2. ELECTROSTATIC INTERPRETATION OF THE ZEROS OF THE BESSEL POLYNOMIAL

$R_n(c;x) \quad (-c \notin \{0,1,\dots\})$,

Let electric point charges p , of strength $(a+1)/2$ and q , of strengh $(c-a)/2$ be located at the points 0 and a^{-1} $(a > 0)$ of the real axis. Assuming a logarithmic potential the field force at $z \in \mathbb{C}$ is :

$$F(x) = \frac{a+1}{2\bar{z}} + \frac{c-a}{2(\bar{z}-a^{-1})} = \frac{1}{2}\ \frac{(c+1)\bar{z}-(1+a^{-1})}{\bar{z}(\bar{z}-a^{-1})} \quad .$$

If $a \longrightarrow \infty$, the point charges become $+\infty$ and $-\infty$ in strength respectively. Their sum remains constant $(c+1)/2$. In the limit we obtain a "generalized dipole" at 0 . The field force $G(z)$ at z is given by

$$G(z) = \frac{(c+1)\bar{z}-1}{2\bar{z}^2} \quad .$$

We now state:

"If n positive free unit charges at the points z_k $(k=1,2,\dots,n)$ are in electrostatic equilibrium in the field $G(z)$, then z_k $(k=1,2,\dots,n)$ are the zeros of the Bessel polynomial $R_n(c;x)$".

We proceed as follows: The conditions for equilibrium are

$$(2.1)\qquad \frac{1}{2}\,\frac{(c+1)\bar{z}_k-1}{\bar{z}_k^2}+\sum_{\substack{j=1\\ j\neq k}}^{n}\frac{1}{\bar{z}_k-\bar{z}_j}=0 \qquad (k=1,2,\dots,n)$$

Putting $y_n(z)=\prod_{j=1}^{n}(z-z_j)$ we can write for (2.1):

$$\frac{1}{2}\,\frac{(c+1)\bar{z}_k-1}{\bar{z}_k^2}+\frac{1}{2}\,\frac{y_n''(\bar{z}_k)}{y_n'(\bar{z}_k)}=0 \qquad (k=1,2,\dots,n)$$

Due to symmetry, bars can be omitted, so we have

$$z_k^2y_n''(z_k)+[(c+1)z_k-1]y_n'(z_k)=0 \qquad (k=1,2,\dots,n)$$

This means: $z^2y_n''(z)+[(c+1)z-1]y_n'(z)$ is a polynomial in z of degree n, divisible by $y_n(z)$, whence for some constant λ_n

$$z^2y_n''(z)+[(c+1)z-1]y_n'(z)-\lambda_ny_n(z)=0\ .$$

Equating to zero the coefficent of z^n yields $\lambda_n=n(c+n)$, so

$$z^2y_n''(z)+[(c+1)z-1]y_n'(z)-n(c+n)y_n(z)=0 \qquad (-c\notin\{0,1,\dots,n\}).$$

Since this differential equation has essentially only one polynomial solution (compare (1.3)) our statement is proved.

Remark 2.1. The electrostatic interpretation of the confluence $a\to\infty$ is, that a generalized dipole at 0 is created and the free positive unit charges, confined to the real line if $c>a>-1$, are flung into the complex plane to take up positions at the zeros of the Bessel polynomial $R_n(c;x)$.

3. CHARGE DISTRIBUTION ON AN m-STAR

An m-star S_m is defined as a collection of rays in the complex plane as follows:

$$S_m = \{z \in \mathbb{C} \mid z = \rho e^{\frac{2\pi k}{m}i} \quad , \quad 0 \leq \rho \leq r \; ; \; k = 0,1,\ldots,m-1\} \; .$$

We pose the following problem:

Positive point charges all of strength q are placed at the endpoints $\rho = r$ of S_m and a non-negative charge of strength p is at the origin. We try to find the positions $z_1, z_2, \ldots, z_n$ of n positive free unit charges $(n > m)$ in electrostatic equilibrium. Assume a logarithmic potential. Only solutions with the rotational symmetry of the regular m-gon are considered.

Solution: Let $\nu \in \{1,2,..,n\}$. Put $\zeta\nu = r \exp(2\pi\nu i/m)$. Equating to zero the resultant force on the positive unit charge at z_ν yields

$$(3.1) \quad \begin{cases} q\left(\frac{1}{\bar z_\nu - \bar\zeta_1} + \ldots + \frac{1}{\bar z_\nu - \bar\zeta_m}\right) + \frac{p}{\bar z_\nu} \\ + \frac{1}{\bar z_\nu - \bar z_1} + \frac{1}{\bar z_\nu - \bar z_2} + \ldots + \frac{1}{\bar z_\nu - \bar z_{\nu-1}} + \frac{1}{\bar z_\nu - \bar z_{\nu+1}} + \ldots + \frac{1}{\bar z_\nu - \bar z_n} = 0 \; . \end{cases}$$

Summing between the brackets and putting $f_n(z) = \prod_{\nu=1}^{n} (z-z_\nu)$, we can for (3.1)

$$(3.2) \qquad \frac{qm\bar z_\nu^{m-1}}{\bar z_\nu^m - r^m} + \frac{p}{\bar z_\nu} + \frac{f''_n(\bar z_\nu)}{2f'_n(\bar z_\nu)} = 0 \qquad (\nu = 1,2,\ldots,n)$$

If $\bar z_\nu$ is a zero of $f_n(z)$, then also z_ν is . From (3.2) it follows that

$$(z^m - r^m)zf''_n(z) + 2\left[(p+qm)z^m - pr^m\right]f'_n(z)$$

is a polynomial in z of degree $\leq m+n-1$, with zeros $z_1, \ldots, z_n$. So this polynomial is divisible by $f_n(z)$ and we have

$$(3.3) \qquad (r^m - z^m)zf''_n(z) - 2\left[(p+qm)z^m - pr^m\right]f'_n(z) + \\ - (c_o^{(n)} + c_1^{(n)}z + \ldots + c_{m-1}^{(n)}z^{m-1})f_n(z) = 0$$

The rotational symmetry entails: z_ν is a zero of $f_n(z) \Rightarrow z_\nu \exp(2\pi ki/m)$ is a zero of $f_n(z)$ $(k=0,1,\ldots,m-1 \; ; \; \nu=1,\ldots,n)$. So we have identically in z :

$$f_n(z) \equiv z^n + a_{n-1}z^{n-1} + \ldots + a_1 z + a_o$$

$$\equiv z^n + a_{n-1} z^{n-1} e^{\frac{2\pi k}{m} i} + \ldots + a_1 z e^{\frac{2\pi k(n-1)}{m} i} + a_o e^{\frac{2\pi k i n}{m}}$$

Assuming no free charge at 0, we must have $n \equiv 0 (\mathrm{mod}\ m)$ i.e. $a_j = 0$ if $j \not\equiv 0 (\mathrm{mod}\ m)$. Hence $f_n(z)$ has the form $z^n + a_{n-m} z^{n-m} + \ldots + a_o$ $(a_o \neq 0)$. Equating the coefficients of z^{m+n-1} in (3.3) yields

$$c_{m-1}^{(n)} = -n(n-1+2p+2qm) .$$

Moreover $c_o^{(n)} = c_1^{(n)} = \ldots = c_{m-2}^{(n)} = 0$ as follows from the structure of f_n. So (3.3) becomes

$$(3.4) \qquad (r^m - z^m) z f_n''(z) = 2\left[(p+qm) z^m - pr^m\right] f_n'(z) + n(n-1+2p+2qm) z^{m-1} f_n(z) = 0 .$$

Hence a solution to our problem is : Place the n positive free unit charges at the zeros of the n-th degree polynomial solution of (3.4).

<u>Remark 3.1</u>. Equation (3.4) has essentially only one polynomial solution. To see this, use the tranformation $w = z^m$ (we may assume $r = 1$). Then (3.4) assumes the form of (1.1), where

$$a = \frac{-1+2p}{m} > -1 \qquad (p \geq 0 , q > 0 , m > 1) .$$

$$c = \frac{-1+2p+2qm}{m} , \qquad c-a = 2q > 0 .$$

So the condition $c > a > -1$ is fullfilled, hence there is essentially one polynomial solution of (3.4).

<u>Special cases</u>

1) Let $p = 0$, $r = 1$, then (3.4) becomes

$$(1-z^m) f_n''(z) - 2qmz^{m-1} f_n'(z) + n(n+2qm-1) z^{m-2} f_n(z) = 0 .$$

This is the differential equation for an orthogonal polynomial belonging to a new class of orthogonal polynomials $P_n(z)$ introduced by R. Smith [1,2] in 1981. Smith's polynomials satisfy the differential equation

$$(1-z^m) z P_n''(z) - (k-1+\delta z^m) P_n'(z) + n(n+\delta-1) z^{m-1} P_n(z) = 0 ,$$

$(0 < k < m$, $\delta > 1-k)$. If $m > 2$ and k and m are mutually prime, each specification (m,k,δ) gives a new class of orthogonal polynomials (Smith [1,2], loc. cit.). Our case corresponds to an $(m,1,2qm)$ specification.

2) $r = 1$, $m = 1$, $z = x \in \mathbb{R}$. Then (3.4) refers to the "Jacobi case" treated first by Stieltjes [4] (in the case of the real interval $[-1,1]$).

3) $r = 1$, $m = 2$, $z = x \in \mathbb{R}$, $p = (2a+1)/2$, $q = (c-a)/2$, $c > a > -1/2$. Then (3.4) takes the form (1.4a). More specifically:

If the points -1 and $+1$ of the real axis, positive charges of strengths $(c-a)/2$ are placed and a positive charge of strength $(2a+1)/2$ at the origin, then $2n$ positive unit charges are at equilibrium at the zeros of the generalized ultraspherical polynomial $P_{2n}(a,c;x)$.

Assuming the unphysical situation that one of the free positive unit charges is placed at 0, we see this amounts to replacing a by $a' = a+1$. In order to keep the charges at the endpoints equal to $(c-a)/2$, we must replace c by $c' = c+1$. So in the new situation we have (see Fig. 1):

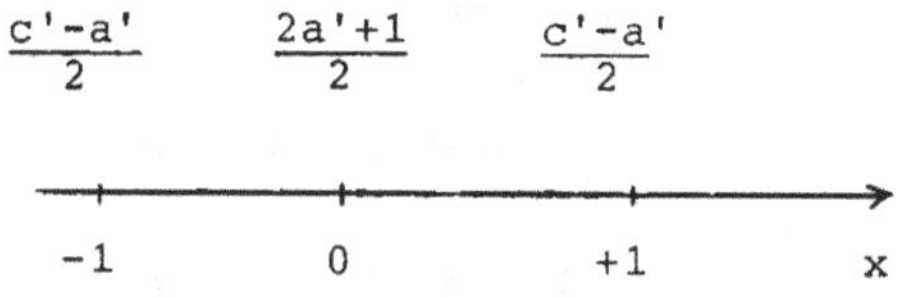

$a' = a+1$

Fig. 1

$c' = c+1$

In the preceding case we had: The $2n$ equilibrium positions were the zeros of $J_n(a,c:x^2)$. In the present case the equilibrium positions of the $2n+1$ unit charges are the zeros of $xJ_n(a+1,c+1,x^2) = P_{2n+1}(a,c;x)$. This can be derived as follows. Rewrite equation (1.4a) for the situation in Fig. 1, i.e.,

$$(3.5) \qquad x(1-x^2)y'' + [2a'+1-(2c'+1)x^2]y' + 2n(2n+2c')xy = 0 .$$

Put $u = xy$, where y is the (essentially unique) polynomial solution of (3.5). A simple calculation leads to the following differential equation for u:

$$x^2(1-x^2)u'' + [2a'-1-(2c'-1)x^2]xu' +$$
$$+ [-2a'+1+(2n+1)(2n-1+2c')]u = 0$$

With $a' = a+1$, $c' = c+1$, this leads to

$$x^2(1-x^2)u'' + [2a+1-(2c+1)x^2]xu' +$$
$$+ [-2a-1+(2n+1)(2c+2n+1)x^2]u = 0 .$$

with polynomial solution $P_{2n+1}(a,c;x)$ (compare 1.4b)).

CONFLUENCES. We consider two types, A and B .

A. At the centre of the m-star S_m , considered in section 3 , we place the charge $p = (1+ma)/2$ and on the rays the charges $q = (c-a)/2$ at $a^{-1/m}\omega_k$ $(k=1,\ldots,m)$ where $a > 0$ and $\omega_1,\ldots,\omega_m$ are the roots of $z^m = 1$. For the field force $F(z)$ we have

$$F(z) = \frac{1+ma}{2} \cdot \frac{1}{\bar{z}} + \frac{c-a}{2} \sum_{k=1}^{m} \frac{1}{\bar{z}-a^{-1/m}\omega_k} = \frac{1}{2} \frac{(1+mc)\bar{z}^m-m-a^{-1}}{\bar{z}^{m+1}-a^{-1}\bar{z}} .$$

We let $a \to \infty$. Then a "generalized (m+1)-pole" at 0 is created. For the field force $G(z)$ we find

$$G(z) = \frac{1}{2} \frac{(1+mc)\bar{z}^m-m}{\bar{z}^{m+1}} . \tag{4.1}$$

If we restrict ourselves to solutions with the rotational symmetry of the regular m-gon, we can find the equilibrium positions $z_1,\ldots,z_n$ of n positive unit charges in the field in (4.1) by deriving a differential equation for $y_n(z) = \prod_{k=1}^{n} (z-z_k)$. We find

$$z^{m+1}y_n''(z) + [(1+mc)z^m-m]y_n'(z) - n(n+mc)z^{m-1}y_n(z) = 0 \tag{4.2}$$

Special cases

1) $m = 1$, then (4.2) yields the Bessel polynomials.
2) $m = 2$, then (4.2) yields the lacunary Bessel polynomials of even degree i.e. $S_{2n}(c;x)$.

Remark 4.1. The result in the case $m = 2$ can be phrased as follows: In the field of a generalized 3-pole at 0 we have: 2n positive unit charges are at equilibrium at the zeros of the lacunary Bessel polynomial $S_{2n}(c;x)$.

Remark 4.2. The zeros of y_n (in (4.2)) are no longer on S_m.

Remark 4.3. We obtained in the case $m = 2$ an electrostatic interpretation of the confluences

$$\lim_{a\to\infty} a^{-n}P_{2n}(a,c;a^{1/2}x) = S_{2n}(c;x) ;$$

$$\lim_{a\to\infty} a^{-n-\frac{1}{2}}P_{2n+1}(a,c;a^{1/2}x) = S_{2n+1}(c;x) .$$

B. Now consider the following set up of charges on S_m :

$$p = (1+ma)/2 \quad \text{at} \quad 0 \quad , \quad q = (c-a)/2 \quad \text{at} \quad c^{1/m}\omega_k \quad (k=1,\ldots,m) ;$$

$\omega_1,\ldots,\omega_m$ are the roots of $z^m = 1$.

For the field force $F(z)$ we have

$$F(z) = \frac{1+ma}{2} \cdot \frac{1}{\bar{z}} + \frac{c-a}{2} \sum_{k=1}^{m} \frac{1}{\bar{z}-c^{1/m}\omega_k} = \frac{1}{2}\,\frac{\bar{z}^m+c\left[m\bar{z}^m-(1+ma)\right]}{\bar{z}^{m+1} - c\bar{z}}$$

We let $c \longrightarrow \infty$ to obtain

$$G(z) = \lim_{c\to\infty} F(z) = -\frac{1}{2}\,\frac{m\bar{z}^m - (1+ma)}{\bar{z}}$$

For $y_n = \prod_{k=1}^{n}(z-z_k)$, where $z_1,\ldots,z_n$ are the equilibrium positions of n positive unit charges we find the differential equation

$$(4.3) \qquad zy_n''(z) - \left[mz^m-(1+ma)\right]y_n'(z) + mnz^{m-1}y_n(z) = 0$$

This is the differential equation of the second type of orthogonal polynomials introduced by R. Smith [2] . See equiation (1.8) in our introduction.

Also in this case we can prove that (4.3) essentially has only one polynomial solution.

Special cases

1) $m = 1$. This leads to Laguerre polynomials.

2) $m = 2$ leads to generalized Hermite polynomials $H_{2n}^{(a)}(x)$.

If we assume (again!) the unphysical situation where one out of $2n+1$ free positive unit charges is placed at the origin we find:

The equilibrium positions of $2n+1$ positive unit charges in the field $G(z)$ are at the zeros of $H_{2n+1}^{(a)}(x)$.

Remark 4.4. Our considerations imply an electrostatic interpretation of the confluence

$$\lim_{c\to\infty} c^n J_n(a,c;c^{-1}x) = L_n^{(a)}(x) .$$

Remark 4.5. The field $G(z)$ in the Laguerre case is

$$G(z) = \frac{\frac{a+1}{2}}{z} - \frac{1}{2}$$

and can be interpreted as the superposition of the field of a single charge $(a+1)/2$ at 0 and a homogenuous field of strength $-1/2$.

REFERENCES

1. R. SMITH, Similarity solutions of a non linear differential equation, IMA Journal of Applied Mathematics, 28 (1982), 149-160.

2. R. SMITH, An abundance of Orthogonal Polynomials, IMA Journal of Applied Mathematics, 28 (1982), 161-167.

3. T.J. STIELTJES, Sur certain polynômes qui verifient une equation différentielle linéair du second order et sur la théorie des fonctions de Lamé, OEuvres I p. 434-439 . Noordhof, Groningen (1914).

4. T.J. STIELTJES, Sur les racines de l'equation $X_n = 0$. OEuvres II, p. 73-88. Noordhof, Groningen (1918).

5. G. SZEGŐ. Orthogonal Polynomials, 4th ed. Amer. Math. Soc., Providence, R.I., 1975.

ASSOCIATED DUAL HAHN POLYNOMIALS

Mourad E.H. ISMAIL

Department of Mathematics,
Arizona State University, Tempe
Arizona 85287, USA

Jean LETESSIER and Galliano VALENT

Laboratoire de Physique Theorique
et Hautes Energies,
Université Paris VII, Tour 24-14
5eme étage, 3, place Jussieu,
75251 Paris Cedex 05 - FRANCE

ABSTRACT

A generating function, the spectral measure and two explicit forms are obtained for each of the two families of associated continuous dual Hahn polynomials.

We give a short account of the results obtained in [1] .

The dual Hahn polynomials are defined by the three terms recurrence relation:

$$Q_o(x) = 1 \qquad Q_1(x) = \frac{\lambda_o+\mu_o-x}{\mu_1}$$

$$-xQ_n(x) = \mu_{n+1}Q_{n+1}(x) + \lambda_{n-1}Q_{n-1}(x) - (\lambda_n+\mu_n)Q_n(x) \qquad n\geq 1$$

with:

$$\lambda_n = (n+a-\beta)(n+b-\beta) \qquad \mu_n = n(n+\alpha-\beta) \qquad n\geq 0$$

For these polynomials there exist a unique measure [4] $d\mu(x)$ such that

$$\pi_n \int_o^\infty d\mu(x)Q_n(x)Q_p(x) = \delta_{np}$$

where

$$\pi_o = 1 \qquad \pi_n = \frac{\lambda_o \ldots \lambda_{n-1}}{\mu_1 \ldots \mu_n} = \frac{(a)_n \, (b)_n}{(1+\alpha)_n (1+\beta)_n}$$

Associated dual Hahn are obtained by the change

$$n \longrightarrow n+\beta \qquad \beta > 0$$

in the recurrence relation. This corresponds to

$$\lambda_n = (n+a)(n+b) \qquad n \geq 0 \qquad \mu_n = (n+\alpha)(n+\beta)$$

However such an extension is not unique, as emphasized in [2], because we may take:

Case I: $\mu_n = (n+\alpha)(n+\beta) \qquad n \geq 0$ (define $\eta = 0$ is this case)

Case II: $\begin{cases} \mu_n = (n+\alpha)(n+\beta) & n \geq 1 \\ \mu_o = 0 & \end{cases}$ (define $\eta = 1$ in this case)

So we obtain two different families of associated polynomials. Let us point out that this twofold structure is completely general and not linked to the fact that λ_n and μ_n are quadratic polynomials. In what follows we use:

$$\gamma = \frac{1}{2}(1+\alpha+\beta-a-b) \qquad s = \sqrt{\gamma^2 - x}$$

The generating function:

$$G(x,z) = \sum_{n=o}^{\infty} \pi_n z^n Q_n(x)$$

is the solution of the inhomogeneous second order differential equation:

$$z(1-z)^2 \frac{d^2G}{dz^2} + (1-z)\left[1+\alpha+\beta-(1+a+b)z\right] \frac{dG}{dz} + (1-z)\left(\frac{\alpha\beta}{z} - ab\right)G + xG =$$
$$= \alpha\beta \frac{(1-z)^\eta}{z}$$

$$\begin{cases} G(x,0) = 1 \\ G(x,z) \quad \text{analytic for} \quad |z| < 1 \end{cases}$$

We get for the generating function:

$$G(x;z) = \frac{\alpha\beta}{\alpha-\beta}\Big\{(1-z)^{\gamma-s} M_{\alpha\beta}(z) \int_o^1 dt\, t^{\beta-1}(1-tz)^{\eta-1-\gamma-s} M_{\beta\alpha}(zt)$$
$$-(\text{same with } \alpha \longleftrightarrow \beta \;)\Big\}$$

with

$$M_{\alpha\beta}(z) = {}_2F_1\left(\begin{matrix}\frac{1}{2}(1+\alpha-\beta+a-b)-s\,,\ \frac{1}{2}(1+\alpha-\beta-a+b)-s \\ 1+\alpha-\beta\end{matrix}\,;\ z\right)$$

The analysis of $G(x,z)$ for $z \longrightarrow 1$ gives the asymtotics of

$Q_n(x)$ for large n .

Defining then:

$$Q_n^*(x) = \frac{-1}{(1+\alpha)(1+\beta)} \quad Q_{n-1} \quad (x;a+1,b+1,\alpha+1,\beta+1)$$

we obtain, using a theorem of Stieltjes [3]

$$\lim_{n\to\infty} \frac{Q_n^*(x)}{Q_n(x)} = \int_0^\infty \frac{d\mu(t)}{x-t}$$

This Stieltjes transform can be inverted to get the spectral measure:

$$\frac{d\mu(x)}{dx} = \frac{A\ \sigma}{(\sigma^2+\gamma^2)^\eta} \left| \frac{\Gamma(\eta+\alpha-\gamma+i\sigma)\Gamma(\eta+\beta-\gamma+ i\sigma)\Gamma(1-\eta+\gamma+i\sigma)}{\Gamma(1+2_i\sigma)} \right|^2 \times$$

$$\times \left| {}_3F_2 \left(\begin{matrix} a+\eta-1\ , & b+\eta-1\ , & \eta-\gamma+i\sigma \\ \eta+\alpha-\gamma+i\sigma\ , & \eta+\beta-\gamma+i\sigma & \end{matrix} ; 1 \right) \right|^{-2}$$

$$A^{-1} = \pi\Gamma(a)\Gamma(b)\Gamma(\alpha+1)\Gamma(\beta+1) \qquad \sigma = \sqrt{x-\gamma^2} \qquad x \geq \gamma^2$$

The symmetries $\alpha \longleftrightarrow \beta$ and $a \longleftrightarrow b$ are manifest.

It can be checked that for α or $\beta \longrightarrow 0$ we recover the orthogonality measure of dual Hahn polynomials:

$$\frac{d\mu}{dx} = \frac{\sigma}{\pi\Gamma(a)\Gamma(b)\Gamma(1+\alpha)} \left| \frac{\Gamma(-\gamma+i\sigma)\Gamma(a+\gamma+i\sigma)\Gamma(b+\gamma+i\sigma)}{\Gamma(1+2_i\sigma)} \right|^2$$

$$Q_n(x) = {}_3F_2 \left(\begin{matrix} -n\ , & -\gamma+i\sigma\ , & -\gamma-i\sigma \\ & a\ ,\ b & \end{matrix} ; 1 \right)$$

For $\eta = 0$ and a or $b \longrightarrow 1$ we recover the dual Hahn polynomials but with different parameters:

$$Q_n(x) = {}_3F_2 \left(\begin{matrix} -n\ , & 1+\frac{\alpha+\beta-a}{2}+i\sigma\ , & 1+\frac{\alpha+\beta-a}{2}-i\sigma \\ & 1+\alpha\ ,\ 1+\beta & \end{matrix} ; 1 \right)$$

For $\eta = 1$ the limit a or $b \longrightarrow 0$ is singular since it leads to $\lambda_0 = 0$.

Using the generating function we obtain two different explicit forms for the associated dual Hahn polynomials. The first one is:

$$\pi_n Q_n(x) = \frac{(b)_n}{n!} \sum_{k=0}^{n} \frac{(-n)_k (b+\gamma+i\sigma)_k (b+\gamma-i\sigma)_k}{(b)_k (1+\alpha)_k (1+\beta)_k} \; {}_3F_2\left(\begin{matrix} \alpha, \beta, b+\eta-1 \\ b+\gamma+i\sigma, \; b+\gamma-i\sigma \end{matrix} ; 1\right)$$

where

$${}_3F_2\left(\begin{matrix} \alpha, \beta, \; b+\eta-1 \\ b+\gamma+i\sigma \; , \; b+\gamma-i\sigma \end{matrix} ; 1\right)_k = \sum_{l=0}^{k} \frac{(\alpha)_l (\beta)_l (\eta-1+b)_l}{l! \, (b+\gamma+i\sigma)_l (b+\gamma-i\sigma)_l}$$

on which it is manifest that the Q_n are polynomials in x, symmetric under the exchange $\alpha \leftrightarrow \beta$.

Let us observe that either for $\beta \longrightarrow 0$ or $\eta = 0$ and $b = 1$ the partial sum of ${}_3F_2$ trivializes to 1 and we get directly the dual Hahn polynomials.

The second form for the $Q_n(x)$'s is

$$\pi_n Q_n(x) = \sum_{k=0}^{n} \frac{(-\gamma+i\sigma)_{n-k} (\gamma+a-i\sigma)_k (\gamma+b-i\sigma)_k}{(n-k)! \; (1+\alpha)_k (1+\beta)_k} \; {}_3F_2\left(\begin{matrix} \alpha, \; \beta, \; \gamma-\eta-i\sigma \\ \gamma+a-i\sigma, \gamma+b-i\sigma \end{matrix} ; 1\right)_k$$

on which both symmetries $\alpha \longleftrightarrow \beta$ and $a \longleftrightarrow b$ are obvious, as well as the α or $\beta \longrightarrow 0$ limit. The case $\eta = 0$ and $b = 1$ seems to give a new summation formula.

REFERENCES

1. M.E.H. ISMAIL, J. LETESSIER, G. VALENT, in preparation.
2. M.E.H. ISMAIL, J. LETESSIER, G. VALENT, "Linear birth and death models and associated Laguerre and Meixner polynomials". Submitted for publication.
3. T.J. STIELTJES, Ann. Sci. Toulouse 8 (1894) J1-122 ; 9 (1895) A1-47 , Oeuvres, vol. 2, 398-566.
4. S. KARLIN, J. McGREGOR, Trans. Amer. Math. Soc. 85 (1957), 489-546.

Szego's Theorem for polynomials orthogonal with respect to varying measures.

G. López

Math. Dept., Hav. Univ.

Havana, Cuba

1. Introduction

Let $d\sigma$ be a finite positive Borel measure on the interval $[0,2\pi]$ whose support contains an infinite set of points. We denote $d\theta$, Lebesgue's measure on $[0,2\pi]$ and $\sigma'=\sigma'(\theta)=d\sigma/d\theta$ the Radon-Nykodym derivative of $d\sigma$ with respect to $d\theta$. Let $\{W_n\}$, $n\in\underline{N}$, be a sequence of polynomials such that, for each $n\in\underline{N}$, W_n has degree n ($\deg W_n=n$) and all its zeros $(w_{n,i})$, $1\le i\le n$, lie in $[|w|\le 1]$. In the following the set of natural (resp. complex) numbers will be denoted by $\underline{N}$ (resp. $\underline{C}$). Set

$$d\sigma_n(\theta) = d\sigma(\theta)/|W_n(z)|^2, \quad z=e^{i\theta}.$$

Suppose that:

$$\|d\sigma_n\|= \int d\sigma_n(\theta)<+\infty , \quad n\in\underline{N};$$

The limits of integration with respect to θ will always be 0 and 2π, thus they will not be indicated in the following.

The condition above guarantees that for each pair (n,m) of natural numbers we can construct a polynomial $\phi_{n,m}(z)= \alpha_{n,m}z^m+\dots$, that is uniquely determined by the conditions:

$$1/2\pi\int \bar{z}^j\phi_{n,m}(z)d\sigma_n(\theta)= 0,\ j=0,1,\dots,m-1,\ (z= e^{i\theta});$$

$$1/2\pi\int |\phi_{n,m}(z)|^2 d\sigma_n(\theta)= 1;\ \deg \phi_{n,m}= m \text{ and } \alpha_{n,m}> 0.$$

In order to avoid unnecessary complications, in the following, we will restrict our attention to the case when $w_{n,i}\in[0,1]$ for all $n\in\underline{N}$, $i=1,\dots,n$. Moreover, we will assume that the indexes are taken so that $w_{n,1}\le w_{n,2}\le\dots\le w_{n,n}$.

Let $W_{n,m}(z)=\prod_{j=1}^{m}(z-w_{n,j})$ and

$$c_{n,m}=\int|W_{n,m}(z)|^{-2}d\sigma(\theta),\ n\in\underline{N},\ m=1,\dots,n,\ c_{n,0}=\sigma[0,2\pi]$$

Definition. We say that $(\sigma,\{W_n\},k)$ satisfies the C-condition (after Carleman) on $[0.2\pi]$ if

i) $\|d\sigma_n\|=\int d\sigma_n(\theta)<+\infty$, $n\in\underline{N}$;

ii) $\int\prod_{i=1}^{|k|}|z-w_{n,i}|^{-2}d\sigma(\theta)\leq M<+\infty$, $z=e^{i\theta}$, for $k=-1,-2,\dots$;

iii) $\lim_{n}\sum_{m=1}^{n}(c_{n,m})^{-2m}=+\infty$.

If β_n and β are positive Borel measures on $[0,2\pi]$ we denote by $\beta_n\underset{*}{\rightarrow}\beta$ the weak convergence of β_n to β as $n\rightarrow\infty$. That means, that for every continuous 2π-periodic function f on $[0,2\pi]$

$$\lim_{n}\int f(\theta)d\beta_n(\theta)=\int f(\theta)d\beta(\theta).$$

We will use the following result which is the key to all further considerations:

Theorem 1. Suppose that $(\sigma,\{W_n\},k)$ satisfies the C-condition on $[0,2\pi]$ then

$$d\beta_n(\theta)=(|W_n(z)|/|\emptyset_{n,n+k}(z)|)^2d\theta\underset{*}{\rightarrow}\ d\sigma(\theta),\quad z=e^{i\theta}. \tag{1}$$

The proof of (1) is rather extensive, and will appear in a forthcoming paper submitted to Mat. Sb. [7] so we will limit ourselves to some remarks. The proof is divided into two cases. In the first, we suppose that $\lim_{n}\sum_{i=1}^{n}(1-|w_{n,i}|)=+\infty$. This case was actually proved in [6], a simplified version appeared in [5], a shorter proof is provided in [7]. The second case considered is when $\lim_{n}\sum_{i=1}^{n}(1-|w_{n,i}|)<+\infty$. Here, we use a generalization of a method due to T.Carleman for uniqueness of analytic functions (see [1], Chap.3). This idea already proved to be effective in two previous papers of ours (see [3,4]).

In §2, we use theorem 1 to prove an extension of Szego's classical result (theorem 12.1.1, [8]), for sequences of orthogonal polynomials with respect to varying measures on the unit circle. Using standard techniques a version of Szego's result for finite intervals of the real line (theorem 12.1.2, [8]) can also be obtained (see [7]).

2. On Szego's Theorem

If P_n is a polynomial of degree exactly equal to n, we denote as usual $P_n^*(w) = w^n\overline{P_n}(1/w)$.

Theorem 2. Suppose that $(\sigma, \{W_n\}, k)$ satisfies the C-condition on $[0,2\pi]$ and suppose that $\log \sigma' \in L_1[0,2\pi]$. Then

$$\lim_n \quad (W_n^*/\phi_{n,n+k}^*)(w) \quad = \quad S(w), \tag{2}$$

where

$$S(w) = \exp[(1/4\pi)\int[(z+w)/(z-w)]\log\sigma'(\theta)d\theta], \ z=e^{i\theta},$$

and the convergence in (2) is uniform on each compact subset of $[|w|<1]$.

Proof. Note that if we multiply W_n by a certain constant the corresponding orthonormal polynomials are multiplied by the same constant and the ratio on the left-hand side of (2) remains unaltered. Thus without loss of generality we may suppose in the following that W_n is monic.

First let us show that $\{W_n^*/\phi_{n,n+k}^*\}$, $n\in\underline{N}$, is uniformly bounded on each compact subset of $[|w|<1]$. In fact, since $\phi_{n,n+k}^*$ has no zero in $[|w|\leq 1]$, using Cauchy's formula and Cauchy-Schwartz's inequality we obtain for $|w|<1$

$$|(W_n^*/\phi_{n,n+k}^*)(w)| \leq (1/2\pi)\int|z(W_n^*/\phi_{n,n+k}^*)(z)(z-w)^{-1}|d\theta \leq$$

$$[(1/2\pi)\int|z(z-w)^{-1}|^2d\theta]^{1/2} \ [(1/2\pi)\int|(W_n^*/\phi_{n,n+k}^*)(z)|^2d\theta]^{1/2} \leq$$

$$(\inf\{|w-z|:|z|=1\})^{-1} \ [(1/2\pi)\int|(W_n^*/\phi_{n,n+k}^*)(z)|^2d\theta]^{1/2}.$$

Now, from theorem 1 we have that the last integral tends to $\sigma[0,2\pi]$ as $n\to\infty$, since $|(W_n^*/\phi_{n,n+k}^*)(z)| = |(W_n/\phi_{n,n+k})(z)|$. Thus, if K is a compact subset of $[|w|<1]$ then for each $\varepsilon>0$, and all $n\in\underline{N}$ sufficiently large

$$\sup\{|(W_n^*/\phi_{n,n+k}^*)(w)|: w\in K\} \leq$$

$$(\inf\{|w-z|:|z|=1, w\in K\})^{-1}[(\sigma[0,2\pi]+ \ )/2\pi]^{1/2}.$$

Since $W_n^*/\phi_{n,n+k}^*$ is analytic in $[|w|<1]$ for all $n\in\underline{N}$, our statement is obviously true.

Take, $\{W_n^*/\emptyset_{n,n+k}^*\}$, $n\in\Gamma, \Gamma\subset \underline{N}$, a convergent subsequence whose limit is S_Γ. We will prove that $S_\Gamma\in H_2[|w|<1]$, $S_\Gamma\neq 0$ and moreover,

$$S_\Gamma(0)\geq \exp[(1/4\pi)\int \log\sigma'(\theta)d\theta]. \qquad (3)$$

In fact,

$$\lim_{n\in\Gamma}|W_n^*/\emptyset_{n,n+k}^*(w)|^2=|S_\Gamma(w)|^2,$$

uniformly on each compact subset of $[|w|<1]$. On the other hand, for each $\varepsilon>0$ and all sufficiently large n, $0<r<1$,

$$\int_{|w|=r}|(W_n^*/\emptyset_{n,n+k}^*)(w)|^2d\theta\leq\int_{|w|=1}|(W_n^*/\emptyset_{n,n+k}^*)(w)|^2d\theta\leq\sigma[0,2\pi]+\varepsilon.$$

Thus, taking limits we obtain that

$$\int_{|w|=r}|S_\Gamma(w)|^2d\theta\leq \sigma[0,2\pi]+\varepsilon.$$

Since this inequality holds for each $0<r<1$ we have that $S_\Gamma\in H_2[|w|<1]$.

It is obvious that $W_n^*\neq 0$ in $[|w|<1]$. Hence, either $S_\Gamma\equiv 0$ or $S_\Gamma\neq 0$. But, denoting $\tilde{\emptyset}_{n,n+k}=(\alpha_{n,n+k})^{-1}\emptyset_{n,n+k}$ and using Jensen's inequality, we obtain

$$[W_n^*(0)/\emptyset_{n,n+k}^*(0)]^2= (\alpha_{n,n+k})^{-2}= (1/2\pi)\int|\tilde{\emptyset}_{n,n+k}(z)/W_n(z)|^2d\sigma(\theta)\geq$$

$$(1/2\pi)\int|\tilde{\emptyset}_{n,n+k}^*(z)/W_n^*(z)|^2\sigma'(\theta)d\theta\geq$$

$$\exp[(1/2\pi)\int\log(|\tilde{\emptyset}_{n,n+k}^*(z)/W_n^*(z)|^2\sigma'(\theta))d\theta]= \exp[(1/2\pi)\int\log\sigma'(\theta)d\theta].$$

And so

$$\lim_{n\in\Gamma} W_n^*(0)/\emptyset_{n,n+k}^*(0)= S_\Gamma(0)\geq \exp[(1/4\pi)\int\log\sigma'(\theta)d\theta]> 0.$$

Since tha family is uniformly bounded on each compact subset, it is sufficient to prove that any convergent subsequence tends to S(w). To this end, if $\lim_{n\in\Gamma} [W_n(w)^*(w)/\emptyset_{n,n+k}^*(w)]= S_\Gamma(w)$, it is sufficient to show that

$$\mathrm{Re}[\log S_\Gamma(w)]= \log|S_\Gamma(w)|= \mathrm{Re}[\log S(w)]= (1/4\pi)\int P(w,z)\log\sigma'(\theta)d\theta,$$

where P(w,z) represents Poisson's kernel.

However, using once more Jensen's inequality and theorem 1 we get

$$|S_\Gamma(w)|^2=\lim_{n\in\Gamma}|W_n^*(w)/\emptyset_{n,n+k}^*(w)|^2=$$

$$\lim_{n\in\Gamma}\exp[(1/2\pi)\int P(w,z)\log|W_n^*(z)/\emptyset_{n,n+k}^*(z)|^2d\theta]\leq$$

$$\lim_{n\in\Gamma} (1/2\pi)\int P(w,z)|W_n^*(z)/\emptyset_{n,n+k}^*(z)|^2 d\theta] = (1/2\pi)\int P(w,z)d\sigma(\theta).$$

Taking limits in the above inequality as $r \longrightarrow 1-0$, $w= re^{i\theta}$, using Fatou's Theorem (see [2], pag. 34) and the fact that $S_r \in H_2[|w|<1]$ we obtain that

$$|S_r(z)|^2 \le \sigma'(\theta),$$

a.e. with respect to Lebesgue's measure on $[0,2\pi]$, where $z= e^{i\theta}$.

Hence,

$$\mathrm{Re}[\log S_r(w)] = \log|S_r(w)| = (1/4\pi)\int P(w,z)\log|S_r(z)|^2 d\theta \le$$
$$(1/4\pi)\int P(w,z)\log\sigma'(\theta)d\theta.$$

But according to (3) we have that

$$\mathrm{Re}[\log S_r(0)] = \log S_r(0) \ge (1/4\pi)\int \log\sigma'(\theta)d\theta = (1/4\pi)\int P(0,z)\log\sigma'(\theta)d\theta.$$

Therefore, using the maximum principle for harmonic functions it follows that

$$\mathrm{Re}[\log S_r(w)] = (1/4\pi)\int P(w,z)\log\sigma'(\theta)d\theta,$$

Let us consider two limit cases:

Corollary 1 (Szego's Theorem). If $\log \sigma' \in L_1[0,2\pi]$ then uniformly on each compact subset of $[|w|>1]$

$$\lim_n [\emptyset_n(w)/w^n] = [\bar{S}(1/\bar{w})]^{-1},$$

where $\emptyset_n$ denotes the n-th orthonormal polynomial with respect to σ.

For the proof, it is sufficient to apply theorem 2 to $(\sigma,\{W_n\},0)$, where $W_n=z^n$, $n\in\underline{N}$. Observe that then $W_n^*(w)\equiv 1$.

In case that all the zeros of W_n lie on the opposite extreme point of [0.1] we have:

Corollary 2. Let $W_n(z)= (z-1)^n$. Suppose that $(\sigma,\{W_n\},k)$ satisfies the C-condition and $\log \sigma' \in L_1[0,2\pi]$, then uniformly on each compact subset of $[|w|>1]$

$$\lim_n \emptyset_{n,n+k}(w)/(w-1)^n = w^k[\bar{S}(1/\bar{w})]^{-1}.$$

References

1. Carleman,T., Sur les fonctions quasianalytiques, Paris, 1926.

2. Grenander,U., Szego,G., Toeplitz Forms and their Applications, Univ. of California Press, Berkeley-Los Angeles, 1958.
3. López,G., Conditions for convergence of multipoint Padé approximants for functions of Stieltjes type, Mat. Sb. 107(149), 69-83,1978; Eng. translation in Math. USSR Sb. 35, 363-376, 1979.
4. López,G., On the convergence of Pade approximants for meromorphic functions of Stieltjes type, Mat. Sb. 111(153), 308-316, 1980; Eng. translation in Math. USSR Sb. 38, 281-288, 1981.
5. López,G., On the asymptotics of orthogonal polynomials and the convergence of multipoint Pade approximants, Mat. Sb. 128(170), 216-229, 1985; to appear translated in Math. USSR Sb..
6. López,G., Asymptotics of polynomials orthogonal with respect to varying measures, (submitted to Const. Approx. Theory).
7. López,G., Unified approach to asymptotics of sequences of orthogonal polynomials on finite and infinite intervals. (submitted to Mat. Sb.).
8. Szego,G., Orthogonal Polynomials, Coll. Pub. XX , Providence, R.I..

ASSOCIATED ASKEY-WILSON POLYNOMIALS AS LAGUERRE-HAHN ORTHOGONAL POLYNOMIALS

Alphonse P. Magnus

Institut Mathématique U.C.L.
Chemin du Cyclotron 2
B-1348 Louvain-la-Neuve (Belgium)

ABSTRACT

One looks for [formal] orthogonal polynomials satisfying interesting differential or difference relations and equations (Laguerre -Hahn theory). The divided difference operator used here is essentially the Askey-Wilson operator

$$Df(x) = \frac{E_2 f(x) - E_1 f(x)}{E_2 x - E_1 x} = \frac{f(y_2(x)) - f(y_1(x))}{y_2(x) - y_1(x)}$$

where $y_1(x)$ and $y_2(x)$ are the two roots of $Ay^2 + 2Bxy + Cx^2 + 2Dy + 2Ex + F = 0$.

The related Laguerre-Hahn orthogonal polynomials are then introduced as the denominators $P_0, P_1, \ldots$ of the successive approximants Q_n/P_n of the Gauss-Heine-like continued fraction $f(x) = 1/(x - r_0 - s_1/(x - r_1 - s_2/\ldots))$ satisfying the Riccati equation $a(x)Df(x) = b(x)E_1 f(x) E_2 f(x) + c(x) Mf(x) + d(x)$ where a,b,c,d are polynomials and $Mf(x) = (E_1 f(x) + E_2 f(x))/2$.

In the classical case (degrees $a,b,c,d \leq 2,0,1,0$) , closed-forms for the recurrence coefficients r_n and s_n are obtained and show that we are dealing essentially with the associated Askey-Wilson polynomials.

One finds for P_n difference relations $(a_n + a)DP_n = (c_n - c)MP_n + 2s_n d_n MP_{n-1} - 2bMQ_n$ and a writing in terms of solutions of linear second order difference equations $P_n = (X_n Y_{-1} - Y_n X_{-1})/(X_0 Y_{-1} - Y_0 X_{-1})$.

1. INTRODUCTION. DIFFERENCE OPERATORS AND EQUATIONS

Classical orthogonal polynomials are solutions of remarkable differential relations and equations. This makes them very useful as elements in the representation of solutions of problems of mathematical physics, numerical analysis, etc. (see [N1] Chap. 2, [WP] Chap. 11). Laguerre ([LA] , see also [AT] , [MC]) found a systematic way of generating orthogonal polynomials satisfying differential relations and equations. A new setting of his theory will be given here.

Classical orthogonal polynomials of a discrete variable are also used in the same fields. Now, they satisfy difference relations and equations where the fundamental difference operator is $\Delta f(x) =$

$= f(x+1)-f(x)$. It happens that the difference formulas are very similar to the differential ones ([N1] , Chap. 2, Sect. 12).

Another interesting difference operator is $\Delta_q f(x) = f(qx)-f(x)$ ([HE] p. 99 , [H2]), also used in orthogonal polynomials theory ([H1] , [H2] , [H3]).

By investigating new families of orthogonal polynomials, Wilson considered the following divided difference operator (see [AS] , Sect. 5) : put $x = a+bt^2$, then

$$\Delta_W f(x) = [f(a+b(t+1)^2)-f(a+b(t-1)^2)]/t$$

Finally, Askey and Wilson ([AS] Sect. 5) found a further extension: if $x = a+b(q^m+q^{-m})$, then

$$\Delta_{AW} f(x) = [f(a+b(q^{m+1/2}+q^{-m-1/2}))-f(a+b(q^{m-1/2}+q^{-m+1/2}))]/(q^m-q^{-m}).$$

Each of these operators is an extension of the preceding one, which can be recovered as a special case and/or a limit case, up to a linear transformation of the variable. For instance, $\Delta f(x)$ is the limit when $q \to 1$ of Δ_q acting on X of the function $f(X-1/(q-1))$ with $X = x+1/(q-1)$ ... A general form of operator avoiding these troubles is given now.

We consider the most general divided difference operator of the form

$$Df(x) = \frac{E_2 f(x)-E_1 f(x)}{E_2 x-E_1 x} = \frac{f(y_2(x))-f(y_1(x))}{y_2(x)-y_1(x)} \tag{1.1}$$

leaving a polynomial of degree $n-1$ when f is a polynomial of degree n . The two first non obvious conditions are

$$y_1(x) + y_2(x) = \text{polynomial of degree } 1$$

and $(y_1(x))^2 + y_1(x)y_2(x) + (y_2(x))^2$ = polynomial of degree 2

equivalent to $y_1(x)y_2(x)$ = polynomial of degree ≤ 2 .

This defines y_1 and y_2 as the two roots of an equation of the form

$$Ay^2 + 2Bxy + Cx^2 + 2Dy + 2Ex + f = 0 , \qquad A \neq 0 . \tag{1.2}$$

Then, $y^{n+1} = (y_1+y_2)y^n-y_1y_2y^{n-1}$, i.e., $Dx^{n+1} = (y_1+y_2)Dx^n - y_1y_2Dx^{n-1}$ shows that Dx^n is indeed a polynomial of degree $\leq n$ (see [MR] Section 2 for a similar derivation). Conditions ensuring Dx^n to be of exact degree $n-1$ will be given in a moment.

The most important identities involving y_1 and y_2 are

$$y_1 + y_2 = -2(Bx+d)/a \tag{1.3a}$$

$$y_1 y_2 = (Cx^2 + 2Ex + F)/A \tag{1.3b}$$

$$(y_2-y_1)^2 = 4[(B^2-AC)x^2 + 2(BD-AE)x + D^2-AF]/A^2 \tag{1.3c}$$

$$y_1, y_2 = -(Bx+D)/A \pm \left[(B^2-AC)\left(x + \frac{BD-AE}{B^2-AC}\right)^2 + \frac{A\theta}{B^2-AC}\right]^{1/2}$$

$$\text{if } B^2 \neq AC \tag{1.4}$$

where

$$\theta = \begin{vmatrix} A & B & D \\ B & C & E \\ D & E & F \end{vmatrix} .$$

In general $(B^2 \neq AC$ and $\theta \neq 0)$, we recover Δ_{AW} ; Δ_q if $B^2 \neq AC$ and $\theta = 0$; Δ_W if $B^2 = AC$ and $\theta \neq 0$; Δ if $B^2 = AC$ and $\theta = 0$ (implying $BD = AE$). Only the differential operator d/dx must still be considered as a limit case.

The companion operator to D is

$$Mf(x) = (E_1 f(x) + E_2 f(x))/2 .$$

One has

$$D(fg) = Df\ Mg + Mf\ Dg ,$$
$$M(fg) = Mf\ Mg + (y_2-y_1)^2/4\ Df\ Dg ,$$
$$D(1/f) = -Df/(E_1 f\ E_2 f) .$$

A first-order difference equation is a link between $E_1 f(x) = f(y_1(x))$ and $E_2 f(x) = f(y_2(x))$ or, if we call $X = y_1(x)$, between $f(X)$ and $Ef(X) = f(y_2(y_{-1}(X)))$, where y_{-1} and y_{-2} are the inverse functions of y_1 and y_2 (this requires $C \neq 0$) :

$$y_{-1}(y_1(x)) \equiv x \quad , \quad y_{-2}(y_2(x)) \equiv x \quad , \quad Ef(x) = f(y_2(y_{-1}(x))) ,$$
$$E^{-1}f(x) = f(y_1(y_{-2}(x))) \quad , \quad E_{-1}f(x) = f(y_{-1}(x)) \quad ,$$
$$E_{-2}f(x) = f(y_{-2}(x)) ,$$
$$E = E_{-1}E_2 = (E_1)^{-1}E_2 \quad , \quad E^{-1} = E_{-2}E_1 = (E_2)^{-1}E_1$$

This introduces the adjoint operators

$$D^*f(x) = (E_{-2}f(x) - E_{-1}f(x))/(y_{-2}(x)-y_{-1}(x)) ,$$
$$M^*f(x) = (E_{-1}f(x) + E_{-2}f(x))/2$$

Remark that, as $y_1(x) + y_2(x) = -2(Bx+D)/A$,

$$Ex+x = -2(By_{-1}(x) + D)/A \quad , \quad E^{-1}x+x = -2(By_{-2}(x)+D)/A \quad . \tag{1.5}$$

A difference equation links values on several powers $E^k(X)$, which are situated on the lattice discussed by Nikiforov and Suslov [NI], indeed, consider

$$EX + E^{-1}X = -2(B(y_{-1} + y_{-2})+2D)/A-2X = 2(2\frac{B^2}{AC} - 1)X + \\ + 4(BE-CD)/(AC)$$

and apply E^k : $E^{k+1}X + E^{k-1}X = 2(2\frac{B^2}{AC} -1)E^kX + 4(BE-CD)/(AC)$.

Let q satisfy

$$q + q^{-1} = 2(2B^2/(AC)-1) , \tag{1.6}$$

then, E^kX has the form

$$E^kX = \alpha q^k + \beta q^{-k} + \frac{CD-BE}{B^2-AC} , \quad q \neq 1 ;$$

$$E^kX = \alpha + \beta k + 2\frac{BE-CD}{AC}k^2 , \quad q = 1 \tag{1.7}$$

for a fixed X. These are indeed the forms discussed in [NI].

The relation (1.4) shows that $y_1(x)/x$ and $y_2(x)/x$ have limits $(C/A)^{1/2}q^{1/2}$ and $(C/A)q^{-1/2}$ when $x \to \infty$. Therefore, Dx^n will be of exact degree $n-1$ if $q^n \neq 1$ and $q \neq 1$; if $q = 1$ (i.e. $B^2 = AC$), (1.3a) and (1.3c) show directly that $Dx^n = n(-B/A)^{n-1}x^{n-1} + \ldots$ Remark also that if $q^N = 1$ for some integer N, each lattice $\{E^kx\}$ is reduced to a finite set, this corresponds however to useful orthogonal polynomials [AS].

Second order difference operators link $E^{-1}f(x)$, $f(x)$ and $Ef(x)$. Interesting operators with rational coefficients are (R is a rational function) :

$$D^*(R(x)Df(x)) = \{R(y_{-2}(x))\frac{f(x)-E^{-1}f(x)}{x-E^{-1}x} - \\ - R(y_{-1}(x))\frac{Ef(x)-f(x)}{Ex-x}\} / (y_{-2}(x)-y_{-1}(x)) \tag{1.8}$$

$$M^*[(y_2(x)-y_1(x)) R(x) Df(x)] / (y_{-2}(x)-y_{-1}(x)) = \\ \frac{1}{2}[(x-E^{-1}x) RDf(y_{-2}(x)) + (Ex-x)RDf(y_{-1}(x))]/(y_{-2}(x)-y_{-1}(x)) = \\ = (x+D/A) D^*(R(x)Df(x)) + (B/A) D^*(xR(x)Df(x)) \tag{1.9}$$

2. LAGUERRE-HAHN ORTHOGONAL POLYNOMIALS

Formal orthogonal polynomials related to the sequence of "moments" $\{\mu_n\}$ are the diagonal Padé denominators of

$$f(x) = \sum_{0}^{\infty} \mu_n x^{-n-1} \tag{2.1}$$

This means that the numerator Q_n (of degree $n-1$) and the denominator P_n (of degree n) are normally determined by

$$f(x) - Q_n(x)/P_n(x) = O(x^{-2n-1}) \qquad x \to \infty \quad .$$

Equivalently, Q_n/P_n is the nth approximant of the Jacobi continued fraction of f

$$f(x) = 1/\left[x-r_0-s_1/(x-r_1-s_2/\ldots)\right] \qquad (\mu_0 = 1) \tag{2.2}$$

and P_n and Q_n satisfy the three-term recurrence relation

$$P_{n+1}(x) = (x-r_n)P_n(x) - s_n P_{n-1}(x) \qquad P_0=1 \qquad P_{-1}=0 \tag{2.3}$$

$$Q_{n+1}(x) = (x-r_n)Q_n(x) - s_n Q_{n-1}(x) \qquad Q_1=1 \qquad Q_0 = 0 \qquad s_0 Q_{-1} = -1$$

Laguerre [LA] realized that families of orthogonal polynomials satisfying remarkable differential relations and equations are conveniently described through their generating functions of moment (2.1) (see also [AT] [MC]) . The relevant extension follows .

<u>Definition 2.1</u>. Laguerre-Hahn orthogonal polynomials are diagonal Padé denominators of an expansion (2.1) satisfying a Riccati equation

$$a(x)Df(x) = b(x)E_1f(x)E_2f(x) + c(x)Mf(x) + d(x) \tag{2.4}$$

where a,b,c,d are polynomials .

This will lead to difference relations and equations in Sections 5 and 6 . Conversely, there is increasing evidence [H1] [H3] [BO] [B2] that all orthogonal polynomials satisfying non trivial difference relations and equations are those of definition 2.1. As most of this evidence has been gathered by W. Hahn (see [H1] [H3] and references therein), the corresponding polynomials are called here Laguerre-Hahn polynomials.

Other interesting forms of the Riccati equation (2.4) showing how it links E_1f and E_2f are

$$bE_1fE_2f + \left[(y_2-y_1)^{-1}a + c/2\right]E_1f - \left[(y_2-y_1)^{-1}a - c/2\right]E_2f + d = 0 \tag{2.5}$$

$$(bE_1f - a/(y_2-y_1) + c/2)\ (bE_2f + a/(y_2-y_1) + c/2) =$$
$$= -bd-a^2/(y_2-y_1)^2 + c^2/4 \tag{2.6}$$

Although measures and weights will not be studied throughly here, let us remark that if f is a Stieltjes function

$$f(x) = \int_S (x-u)^{-1} d\sigma(u) \quad , \quad x \notin S$$

with σ increasing on its real support S, $w(x) = \sigma'(x)$ is obtained almost everywhere as limit of $-\pi^{-1}$ Im $f(x+i\varepsilon)$ when $\varepsilon \to 0$ $(\varepsilon > 0)$. It is then easy to show that, if $b \equiv 0$,

$$aDw = cMw , \tag{2.7}$$

equivalently $(a-(y_2-y_1)c/2)E_2w = (a+(y_2-y_1)c/2)E_1w$.

If there is a masspoint of weight $\mu(X)$ at some $X = E_1x$ (pole of f with residue $\mu(X)$), (2.5) tells that, still if $b = 0$, there is another one at $EX = E_2x$ with a weight

$$\mu(EX) = \mu(E_2x) = \frac{a+(y_2-y_1)c/2}{a-(y_2-y_1)c/2} \frac{dy_2}{dy_1} \mu(E_1x) .$$

This shows that there are masspoint on lattices $\{E^kX\}$ interrupted at zeros of $a \pm (y_2-y_1)c/2$ [if $b \equiv 0$]. If $b \neq 0$, it will be shown in section 6 how f can be represented as a ratio of solutions of second order difference equations. The measure can then be expressed in terms of these solutions, as in sect. 5 of [BE]. A very large set of measures, still far from being completely explored, are concerned.

3. FUNDAMENTAL RECURRENCE RELATIONS

The importance of the Riccati form (2.4) is that the same form holds for the equation satisfied by f_1 $(f(x) = 1/(x-r_o-s_1f_1))$. Iteration of this remark ([BA] p. 163) will generate a sequence of Riccati equations whose coefficients will eventually enter difference relations and equations.

Theorem 3.1. Let the Jacobi continued fraction (2.2) satisfy the Riccati equation (2.4) where a,b, c and d are polynomials of degrees $\leq m+2$, m, $m+1$ and m. Then, $f_n(x) =$ $= 1/(x-r_n-s_{n+1}/(x-r_{n+1} - \dots))$ satisfies

$$a_n(x)Df_n(x) = b_n(x)E_1f_n(x)E_2f_n(x) + c_n(x)Mf_n(x) + d_n(x) \tag{3.1}$$

with polynomials a_n b_n c_n d_n of degrees still bounded by $m+2$, m, $m+1$ and m.

Moreover,

$$a_{n+1}(x) = a_n(x) = (y_2-y_1)^2 d_n(x)/2 \tag{3.2a}$$

$$b_{n+1}(x) = s_{n+1}d_n(x) \tag{3.2b}$$

$$c_{n+1}(x) = -c_n(x) - 2(Mx-r_n)d_n(x) \tag{3.2c}$$

$$d_{n+1}(x) = [a_n(x) + b_n(x) + (Mx-r_n)c_n(x) + (y_1-r_n)(y_2-r_n)d_n(x)]/s_{n+1} \tag{3.2d}$$

$(a_o \equiv a\ ,\ b_o \equiv b\ ,\ c_o \equiv c\ ,\ d_o \equiv d)$.

Indeed , put $f_n(x) = 1/(x-r_n-s_{n+1}f_{n+1}(x))$ in (3.1) and (3.2) follows easily by an induction starting with (2.4) at $n = 0$ $(f_o \equiv f)$. Only the degree of d_{n+1} must still be discussed . As $f_n(x) = 1/(x-r_n) + O(x^{-3})$ for large x , (3.1) yields

$-a_nD(x-r_n)^{-1} + b_nE_1(x-r_n)^{-1}\ E_2(x-r_n)^{-1} + c_nM(x-r_n)^{-1} + d_n = "(x^{m-2})$,

or $a_n+b_n+(Mx-r_n)c_n + (y_1-r_n)(y_2-r_n)d_n = "(x^m)$ and this means degree $d_{n+1} \leq m$. #

<u>Remark 3.2.</u> The polynomial

$$\chi(x) = (a_n(x))^2 + (y_2(x)-y_1(x))^2[b_n(x)d_n(x)-(c_n(x))^2/4] \tag{3.3}$$

of degree $2m+4$ is independent of n ;

$$\Sigma_n(x) = b_n(x)d_n(x) - (c_n(x))^2/4 =$$
$$= b_o(x)d_o(x) - (c_o(x))^2/4 + (1/2)(a_o(x)+a_n(x)) \sum_o^{n-1} d_k(x) \tag{3.4}$$

Remark also that χ appears at the right-hand side of (2.6) .

We proceed now with a first exploration of the polynomials a_n , b_n , c_n and d_n :

<u>Lemma 3.3.</u>

The dominant coefficients of $a_n(x) = a_{n,o}x^{m+2}+\ldots$, $c_n(x) = c_{n,o}x^{m+1}+\ldots$ and $d_n(x) = d_{n,o}x^m+\ldots$ have the form

$$a_{n,o} = -(K/2)\left(q^{-n-\mu}+q^{n+\mu}\right) ,$$

$$c_{n,o} = K(A/C)^{1/2}\left(q^{-n-\mu}-q^{n+\mu}\right) / \left(q^{-1/2}-q^{1/2}\right) ,$$

$$d_{n,o} = (KA/C)\left(q^{-n-\mu-1/2}-q^{n+\mu+1/2}\right)/\left(q^{-1/2}-q^{1/2}\right) , \quad \text{if} \quad q \neq 1 ;$$

$$a_{n,o} = -K \ , \quad c_{n,o} = 2K(A/C)^{1/2}(n+\mu) ,$$

$$d_{n,o} = (2KA/C)(n+1/2+\mu) \quad \text{if} \quad q = 1 .$$

Indeed , from (3.2a..d) , (1.3c) and (1.3a) , one has

$$a_{n+1,o} = a_{n,o} - 2A^{-2}(B^2 - AC)d_{n,o} \quad , \quad c_{n+1,o} = -c_{n,o} + (2B/A)d_{n,o}$$

and $a_{n,o} - (B/A)c_{n,o} + (C/A)d_{n,o} = 0$.

After elimination of $a_{n,o}$ and $a_{n+1,o}$:

$$\begin{bmatrix} c_{n+1,o} \\ d_{n+1,o} \end{bmatrix} = \begin{bmatrix} -1 & 2B/A \\ -2B/C & 4B^2/(AC)-1 \end{bmatrix} \begin{bmatrix} c_{n,o} \\ d_{n,o} \end{bmatrix}$$

This matrix has eigenvalues q and q^{-1} (from (1.6)) . The square root of this matrix with eigenvalues $q^{1/2}$ and $q^{-1/2}$ is

$$\begin{bmatrix} 0 & (C/A)^{1/2} \\ -(A/C)^{1/2} & 2B/(AC)^{1/2} \end{bmatrix} \quad , \text{ so that } \quad c_{n+1/2,o} = (C/A)^{1/2} d_{n,o} ,$$

$d_{n+1/2,o} = (q^{1/2} + q^{-1/2})d_{n,o} - d_{n-1/2,0}$. This is solved in terms of powers of q and q^{-1} if $q \neq 1$, of linear functions of n if $q = 1$. The constants K and μ of the theorem are derived from the initial values $c_{o,o}$ and $d_{o,o}$. #

There is no simple rule for the other coefficients of the polynomials, they are related by equations where the further unknowns r_n and s_n are also involved. If the d_n's are supposed to be known, it is possible to eliminate the a's , b's and c's and to end with equations for r_n and s_n . When $m = 0$ (classical case), d_n is a constant known by lemma 3.3 and formulas in closed form can be found. The polynomials studied by Askey and Wilson [AN] [AS] [PE] [WI] will be recovered.

4. RELATIONS WITH ASKEY-WILSON POLYNOMIALS

<u>Theorem 4.1</u>. If the degrees of a,b,c,d are less or equal than 2,0,1,0 , the orthogonal polynomials of definition 2.1 satisfy the recurrence relation (2.3) where

$$s_n = S((n+\mu)^2)/[d_{n-1}(d_{n-1/2})^2 d_n]$$

and where r_n can be written in four different ways as

$$r_n = y_1(x_i) - R_i(n+\mu)/(d_n d_{n-1/2}) - R_i(-n-\mu-1)/(d_n d_{n+1/2})$$

$$= y_2(x_i) - R_i(-n-\mu)/(d_n d_{n-1/2}) - R_i(n+\mu+1)\ (d_n d_{n+1/2})$$

with $R_i(x)R_i(-x) \equiv S(x^2)$, $i=1,\dots,4$. If $q \neq 1$, $S(x^2)$ is a polynomial of degree four in q^x+q^{-x} , each $R_i(x)$ is a Laurent polynomial of degree two in q^x (i.e, a linear combination of q^{-2x}, q^{-x} , 1 , q^x and q^{2x}), and $d_n = \text{const.}(q^{n+\mu+1/2}-q^{-n-\mu-1/2})$; if $q = 1$, S and each R_i are polynomials of degree four and $d_n = \text{const.}\ (n+\mu+1/2)$.

Indeed, from theorem 3.1, r_n and s_n can be obtained if we find completely the expressions of the polynomials a_n , b_n , c_n and d_n . As $m = 0$, d_n is a constant already known (lemma 3.3). From (3.2a) , $a_n(x) = a_o(x)-(1/2)(y_2-y_1)^2 \sum_0^{n-1} d_k$. Let us introduce D_n such that $D_{n+1}-D_n = d_n$, a function of n that will be very useful. From lemma 3.3, $D_n = (KA/C)\ (q^{-1/2}-q^{1/2})^{-2}(q^{-n-\mu}+q^{n+\mu})$ + constant if $q \neq 1$, $D_n = (KA/C)(n+\mu)^2$ + const. if $q = 1$.

Considered as a function of n , $a_n(x)$ is therefore a polynomial of first degree in D_n .

In the polynomial $c_n(x) = c_{n,o}x+c_{n,1}$, only $c_{n,1}$ remains to be determined. This is done by looking at the coefficient of x in (3.4) exhibiting $c_{n,o}c_{n,1}$ as a quadratic polynomial in D_n .

This shows that $c_{n,o}c_n(x)$ as a function of n is a polynomial of second degree in D_n (in (3.3) , the coefficient of x^4 in χ shows that $(c_{n,o})^2$ is quadratic in D_n) .

We have now all the material needed for the recurrence coefficients : from (3.2c) , $r_n = Mx + (c_n(x) + c_{n+1}(x))/(2d_n)$ for any choice of x ; in (3.3) , everything is known but $b_n = s_n d_n d_{n-1}$ (from (3.2b)) , so that $s_n = [(y_2(x)-y_1(x))^{-2}(\chi(x)-(a_n(x))^2) + (c_n(x))^2/4]/(d_{n-1}d_n)$ for any choice of x . By multiplying the numerator and the denominator by $(c_{n,o})^2$, s_n appears indeed as a quartic polynomial in D_n divided by $d_{n-1}(c_{n,o})^2 d_n$ and the first part of the theorem follows from lemma 3.3 $(c_{n,o} = (C/A)^{1/2} d_{n-1/2,o})$. There is a connection between r_n and s_n which appears when one chooses x as one of the four roots of $\chi(x) = 0$ (whence the four possibilities announced in the theorem). Then

$$\left([c_n(x)+2(y_2(x)-y_1(x))^{-1}a_n(x)]/(2d_n)\right)\left([c_n(x)-2(y_2(x)-y_1(x))^{-1}a_n(x)]/(2d_{n-1})\right)$$

and

$$\left([c_n(x)-2(y_2(x)-y_1(x))^{-1}a_n(x)]/(2d_n)\right)\left([c_n(x)+2(y_2(x)-y_1(x))^{-1}a_n(x)]/(2d_{n-1})\right)$$

are two different factorizations of s_n . After multiplication of the numerators and the denominators by $(1/2)d_{n-1/2}$, each of the numerators appears as an even function of $n+\mu$ plus or minus an odd function of $n+\mu$: $-R_i(n+\mu)$ and $-R_i(-n-\mu)$. Finally, the sum of the factors of the first factorization (where n is replaced by $n+1$ in the second factor) gives (use (3.2a) and (3.2.c)) $r_n - Mx + (y_2(x)-y_1(x))/2 = r_n-y_1(x)$. The second factorization gives $r_n-y_2(x)$. #

Problem 4.2. Find a geometric description of the roots of $\chi(x) = 0$.

Theorem 4.1. Tells that we are dealing with a family of orthogonal polynomials depending on 6 essential parameters (q , μ , and the 4 zeros of S) and 2 inessential parameters wich are the dominant coefficient of S (dilation of the variable) and one of the $y_i(x_j)$'s (translation). Apparently, the conic (1.2) defining y_1 and y_2 requires 5 parameters and a,b,c,d of degrees $2,0,1,0$ ask for another 7 of whom we subtract 1 for homogeneity and 1 for the constraint $\mu_o = 1$, leaving 10 parameters. Several operators E_1 and E_2 will therefore lead to the same family of orthogonal polynomials. The two remaining degrees of freedom can be used to make $E_1 x \equiv x$ (possible only if $\theta = 0$ (see (1.4))) or to have a convenient symmetry $A = C$ and $D = E$ implying $E_{-1} = E_2 = E^{1/2}$, $E_{-2} = E_1 = E^{-1/2}$ (Askey and Wilson's choice : [AS] Section 5).

Theorem 4.3. Under the conditions of Theorem 4.1 and if $b = 0$, the orthogonal polynomials of definition 2.1 are the Askey-Wilson polynomials [AN] [AS] [WI] . If $b \neq 0$, they are the associated Askey-Wilson polynomials.

Indeed, if we look at different recurrence relation writing for the Askey-Wilson polynomials, such as [AN] p. 56 or [AS] p. 5 , we obtain the forms of Theorem 4.1 (our s_n is the $A_{n-1}C_n$ of [AN] and [AS]). However, there is a constraint $S(\mu^2) = 0$ so that $s_o = 0$ in [AN] and [AS]. This happens indeed if $b=0$. If $b\neq 0$, let ν be such that $R_i(\nu)=0$ and let us write $r_n=r_{n+\nu-\mu}$, $s_n=s_{n+\nu-\mu}$. Then r_n and s_n are the recurrence coefficients of a valid set of Askey-Wilson polynomials with ν insted of μ . This shows that we are dealing with the associated Askey-Wilson polynomials.

Wilson [WI] and Askey-Wilson [AN] [AS] polynomials have been introduced through contiguity relations for ${}_4F_3$ and ${}_4\varphi_3$ hypergeometric functions. These hypergeometric expansions can also be recovered here:

Theorem 4.4. With r_n and s_n given by Theorem 4.1, the recurrence relation

$$Z_{n+1} = (x-r_n)Z_n - s_n Z_{n-1} \tag{4.1}$$

has special solutions of the form

$$Z_n = \prod^{n} \frac{R_i(\mp(m+\mu))}{d_{m-1}d_{m-1/2}} \sum_{m=0}^{\infty} \rho^{-m} \prod_{k=1}^{m} \frac{\delta(n+\mu\mp\nu_1-k+1)}{\delta(k)\delta(k\pm\nu_1\pm\nu_2)} \frac{\delta(n+\mu\pm\nu_1+k)}{\delta(k\pm\nu_1\pm\nu_3)} \frac{(x-E^{\mp(k-1)}y_j(x_i))}{\delta(k\pm\nu_1\pm\nu_4)} \tag{4.2}$$

where the upper signs correspond to $j=1$ and the lower sings to $j=2$ and where $\delta(n) = d_{n/2-\mu-1/2} = (KA/C)(q^{-n/2}-q^{n/2})/(q^{-1/2}-q^{1/2})$ if $q \neq 1$; $\delta(n) = 2KAn/C$ if $q = 1$.

Indeed, for one of the values of $i = 1,\ldots,4$ allowed by Theorem 4.1, (4.1) becomes

$$Z_{n+1} = \left\{x-y_j(x_i) + \frac{R_i(\pm(n+\mu))}{d_n d_{n-1/2}} + \frac{R_i(\mp(n+\mu+1))}{d_n d_{n+1/2}}\right] Z_n - \frac{R_i(\pm(n+\mu))R_i(\mp(n+\mu))}{d_{n-1}(d_{n-1/2})^2 d_n} Z_{n-1} .$$

If $x = y_j(x_i)$, (4.1) has the simple particular solution $\prod^{n} \frac{R_i(\mp(m+\mu))}{d_{m-1}d_{m-1/2}}$. Writing $Z_n = \prod^{n} \frac{R_i(\mp(m+\mu))}{d_{m-1}d_{m-1/2}} Z_n$, the recurrence relation becomes

$$\frac{R_i(\mp(n+\mu+1))}{d_n d_{n+1/2}})(\tilde{Z}_{n+1}-\tilde{Z}_n) = (x-y_j(x_i))\tilde{Z}_n + \frac{R_i(\pm(n+\mu))}{d_n d_{n-1/2}}(\tilde{Z}_n-\tilde{Z}_{n-1}) \tag{4.3}$$

Consider $\tilde{Z}_n$ as a function of $n+\mu+1/2$. Then this recurrence accepts solutions which are even functions of $n+\mu+1/2$ (d_n is an odd function of $n+\mu+1/2$) . Wether there are interesting forms of $\tilde{Z}_n$ as and odd function of $n+\mu+1/2$ is not investigated here. Furthermore, from Theorem 4.1, $R_i(t)$ can be factorized as $R_i(t) = \rho\delta(t-\nu_1)\delta(t-\nu_2)\,\delta(t-\nu_3)\delta(t-\nu_4)$. So if $\pm(n+\mu) = \nu_1$, i.e. $n = -\mu\pm\nu_1$, there is a simple relation between $\tilde{Z}_n$ and $\tilde{Z}_{n+1}$. This suggests for $\tilde{Z}_n$ an expression containing $\delta(n+\mu\mp\nu_1)$, $\delta(n+\mu\mp\nu_1)\delta(n+\mu\mp\nu_1-1),\ldots$ and as it must be even in $n+\mu+1/2$, the following is proposed:

$$\tilde{Z}_n = \sum_{m=0}^{\infty} \xi_m \prod_{k=1}^{m} \delta(n+\mu\mp\nu_1-k+1)\delta(n+\mu\pm\nu_1+k) . \tag{4.4}$$

Expanding $\tilde{Z}_n-\tilde{Z}_{n-1}$, one encounters $\delta(n+\mu\mp\nu_1)\delta(n+\mu\pm\nu_1+m)$ -

$\delta(n+\mu\mp\nu_1-m)\delta(n+\mu\pm\nu_1)$ which is neatly factorized as $\delta(m)\delta(2n+2\mu) = \delta(m)d_{n-1/2}$. After a little work on (4.3), one must satisfy the identity

$$\sum_{m=0}^{\infty} \xi_m \prod_{k=1}^{m-1} \delta(n+\mu\mp\nu_1-k+1)\delta(n+\mu\pm\nu_1+k)(X_{n,m}-(x-y_j(x_i))d_n\delta(n+\mu\mp\nu_1-m+1)\delta(n+\mu\pm\nu_1+m)) = 0,$$

where $X_{n,m} = \rho\delta(m)\left[\delta(n+\mu\pm\nu_1+m)\prod_2^4 \delta(n+\mu+1\pm\nu_k) - \right.$ $\left.- \delta(n+\mu\mp\nu_1-m+1)\prod_2^4 \delta(n+\mu\mp\nu_k)\right]$ is an odd function of $n+\mu+1/2$ and can be factorized as $d_n\left[\alpha_m + \beta_m\delta(n+\mu\mp\nu_1-m+1)\delta(n+\mu\pm\nu_1+m)\right]$. Then $\alpha_m\xi_m + (\beta_{m-1}-x+y_j(x_i))\xi_{m-1} = 0$ follows, so that (4.4) becomes

$$\tilde{Z}_n = \sum_{m=0}^{\infty} \prod_{k=1}^{m} (\alpha_k)^{-1}\delta(n+\mu\mp\nu_1-k+1)\delta(n+\mu\pm\nu_1+k)(x-y_j(x_i)-\beta_{k-1}) . \quad (4.5)$$

α_m is easily found by taking $n = -\mu+m-1\pm\nu_1$ as $\rho\delta(m)\prod_2^4 \delta(m\pm\nu_1\pm\nu_k)$; from the highest powers of q^n and q^{-n} in $X_{n,m}$ and $\beta_o = 0$, one finds $\beta_m = \rho\delta(m)\delta(m+1\pm\sum_1^4 \nu_k)$. So β_m is a combination of q^m, q^{-m} and a constant. The coefficients of q^m and q^{-m} are the same as in $[C(q^{-1/2}-q^{1/2})/KA]^2 R_i(\mp(m/2+1/4))$, from the factorization of R_i. In order to see the link between β_m and $E^{\mp m}\, y_j(x_i)$, one returns to the definition of R_i in theorem 4.1 in terms of $a_n(x_i)$ and $c_n(x_i)$ to get finally $\beta_m = (-1/2)(C/A)^{1/2}$

$$\left[x_i + \frac{BD-AE}{B^2-AC}\right](q^{-m-1/2}+q^{m+1/2}) \mp (1/2)(y_2-y_1)\frac{q^{-m-1/2}-q^{m+1/2}}{q^{-1/2}-q^{1/2}} +$$

+ constant, where the constant is such that $\beta_o = 0$. It then happens that $y_j(x_i)+\beta_m$ is of the form (1.7). Checking that $2y_j(x_i)+\beta_o+\beta_{-1} = -2(Bx_i+D)/A$ and comparing with (1.5) yields indeed $y_j(x_i)+\beta_m = E^{\mp m}\, y_j(x_i)$ and (4.2) follows. #

5. DIFFERENCE RELATIONS

We return to the general theory of sections 2 and 3 in order to show how the polynomials of definition 2.1 satisfy difference relations involving the polynomials a_n, b_n, c_n and d_n of section 3. Before that, the following functions will be found useful:

$$\tilde{a}_n(x) = -s_1\ldots s_{n-1}\,(a_n(x)/(y_2-y_1) + c_n(x)/2)$$
$$\tilde{c}_n(x) = s_1\ldots s_{n-1}\,(a_n(x)/(y_2-y_1) - c_n(x)/2)$$
$$\tilde{b}_n(x) = s_1\ldots s_{n-1}\, b_n(x)/s_n$$
$$\tilde{d}_n(x) = s_1\ldots s_n\, d_n(x)$$

Remark that the Riccati forms (2.5) and (2.6) (with indexes n) can be written

$$(s_n)^2\tilde{b}_n E_1 f_n E_2 f_n - s_n \tilde{a}_n E_1 f_n - s_n \tilde{c}_n E_2 f_n + \tilde{d}_n = 0 \tag{5.1}$$

and

$$(s_n \tilde{b}_n E_1 f_n - \tilde{c}_n)(s_n \tilde{b}_n E_2 f_n - \tilde{a}_n) = -(s_1 \ldots s_{n-1})^2 \chi/(y_2-y_1)^2 \tag{5.2}$$

<u>Lemma 5.1</u>. Let X_n and Y_n be two independent solutions of $Z_{n+1} = (x-r_n)Z_n - s_n Z_{n-1}$, then there are functions $\alpha,\beta,\gamma,\delta$ of x, independent of n, such that

$$\tilde{a}_n = \alpha(E_1X_{n-1})(E_2X_n) + \beta(E_1Y_{n-1})(E_2X_n) + \gamma(E_1X_{n-1})(E_2Y_n) + \delta(E_1Y_{n-1})(E_2Y_n)$$
$$\tilde{c}_n = \alpha(E_1X_n)(E_2X_{n-1}) + \beta(E_1Y_n)(E_2X_{n-1}) + \gamma(E_1X_n)(E_2Y_{n-1}) + \delta(E_1Y_n)(E_2Y_{n-1})$$
$$\tilde{b}_n = \alpha(E_1X_{n-1})(E_2X_{n-1}) + \beta(E_1Y_{n-1})(E_2X_{n-1}) + \gamma(E_1X_{n-1})(E_2Y_{n-1}) + \delta(E_1Y_{n-1})(E_2Y_{n-1})$$
$$\tilde{d}_n = \alpha(E_1X_n)(E_2X_n) + \beta(E_1Y_n)(E_2X_n) + \gamma(E_1X_n)(E_2Y_n) + \delta(E_1Y_n)(E_2Y_n)$$

$$\tilde{a}_n E_1X_n - \tilde{d}_n E_1X_{n-1} = (\beta E_2X_n + \delta E_2Y_n)\, E_1\Gamma_n \tag{5.3a}$$
$$\tilde{c}_n E_2X_n - \tilde{d}_n E_2X_{n-1} = (\gamma E_1X_n + \delta E_1Y_n)\, E_2\Gamma_n \tag{5.3b}$$

$$\tilde{a}_n E_1Y_n - \tilde{d}_n E_1Y_{n-1} = -(\gamma E_2Y_n + \alpha E_2X_n)\, E_1\Gamma_n \tag{5.4a}$$
$$\tilde{c}_n E_2Y_n - \tilde{d}_n E_2Y_{n-1} = -(\beta E_1Y_n + \alpha E_1X_n)\, E_2\Gamma_n \tag{5.4b}$$

where Γ_n is the Casorati determinant $X_nY_{n-1}-X_{n-1}Y_n$ satisfying $\Gamma_{n+1} - s_n\Gamma_n$.

$$\begin{aligned}\tilde{a}_n\tilde{c}_n - \tilde{b}_n\tilde{d}_n &= (s_1\ldots s_{n-1})^2(-(a_n)^2/(y_2-y_1)^2+(c_n)^2/4-b_nd_n) \\ &= -(s_1\ldots s_{n-1})^2\chi/(y_2-y_1)^2 = (\beta\gamma-\alpha\delta)E_1\Gamma_n\, E_2\Gamma_n\,.\end{aligned} \tag{5.5}$$

Indeed, from (3.2), the recurrence relations for $\tilde{a}_n,\ldots,\tilde{d}_n$ can be written

$$\begin{bmatrix}\tilde{a}_{n+1}\\ \tilde{c}_{n+1}\\ \tilde{b}_{n+1}\\ \tilde{d}_{n+1}\end{bmatrix} = \begin{bmatrix}0 & -s_n & 0 & y_2-r_n\\ -s_n & 0 & 0 & y_1-r_n\\ 0 & 0 & 0 & 1\\ -s_n(y_2-r_n) & -s_n(y_1-r_n) & (s_n)^2 & (y_1-r_n)(y_2-r_n)\end{bmatrix}\begin{bmatrix}\tilde{a}_n\\ \tilde{c}_n\\ \tilde{b}_n\\ \tilde{d}_n\end{bmatrix}$$

Now, if $\Xi_{n+1} = (y_2-r_n)\,\Xi_n - s_n\Xi_{n-1}$ and $\Upsilon_{n+1} = (y_1-r_n)\Upsilon_n - s_n\Upsilon_{n-1}$, that is exactly the recurrence relation for $[\Xi_n\Upsilon_{n-1}, \Xi_{n-1}\Upsilon_n, \Xi_{n-1}\Upsilon_{n-1}, \Xi_n\Upsilon_n]^T$. A basis of solutions of this recurrence is obtained by the four choices $\Xi_n = E_2X_n$ or E_2Y_n, $\Upsilon_n = E_1X_n$ or E_1Y_n (tensorial product). (5.3) and (5.4), which are the actual difference relations, follow easily.

<u>Theorem 5.2</u>. The monic orthogonal polynomials defined in Section 2

satisfy the difference relation

$$(a_n+a)DP_n = (c_n-c)MP_n + 2s_n d_n MP_{n-1} - 2bMQ_n \qquad (5.6)$$

Indeed, with $X_n = P_n$ and $Y_n = Q_n$, α, β, γ and δ of lemma 5.1 are readily found to be (take $n = 0$ and use $P_{-1} = 0$, $P_o = 1$, $Q_{-1} = -1/s_o$, $Q_o = 0$), $\alpha = \tilde{d}_o = d$, $\beta = -s_o\tilde{a}_o = a/(y_2-y_1) + c/2$, $\gamma = -s_o\tilde{c}_o = -a/(y_2-y_1)+c/2$, $\delta = (s_o)^2\tilde{b}_o = b$. Now, adding (5.3a) and (5.3b) :

$$s_1 \ldots s_{n-1}\, [a_n(E_2P_n-E_1P_n)/(y_2-y_1)-c_n(E_1P_n+E_2P_n)/2 -$$

$$- s_n d_n(E_1P_{n-1}+E_2P_{n-1})] = [a(E_2P_n-E_1P_n)/(y_2-y_1) +$$

$$+ c(E_1P_n+E_2P_n)/2 + b(E_1Q_n+E_2Q_n)]\Gamma_n$$

and this is exactly (5.6) after using $\Gamma_n = s_o \ldots s_{n-1}\Gamma_o = -s_1 \ldots s_{n-1}$. #

6. DIFFERENCE EQUATIONS

<u>Theorem 6.1</u>. If $b \equiv 0$ in (2.4), the orthogonal polynomials defined in section 2 satisfy the second-order difference equation with rational coefficients

$$D^*\left(\frac{a}{d_n}DP_n\right) - (y_{-2}-y_{-1})^{-1}M^*\left((y_2-y_1)\frac{c}{d_n}DP_n\right) = \qquad (6.1)$$

$$\left[(y_{-2}-y_{-1})^{-1}M^*\frac{2(a-a_n)}{(y_2-y_1)d_n} + D^*\frac{c_n-c}{2d_n}\right]P_n$$

Other interesting forms are

$$D^*\left(\frac{\tilde{w}}{d_n}DP_n\right) = \left[(v_{-2}-y_{-1})^{-1}M^*\left((y_2-y_1)\frac{\sum_0^{n-1} d_k}{d_n}\right) + D^*\frac{c_n-c}{2d_n}\right]wP_n \qquad (6.2)$$

with w form (2.7) and $\tilde{w} = (a+(y_2-y_1)c/2)E_1w = (a-(y_2-y_1)c/2)E_2w$,

$$D^*\left(\frac{\chi^{1/2}}{d_n}D(w^{1/2}P_n)\right) = \left[D^*\frac{c_n}{2d_n} + 2(y_{-2}-y_{-1})^{-1}M^*\left(\frac{\chi^{1/2}-a_n}{(y_2-y_1)d_n}\right)\right]w^{1/2}P_n \qquad (6.3)$$

Indeed, as $b \equiv 0$, choosing $X_n = X_oP_n$, $Y_n = -s_oY_{-1}Q_n$ in lemma 5.1, one finds $\delta \equiv 0$. Therefore (5.3) involves only X_n and X_{n-1}, with $\beta = \tilde{a}_o/(E_1Y_{-1}E_2X_o)$, $\gamma = \tilde{c}_o/(E_1X_oE_2Y_{-1})$.

Elimination of X_{n-1} yields

$$E_{-1}(\beta/\tilde{d}_n)\Gamma_nEX_n - E_{-2}(\gamma/\tilde{d}_n)\Gamma_nE^{-1}X_n =$$

$$= [E_{-1}(\tilde{a}_n/\tilde{d}_n) - E_{-2}(\tilde{c}_n/\tilde{d}_n)]X_n$$

which can be rewritten using (1.8) and (1.9) as

$$D^*\left[\frac{\beta-\gamma}{2d_n}(y_2-y_1)DX_n\right] - (y_{-2}-y_{-1})^{-1} M^*\left[\frac{\beta+\gamma}{d_n}(y_2-y_1)DX_n\right] =$$

$$\left[\frac{2}{s_o\Gamma_o(y_{-2}-y_{-1})}M^*\frac{a_n}{(y_2-y_1)d_n} - \frac{1}{2s_o\Gamma_o}D^*\frac{c_n}{d_n} - \frac{1}{2}D^*\frac{\gamma+\beta}{d_n} - \frac{1}{y_{-2}-y_{-1}}M^*\frac{\gamma-\beta}{d_n}\right]X_n .$$

With $X_o = Y_{-1} = 1$, β and $\gamma = [-c_o/2\mp a_o/(y_2-y_1)]/s_o$, $\Gamma_o = 1$ and this gives (6.1) which is the extension of Laguerre differential equation [AT] [H1] [LA] [MA] . For the two other forms, $\beta = -\gamma$, possible if $\tilde{a}_o E_1(X_o/Y_{-1}) = -\tilde{c}_o E_2(X_o/Y_{-1})$: $X_o/Y_{-1} = w$. With $Y_{-1} = 1/w$, $X_o = 1$, one has (6.2) , which is the form used by Askey and Wilson ([AS] Section 5). Remark also that, in the

classical case (degrees a,b,c,d = 2,0,1,0) , the right-handsides of (6.1) and (6.2) have the form $\lambda_n P_n$ and $\lambda_n w P_n$: d_n is independent on x , c_p is degreee 1 and use (1.9) [and (3.2a)]. With $Y_{-1} = w^{-1/2}$, $X_o = w^{1/2}$, and using (5.5) , one obtains (6.3).

#

<u>Theorem 6.2</u>. The orthogonal polynomials of definition 2.1 satisfy a linear fourth-order difference equation whose coefficients depend on a,b,c,d and a_n , b_n , c_n and d_n .

Indeed, we consider again lemma 5.1 with $X_n = P_n$ and $Y_n = Q_n$ and α , β , γ and δ determined from $\tilde{a}_o$, $\tilde{b}_o$, $\tilde{c}_o$ and $\tilde{d}_o$ i.e. from a , b , c and d , but now $\delta \neq 0$ in general. Elimination of $X_{n-1} = P_{n-1}$ and $Y_{n-1} = Q_{n-1}$ from (5.3) and (5.4) yields two equations involving P_n , EP_n , $E^{-1}P_n$, Q_n , EQ_n and $E^{-1}Q_n$. Applying the operators E and E^{-1} to the two equivalent equations involving only Q_n and $E^{-1}Q_n$ for the first, and Q_n and EQ_n for the second, we get two new equations. Final elimination of Q_n , EQ_n and $E^{-1}Q_n$ from the four resulting equations leaves a single relation involving $E^{-2}P_n$, $E^{-1}P_n$, P_n , EP_n and E^2P_n , which is the fourth-order difference equation.

The complete writing of this equation would be very tedious. It is much more interesting to see how its solutions are related to solutions of second order difference equations ([H1] eq. 17) :

<u>Theorem 6.3</u>. The orthogonal polynomials of definition (2.1) can be written

$$P_n = (X_n Y_{-1} - Y_n X_{-1})/(X_o Y_{-1} - Y_o X_{-1}) , \tag{6.4}$$

where the functions of x and n X_n and Y_n satisfy simultaneously the three-terms recurrence relation

$$Z_{n+1} = (x-r_n)Z_n - s_n Z_{n-1} , \tag{6.5}$$

the second order difference equation

$$D^*\left(\frac{\chi^{1/2}}{d_n} DZ_n\right) = \left[D^*\frac{c_n}{2d_n} + 2(y_{-2}-y_{-1})^{-1} M^*\left(\frac{\chi^{1/2}-a_n}{(y_2-y_1)d_n}\right)\right] Z_n , \tag{6.6}$$

and the difference relation

$$(\chi^{1/2}+a_n)DZ_n = c_n MZ_n + 2s_n d_n MZ_{n-1} \tag{6.7}$$

Indeed, let us start with X_n and Y_n , two independent solutions of (6.5). Then, (6.4) is obviously a solution which is a polynomial of degree n in x , and this is P_n . We are still free to choose the four functions of x X_o , X_{-1} , Y_o and Y_{-1} . Here is a way to have $\alpha = \delta = 0$ in Lemma 5.1 , which will lead to considerable simplification : it is based on the fact that if $Z_o/(s_o Z_{-1})$ is a solution of the Riccati equation (2.4) , and if the recurrence relation (6.5) holds , then $Z_n/(s_n Z_{n-1})$ is a solution of the Riccati equation (3.1) . This is a consequence of the construction of (3.1) . So let us choose a solution g of (2.4) and fix the ratio $X_o/(s_o X_{-1}) = g$ (of course, g could be f itself, and this will indicate a representation of f as a ratio of solutions of second-order differences equations), and let $g_n = X_n/(s_n X_{n-1})$. We multiply now the four lines of lemma 5.1 by $-s_n E_1 g_n$, $-s_n E_2 g_n$, $(s_n)^2 E_1 g_n E_2 g_n$, 1 and we make the sum . The left-hand side vanishes, as we have reconstructed the Riccati equation of g_n in the form (5.1) . In the right-hand side, we find various products with the vanishing factor $s_n g_n X_{n-1} - X_n$ and the single product left $\delta E_1(s_n g_n Y_{n-1} - Y_n) E_2(s_n g_n Y_{n-1} - Y_n)$. Therefore, $\delta = 0$. Fixing $Y_o/(s_o Y_{-1}) = h$, another solution of (2.4) , we have also $\alpha = 0$.

Now, eliminating X_{n-1} and Y_{n-1} from (5.3) and (5.4) :

$$\begin{aligned} \Gamma_n E_{-1}(\beta/\tilde{d}_n)\, EX_n - \Gamma_n E_{-2}(\gamma/\tilde{d}_n)\, E^{-1}X_n &= \left[E_{-1}(\tilde{a}_n/\tilde{d}_n) - E_{-2}(\tilde{c}_n/\tilde{d}_n)\right] X_n \\ -\Gamma_n E_{-1}(\gamma/\tilde{d}_n)\, EY_n + \Gamma_n E_{-2}(\beta/\tilde{d}_n)\, E^{-1}Y_n &= \left[E_{-1}(\tilde{a}_n/\tilde{d}_n) - E_{-2}(\tilde{c}_n/\tilde{d}_n)\right] Y_n \end{aligned} \tag{6.8}$$

with $\beta\gamma = -(\tilde{a}_o\tilde{c}_o - \tilde{b}_o\tilde{d}_o)/(E_1\Gamma_o E_2\Gamma_o) = -\chi/[s_o{}^2 (y_2-y_1)^2 E_1\Gamma_o E_2\Gamma_o]$. These equations are the some if $\gamma = -\beta$. With the choice $\Gamma_o = 1$, we proceed as in the proof of Theorem 6.1 and we get (6.6), which has indeed the same form as (6.3) .

Now, let us define

$$W_n = E_{-1}\frac{\chi^{1/2}}{(y_2-y_1)d_n} EZ_n - E_{-1}\frac{1}{d_n}\left(\frac{a_n}{y_2-y_1}+\frac{c_n}{2}\right) Z_n$$
$$= -E_{-2}\frac{\chi^{1/2}}{(y_2-y_1)d_n} E^{-1}Z_n + E_{-2}\frac{1}{d_n}\left(\frac{a_n}{y_2-y_1}-\frac{c_n}{2}\right) Z_n , \qquad (6.9)$$

the second equality coming from (6.8) . Applying E and E^{-1} to (6.9) and eliminating Z_n , EZ_n and $E^{-1}Z_n$, it appears that W_n satisfies (6.8) with n-1 instead of n . Writing $W_n = s_n Z_{n-1}$, (6.9) is a difference relation which can be put in the form (6.7) , and (6.5) can be checked using (3.2). #

In the classical case, it would be interesting to get expressions of P_n and $Q_n = (X_n Y_o - Y_n X_o)/(X_1 Y_o - Y_1 X_o)$ in terms of hypergeometric expansions, as in [HE] p. 281 and 285.

Acknowledgements. It is a pleasure to thank M. Ismail and P. Maroni for their informations and kind interest.

REFERENCES

[AN] G.E. ANDREWS, R. ASKEY : Classical orthogonal polynomials, pp. 36-62 in "Polynômes Orthogonaux et Applications, Proceedings, Bar-le-Duc 1984" , Lecture Notes Math. 1171 (C. BREZINSKI et al. editors), Springer, Berlin 1985.

[AS] R. ASKEY , J. WILSON : Some basic hypergeometric orthogonal polynomials that generalize Jacobi polynomials, Memoirs AMS vol. 54 nr . 319 , AMS, Providence, 1985.

[AT] F.V. ATKINSON, W.N. EVERITT : Orthogonal polynomials which satisfy second order differential equations, pp. 173-181 in "E.B. Christoffel", (P.L. BUTZER and F. FEHER, editors), Birkhäuser, Basel 1981.

[BA] G.A. BAKER, Jr , P. GRAVES-MORRIS Padé Approximants II, Addison Wesley, Reading 1981.

[BE] V. BELEVITCH : The Gauss hypergeometric ratio as a positive real function, SIAM J. Math. An. 13 (1982) 1024-1040.

[BO] S.S. BONAN, P. NEVAI : Orthogonal polynomials and their derivatives I, J. Approx. Theory 40 (1984) 134-147.

[B2] S.S. BONAN, D.S. LUBINSKY, P. NEVAI : Orthogonal polynomials and their derivatives II, Preprint.

[H1] W. HAHN : On differential equations for orthogonal polynomials, Funk. Ekv. 21 (1978) 1-9.

[H2] W. HAHN :Lineare geometrische Differenzengleichungen, Bericht Nr. 169 (1981), Math.-Stat. Sektion Forschungszentrum Graz.

[H3] W. HAHN : Über Orthogonalpolynome, die linearen Functionalgleichungen genügen, pp. 16-35 in "Polynômes Orthogonaux et Applications, Proceedings, Bar-le-Duc 1984", Lecture Notes Math.

1171 (C. BREZINSKI et al., editors), Springer, Berlin 1985.

[HE] E. HEINE : Handbuch der Kugelfunctionen I, 2nd edition, G. Reimer, Berlin 1978.

[LA] E. LAGUERRE : Sur la réduction en fractions continues d'une fraction qui satisfait à une équation différentielle linéaire du premierordre dont les coefficients sont rationnels, J. de Math. 1 (1885) 135-165 - Oeuvres II 685-711, Chelsea 1972.

[MA] A.P. MAGNUS : Riccati acceleration of Jacobi continued fractions and Laguerre-Hahn orthogonal polynomials, pp. 213-230 in "Padé Approximation and its Applications, Bad Honnef 1983" , Lecture Notes Math. 1071 (H. WERNER and H.J. BUNGER, editors), Springer Berlin 1984.

[MR] P. MARONI : Introduction à l'étude des δ-polynômes orthogonaux semiclassiques,

[MC] J. McCABE : Some remarks on a result of Laguerre concerning continued fraction solutions of first order linear differential equations, pp. 373-379 in "Polynômes Orthogonaux et Applications, Proceedings, Bar-le-Duc 1984", Lecture Notes Math. 1171 (C. BREZINSKI et all., editors), Springer, Berlin 1985.

[N1] A. NIKIFOROV, V. OUVAROV : Fonctions spéciales de la physique mathématique, Mir, Moscou 1983.

[NI] A.F. NIKIFOROV, S.K. SUSLOV : Classical orthogonal polynomials of a discrete variable on nonuniform lattices, Letters Math. Phys. 11 (1986) 27-34.

[PE] M. PERLSTADT : A property of orthogonal polynomial families with polynomial duals, SIAM J. Math. An. 15 (1984), 1043-1054.

[WI] J.A. WILSON Three-term contiguous relations and some new orthogonal polynomials, pp. 227-232 in "Padé and Rational Approximation" (E.B. SAFF and R.S. VARGA, editors), Ac. Press, N. Y. 1977.

[WP] J. WIMP : Computations with Recurrence Relations, Pitman, Boston 1984.

LE CALCUL DES FORMES LINEAIRES ET LES POLYNÔMES ORTHOGONAUX SEMI-CLASSIQUES

Pascal MARONI
Université P. et M. Curie
Laboratoire d'Analyse Numérique
4, Place Jussieu
75252 Paris Cedex 05

INTRODUCTION

Dans l'étude des suites de polynômes, la forme linéaire par rapport à laquelle une suite est orthogonale, joue un rôle essentiel et il est utile de déterminer ses propriétés indépendamment de toute représentation. Par example, la donnée d'une forme linéaire est équivalente a la donnée de ses moments ; très souvent, la suite des moments vérifie une relation de récurrence. A l'aide d'un cadre algébrique convenable cela se traduit par une équation vérifiée par la forme : c'est une propriété intrinsèque.

L'objet de cet article est double: formaliser un certain nombre d'opérations habituelles sur les formes linéaires et appliquer le formalisme obtenu aux polynômes orthogonaux semi-classiques.

Dans le §1, on introduit le cadre topologique adéquat [7] de manière à définir systématiquement, par transposition, des opérations sur les formes linéaires à partir d'opérations sur les polynômes. De ce point de vue, l'espace vectoriel des séries formelles apparait comme un miroir de l'espace vectoriel des formes linéaires. On introduit un produit multiplicatif de deux formes linéaires, qui intervient naturellement dans la détermination de la forme canonique de la suite associée a une suite orthogonale.

Dans le §2 , on introduit la transformée de Fourier d'une forme linéaire et le produit de convolution de deux formes linéaires [2] , [8].

On établit en passant un résultat équivalent au théorème de Borel sur les séries formelles ;une forme linéaire peut toujours être représentée par une ultradistribution à décroissance rapide.

Enfin dans le §3 , on indique la définition et les diverses caractérisations des suites orthogonales semi-classiques [3] . En particulier, on montre l'identité de l'ensemble des suites semi-classiques et de l'ensemble des suites dont la fonction de Stieltjes formelle

$S(z)$ vérifie l'équation

$$AS' = CS + D \ , \qquad A,C,D \ \text{polynômes} \ .$$

On donne l'équation vérifiée par la forme associée a une suite de Laguerre-Hahn.

Faute de place, on n'a pas pu donner les démonstrations completes des propositions énoncées.

§1. LE CADRE TOPOLOGIQUE

1. Notons $\mathcal{P}$ l'espace vectoriel des fonctions polynomiales d'une variable réelle à valeurs dans $\mathbb{C}$, qu'on munit de sa topologie naturelle de L.F., limite inductive stricte des espaces $\mathcal{P}_n$ avec $\mathcal{P}_n \subset \mathcal{P}_{n+1}$, $n \geq 0$, $\mathcal{P} = \bigcup_{n \geq 0} \mathcal{P}_n$.

Soit $\mathcal{P}'$ le dual topologique de $\mathcal{P}$. Pour $u \in \mathcal{P}'$, on note $u_n = \langle u, x^n \rangle$, $n \geq 0$ les moments de u par rapport à la suite $\{x_n\}_{n \geq 0}$. La topologie duale forte de $\mathcal{P}'$ est définie par le système de semi-normes:

$$|u|_n = \sup_{\nu \leq n} |u_\nu| \quad , \quad n \geq 0 \ ,$$

car elle coïncide avec la topologie duale faible. De plus, $\mathcal{P}'$ est un Fréchet. Soit $\mathcal{E}$ l'espace vectoriel $C^\infty(\mathbb{R})$ muni de sa topologie naturelle; on a:

$$\mathcal{P} \hookrightarrow \mathcal{E} \qquad \text{et} \qquad \overline{\mathcal{P}} = \mathcal{E}$$

ou le signe $\hookrightarrow$ indique qu'il s'agit d'une injection continue. On en déduit:

$$\mathcal{E}' \hookrightarrow \mathcal{P}'$$

Soit Δ le sous-espace vectoriel de $\mathcal{E}'$ engendré par les éléments $D^n\delta$ $n \geq 0$, $(\langle \delta , \varphi \rangle = \varphi(0)$, $\varphi \in \mathcal{E})$ muni de sa topologie de L.F. On a:

$$\Delta \hookrightarrow \mathcal{E}' \qquad \text{et} \qquad \overline{\Delta} = \mathcal{P}'$$

L'application linéaire $F : \Delta \to \mathcal{P}$, définie par:

$$d = \sum_{\nu=0}^{n} d_\nu \frac{(-1)^\nu}{\nu!} D^\nu\delta \longrightarrow F(d) = \sum_{\nu=0}^{n} d_\nu x^\nu$$

vérifie: $F \in \text{Isom}(\Delta, \mathcal{P})$ et donc ${}^tF \in \text{Isom}(\mathcal{P}', \Delta')$.
Or ${}^tF = F$ sur Δ, donc sur $\mathcal{P}'$, de sorte que:

$$\langle F(u), d \rangle = \langle u, F(d) \rangle \quad , \quad u \in \mathcal{P}' \quad , \quad d \in \Delta$$

Le dual Δ' apparait ainsi comme un "miroir" de $\mathcal{P}'$. En fait, la suite $\{\frac{(-1)^n}{n!} D^n\delta\}_{n\geq 0}$ est une base universelle de $\mathcal{P}'$, car si on note

$$S_m(u) = \sum_{\nu=0}^{m} u_\nu \frac{(-1)^\nu}{!} D^\nu\delta \quad ,$$

de $<\frac{(-1)^n}{n!} D^n\delta , x^\nu> = \delta_{n,\nu}$, on a :

$$<S_m(u),x^n> = \begin{cases} u_n & n \leq m \\ 0 & n \geq m+1 \end{cases}$$

et donc $S_m(u) \to u$ dans $\mathcal{P}'$ lorsque $m \to +\infty$. Il en résulte la représentation: [1]

$$u = \sum_{\nu\geq 0} u_\nu \frac{(-1)^\nu}{\nu!} D^\nu\delta \quad , \ u \in \mathcal{P}'$$

qui fournit:

$$F(u) = \sum_{\nu\geq 0} u_\nu z^\nu .$$

Le dual Δ' est ainsi l'espace vectoriel des séries formelles; on a:

$$\mathcal{P} \subsetneq \Delta' \qquad\qquad \text{et} \qquad \overline{\mathcal{P}} = \Delta'$$

Remarque: Il existe une infinité de bases universelles. Soit $\{B_n\}_{n\geq 0}$ une suite libre de $\mathcal{P}$ et soit $\{\mathcal{L}_n\}_{n\geq 0}$ la suite duale, définie par:

$$\mathcal{L}_m(B_n) = <\mathcal{L}_m, B_n> = \delta_{m,n}$$

Alors $u = \sum_{n\geq 0} u_n \mathcal{L}_n$ avec ici $u_n = <u,B_n>$, $n \geq 0$. [2]

On appelle $\mathcal{L}_0$ la forme canonique de la suite $\{B_n\}_{n\geq 0}$.

2. Les applications suivantes son dans $\mathcal{L}(\mathcal{P},\mathcal{P})$:

$$p \to fp(x) = f(x)p(x) \quad , \quad f \in \mathcal{P}$$

$$p \to Dp(x) = p'(x)$$

$$p \to \tau_b p(x) = p(x-b) \quad , \quad b \in \mathbb{C}$$

$$p \to h_a p(x) = p(ax) \quad , \quad a \in \mathbb{C} - \{0\}$$

Par transposition, les applications suivantes sont dans $\mathcal{L}(\mathcal{P}',\mathcal{P}')$:

$$u \to fu = u \circ f$$

$$u \to Du = - u \circ D$$

$$u \to \tau_b u = u \circ \tau_{-b}$$

$$u \to h_a u = u \circ h_a \qquad (h_{-1}u = \check{u})$$

On a:

$$(fu)_n = f(u_n) \overset{\text{def}}{=} \sum_{\nu=0}^{p} a_\nu u_{\nu+n} \quad , \quad n \geq 0$$

$$(Du)_n = -n\, u_{n-1} \quad , \qquad n \geq 0$$

$$(\tau_b u)_n = \sum_{\nu+\mu=n} \frac{u_\nu}{\nu!} \frac{b^\mu}{\mu!} \quad , \qquad n \geq 0$$

$$(h_a u)_n = u_n a^n \quad , \qquad n \geq 0$$

avec

$$f(x) = \sum_{\nu=0}^{p} a_\nu x^\nu \quad .$$

3. Un produit multiplicatif dans $\mathcal{P}'$. Considérons le produit à droite d'une forme linéaire par un polynôme défini par :

$$(u,f) \to uf(x) = \sum_{n=0}^{p} \left(\sum_{\nu=n}^{p} a_\nu u_{\nu-n}\right) x^n$$

Cette application est dans $\mathcal{L}(\mathcal{P}' \times \mathcal{P}, \mathcal{P})$ car :

$$||uf||_p \leq p||f||_p \cdot |u|_p$$

La transposée de $f \to uf$ permet de définir le produit multiplicatif de deux formes linéaires:

$$\langle vu,f\rangle = \langle v,uf\rangle \quad , \qquad v,u \in \mathcal{P}' \quad , \quad f \in \mathcal{P}$$

et on a

$$(vu)_n = \sum_{\nu+\mu=n} v_\nu u_\mu \quad , \qquad n \geq 0$$

On a les proprietés:

$\delta f = f$; $v(uf) = (vu)f$

$\delta u = u$; $vu = uv$; $(uv)w = u(vw)$

$(v,u) \to vu \in \mathcal{L}(\mathcal{P}' \times \mathcal{P}', \mathcal{P}')$

La forme u est inversible si et seulement si $u_o \neq 0$; en particulier, une forme régulière est inversible.

On a facilement la formule de dérivation:

$$D(uf) = (Du)\, f + u\, Df + u\, \theta_o\, f$$

avec

$$\theta_o f(x) = \frac{f(x)-f(0)}{x}$$

La transposée de θ_o se définit par $u \to x^{-1} u$ et on a $(x^{-1}u)_n = u_{n-1}$, $n \geq 0$ avec la convention $u_n = 0$ si $n < 0$.

On en déduit la dérivée d'un produit:

$$D(uv) = (Du)v + u(Dv + x^{-1}(uv)$$

D'autre part:

$$x^{-1}(uv) = (x^{-1}u)v = u(x^{-1}v)$$
$$x^{-n}u = (-1)^n (D\delta)^n\ u, n \geq 0 \qquad \text{avec} \qquad x^{-1}(x^{-n}u) = x^{-(n+1)}u$$
$$D(x^{-1}u) = x^{-1}\ Du - x^{-2}\ u$$
$$Du^{-1} = -(Du)u^{-2} - 2x^{-1}u^{-1}$$

Enfin, contrairement au produit par x^{-1}, on a :

$$x(uv) = (xu)v + u_o(xv)$$

et $x^{-1}(xu) = u - u_o\delta\ ;\ x(x^{-1}u) = u$.

4. Les opérations dans Δ et Δ' . La dualité entre Δ et Δ' s'exprime par

$$\langle w,d\rangle = \sum_{\nu=o}^{n} w_\nu\ d_\nu \qquad \text{si :}$$

$$w = \sum_{\nu\geq o} w_\nu\ z^\nu \qquad \text{et} \qquad d = \sum_{\nu=o}^{n} d_\nu \frac{(-1)^\nu}{\nu!} D^\nu\delta$$

On part du résultat suivant:

$$(\varphi, d) \to \varphi d \in \mathcal{L}(\mathcal{E} \times \Delta, \Delta)$$

avec

$$\varphi d = \sum_{\nu=o}^{n} \left(\sum_{\mu=\nu}^{n} d_\mu \frac{\varphi^{(\mu-\nu)}(0)}{(\mu-\nu)!} \right) \frac{(-1)^\nu}{\nu!} D^\nu\delta$$

et par transpositions, on obtient:

$$dw = \sum_{\nu=o}^{n} \left(\sum_{\mu=\nu}^{n} d_\mu\ w_{\mu-\nu} \right) \frac{(-1)^\nu}{\nu!} D^\nu\delta \in \Delta$$

$$ww' = \sum_{n\geq o} \left(\sum_{\mu+\nu=n} w_\mu w'_\nu \right) z^n \in \Delta' \quad , w,w' \in \Delta'$$

$$\varphi.w = \sum_{n\geq o} \left(\sum_{\mu+\nu=n} \frac{\varphi^{(u)}(0)}{\mu!} w_\nu \right) z^n \in \Delta' , \quad \varphi \in \mathcal{E}$$

Le dual Δ' est un $\mathcal{E}$-module.

On peut égalment définir un produit de convolution sur $\Delta \times \Delta'$; de

$$(d,q) \to d * q \in \mathcal{L}(\Delta \times \Delta, \Delta) \text{ , on a :}$$

$$d * w = \sum_{m\geq o} \left(\sum_{\nu=o}^{n} c^\nu_{m+\nu} (-1)^\nu\ d_\nu\ w_{m+\nu} \right) z^m$$

Exemple: $D\delta * w = \sum_{m\geq o} (m+1)\ w_{m+1}\ z^m = Dw$.

5. Application a la suite associee d'une suite libre

Soit $\{B_n\}_{n\geq 0}$ une suite normalisée $(B_n(x) = x^n + \ldots)$ et soit $u \in \mathcal{P}'$ telle que $u_0 \neq 0$. On définit la suite associée $\{B_n^{(1)}(u)\}_{n\geq 0}$ à la suite $\{B_n\}_{n\geq 0}$ relativement à u par :

$$u_0 \, B_n^{(1)}(x;u) = u\left(\frac{B_{n+1}(x)-B_{n+1}(\xi)}{x-\xi}\right), \qquad n \geq 0$$

Alors, on a aussi :

$$u_0 B_n^{(1)}(x;u) = u \, \theta_0 \, B_{n+1}(x)$$

On cherche à déterminer la forme canonique $u_0^{(1)}(L)$ de la suite $\{B_n^{(1)}(u)\}_{n\geq 0}$ où L désigne la suite duale $\{\mathcal{L}_n\}_{n\geq 0}$ de $\{B_n\}_{n\geq 0}$.

On a par définition:

$$\langle u_0^{(1)}(L), 1\rangle = 1 \quad ; \quad \langle u_0^{(1)}(L) \, , \, B_n^{(1)}(u)\rangle = 0 \, , \qquad n \geq 1$$

donc

$$\langle u_0^{(1)}(L) \, , \, u\theta_0 \, B_{n+1}\rangle = 0 \, , \quad n \geq 1$$

ou

$$\langle u_0^{(1)}(L)u \, , \, \theta_0 \, B_{n+1}\rangle = 0 \, , \qquad n \geq 1$$

donc

$$\langle x^{-1}(u_0^{(1)}(L)u), B_n\rangle = 0 \, , \qquad n \geq 2$$

Soit $\delta_0^{(1)}(L)$ la forme canonique de $\{\theta_0 \, B_{n+1}\}_{n\geq 0} = \{B_n^{(1)}(\delta)\}_{n\geq 0}$;

alors

$$u_0 \delta_0^{(1)}(L) = u_0^{(1)}(L)u$$

Ensuite, on a facilement $\delta_0^{(1)}(L) = x\mathcal{L}_1$ à l'aide de la forme linéaire $\mathcal{F} = x^{-1}\delta_0^{(1)}(L) + \alpha_0 \mathcal{L}_0 + \alpha_1 \mathcal{L}_1$ où on peut déterminer α_0 et α_1 pour que $\mathcal{F} = 0$ sur $\mathcal{P}$. D'où le résultat :

$$u_0^{(1)}(L) = u_0(x\mathcal{L}_1)u^{-1}$$

Lorsque $\{B_n\}_{n\geq 0}$ est faiblement orthogonale d'index (1,2) par rapport à $\mathcal{L}_0$ (voir §3), on a :

$$x\mathcal{L}_1 = \frac{1}{\chi}\left(x^2 \mathcal{L}_0 - (\mathcal{L}_0)_1 \, x\mathcal{L}_0\right)$$

où

$$\chi = (\mathcal{L}_0)_2 - (\mathcal{L}_0)_1^2 \neq 0$$

Lorsqu'on a de plus : $u = u_0 \mathcal{L}_0$, alors :

$$u^{(1)} \overset{\text{def}}{=} \chi u_o^{(1)}(L) = -u_o x^2 u^{-1}$$

C'est le résultat classique qui est connu depuis longtemps lorsque la suite $\{B_n\}_{n \geq o}$ est orthogonale par rapport à u régulière.

§2. LA TRANSFORMEE DE FOURIER D'UNE FORME LINEAIRE

1. L'opérateur de Fourier est un autre isomorphisme topologique de Δ sur $\mathcal{P}$. Si $\mathcal{F}$ est défini par :

$$\mathcal{F}(f)(Z) = \int_{-\infty}^{+\infty} e^{-2i\pi Zx} f(x)dx, \quad f \in L_1 \, , \; Z \in \mathbb{R}$$

alors $\mathcal{F}$ vérifie $\mathcal{F}(D^n\delta) = (2i\pi)^n \, Z^n$, $n \geq 0$, c'est à dire $\mathcal{F} \in \text{Isom}\ (\Delta, \mathcal{P})$ et donc ${}^t\mathcal{F} \in \text{Isom}\ (\mathcal{P}', \Delta')$; de plus ${}^t\mathcal{F} = \mathcal{F}$ sur Δ , donc sur $\mathcal{P}'$ et on a :

$$<\mathcal{F}(u), d> = <u, \mathcal{F}(d)> \quad , \quad u \in \mathcal{P}' \quad , \quad d \in \Delta \; .$$

Lorsque

$$u = \sum_{n \geq o} u_n \frac{(-1)^n}{n!} D^n\delta \quad , \text{ on a :}$$

$$\mathcal{F}(u) = \sum_{n \geq o} \frac{(-1)^n}{n!} u_n (2i\pi)^n Z^n$$

Exemple. Soit $u(x) = e^{-\pi x^2}$. On a $u_n = \int_{-\infty}^{+\infty} e^{-\pi x^2} x^n dx$

c'est à dire $u_{2n+1} = 0$, $n \geq 0$ et $u_{2n} = \frac{1}{\pi^{n+\frac{1}{2}}} \Gamma(n + \frac{1}{2})$, $n \geq 0$

Donc:

$$\hat{u} = \sum_{n \geq o} \frac{(-1)^n}{n!} u_n (2i\pi)^n Z^n = \sum_{n \geq o} \frac{(2n-1)!!}{(2n)!} (-1)^n Z^{2n} =$$

$$= \sum_{n \geq o} (-1)^n \frac{\pi^n Z^{2n}}{n!} = e^{-\pi Z^2}$$

2. On peut introduire un produit de convolution dans $\mathcal{P}'$ de la façon suivante: partons du fait que

$$(\varphi, f) \to \varphi * f \in \mathcal{L}(S \times \mathcal{P}, \mathcal{P})$$

car

$$||\varphi \times f||_n \leq en! \; ||f||_n \; \cdot \; q_{n,o}(\varphi)$$

où S désigne l'espace vectoriel des fonctions $C^\infty(\mathbb{R})$ à décroissance rapide muni du système de semi-normes:

$$q_{n,m}(\varphi) = \sup_{\alpha \leq n} \sup_{\beta \leq m} \int_{-\infty}^{+\infty} |x|^\alpha \, |\varphi^{(\beta)}(x)| dx \, , \qquad n,m \geq 0$$

On en déduit:

$$\langle v * \varphi, f\rangle = \langle v, \check{\varphi} * f\rangle \quad , \quad v \in \mathcal{P}' \quad , \quad \varphi \in S \quad , \quad f \in \mathcal{P}$$

$$\langle v * f, \varphi\rangle = \langle v, \check{f} * \varphi\rangle$$

On peut préciser le produit $v * f$; en fait, on a :

$$v * f(x) = \langle v, f(x-y)\rangle \quad , \quad x \in \mathbb{R}$$

et

$$(v,f) \to v * f \in \mathcal{L}(\mathcal{P}' \times \mathcal{P}, \mathcal{P})$$

car

$$||v * f||_n \leq en! \; ||f||_n \; |v|_n$$

D'où le produit de convolution de deux éléments de $\mathcal{P}'$:

$$\langle u * v, f\rangle = \langle u, \check{v} * f\rangle \quad , \quad u,v \in \mathcal{P}' \quad , \quad f \in \mathcal{P}$$

On a $\quad u * \delta = \delta * u = u \quad , \quad u \in \mathcal{P}'$

Associativité:

$$u * (f * \varphi) = (u * f) * \varphi$$

$$u * (v * f) = (u * v) * f$$

$$w * (u * v) = (w * u) * v$$

Les moments:

$$\langle u * v, \frac{x^n}{n!}\rangle = \sum_{\mu+\nu=n} \langle u, \frac{x^\mu}{\mu!}\rangle \langle v, \frac{x^\nu}{\nu!}\rangle \quad , \quad n \geq 0$$

egalité qui est prise pour définition du produit de deux fonctions dans [2] . D'où :

$$u * v = v * u$$

et

$$|u * v|_n \leq 2^n |u|_n |v|_n$$

ce qui implique $(u,v) \to u * v \in \mathcal{L}(\mathcal{P}' \times \mathcal{P}', \mathcal{P}')$.
L'application $f \to u * f$ permite avec la translation. Réciproquement si $U \in \mathcal{L}(\mathcal{P},\mathcal{P})$ et permute avec la translation, alors il existe $u \in \mathcal{P}'$ telle que

$$U(f) = u * f$$

Exemple: $Df = D\delta * f$ et donc $Du = D\delta * u$.
On a maintenant les relations habituelles:

$$\mathcal{F}(v * \varphi) = \hat{\varphi} \, . \, \hat{v} \quad , \quad v \in \mathcal{P}' \quad , \quad \varphi \in S \, .$$

$$\mathcal{F}(v * f) = \hat{f} \, \hat{v} \quad , \quad v \in \mathcal{P}' \quad , \quad f \in \mathcal{P} \, .$$

$$\mathcal{F}(u * v) = \hat{u} \, \hat{v} \quad , \quad u,v \in \mathcal{P}' \, .$$

$\mathcal{F}(fu) = \hat{f} * \hat{u}$, $f \in \mathcal{P}$, $u \in \mathcal{P}'$.

$\mathcal{F}(D^n u) = (2i\pi z)^n \hat{u}$, $u \in \mathcal{P}'$ $n \geq 0$.

Soit $A : f \rightarrow A(f)(x) = \sum_{\nu=o}^{p} (-1)^{\nu} \nu! \frac{a_{\nu}}{(2i\pi)^{\nu}} x^{\nu}$.

La transposée de A s'exprime par :

$$A(u) = \sum_{\nu \geq o} \frac{u_{\nu}}{(2i\pi)^{\nu}} D^{\nu}\delta \quad , \quad u \in \mathcal{P}'$$

et on a

$$F = \mathcal{F} \circ A$$

$$A(uv) = A(u) * A(v)$$

et donc

$$F(uv) = F(u)F(v).$$

3. Une interprétation du théorème de Borel. Celui-ci dit que l'application $\varphi \rightarrow \varphi.1$ est surjective (en effet, sa transposée qui est l'identité est trivialement injective et son image est *.faiblement fermée dans $\mathcal{E}'$).

Soit $u \in \mathcal{P}'$, alors il existe $\varphi \in \mathcal{E}$ telle que $\varphi.1 = \hat{u}$, donc il existe $v \in \mathcal{O}'$ ($\mathcal{O}'$: espace vectoriel des ultradistributions à décroissance rapide) telle que $\hat{v}.1 = \hat{u}$. Mais on a $\hat{v}.1 = r(\hat{v})$ où $r(v)$ désigne la restriction à $\mathcal{P}$ de v et donc:

$$u = r(v)$$

Ainsi, toute forme linéaire sur $\mathcal{P}$ est la restriction à $\mathcal{P}$ d'une ultradistribution à décroissance rapide. De plus, $\mathcal{P}'$ est isomrphe algébriquement et topologiquement au quotient $\mathcal{O}'/Ker(r)$.

§3. LE POLYNÔMES ORTHOGONAUX SEMI-CLASSIQUES.

Rappelons les différentes caractérisations des suites orthogonales semi-classiques. [3] , [4] . Soit $\{p_n\}_{n\geq o}$ une suite normalisée, orthogonale par rapport à u régulière. On note $Q_n(x) = \frac{1}{n+1} p'_{n+1}(x)$, $n \geq 0$.

On peut prendre comme définition la propriété suivante:

a) La suite $\{p_n\}_{n\geq o}$ est dite semi-classique, de classe s, s'il existe une forme linéaire $\tilde{u}$ telle que la suite $\{Q_n\}_{n\geq o}$ soit quasi-orthogonale d'ordre s par rapport à $\tilde{u}$. Autrement dit, lorsque la suite $\{Q_n\}_{n\geq o}$ vérifie :

$$<\tilde{u}, x^m Q_n(x)> = 0 \ , \quad 0 \leq m \leq n-s-1 \quad , \qquad n \geq s+1$$

Il existe $r = s$ tel que : $<u, x^{r-s} Q_r(x)> \neq 0$

Lorsque $s = 0$, la suite $\{p_n\}_{n=o}$ est une suite classique (Jacobi, Bessel, Laguerre, Hermite).

Les propositions suivantes son équivalentes a la proposition a).

b) Il existe $s \geq 0$ et $0 \leq t \leq s+2$ entiers et un polynôme ϕ de degré t tels que :

$$\phi(x)Q_n(x) = \sum_{\nu=n-s}^{n+t} \theta_{n,\nu} P_\nu(x) \ , \qquad n \geq s$$

$$\exists \sigma \geq s \quad \text{tel que} \quad \theta_{\sigma,\sigma-s} \neq 0$$

c) Il existe un polynôme ψ de degré $p \geq 1$ et un polynôme Λ de degré $q \geq 1$ tels que la forme u vérifie l'équation:

$$(*) \qquad \psi u + D(\phi u) = 0$$

avec $\phi(x) = x\psi(x) - \Lambda(x)$. La classe s est donnée par

$$1 + s = \max\,(p, q-1)$$

Toute solution u régulière de l'équation ci-dessus sera appelée une forme semiclassique.

La suite $\{R_n\}_{n\geq o}$ définie par $R_n(x) = a^{-n} h_a \circ \tau_{-b}\, p_n(x)$, $n \geq 0$ est encore une suite orthogonale semi-classique de classe s , car sa forme canonique $u_1 = (h_{\frac{1}{a}} \circ \tau_{-b})u$ vérifie l'équation.

$$\psi_1 u_1 + D(\phi_1 u_1) = 0$$

où $\psi_1(x) = a^{1-t}\,\psi(ax+b)$, $\phi_1(x) = a^{-t}\,\phi(ax+b)$, $t = \deg \phi$.

Posant $\psi(x) = \sum_{\nu=o}^{p} a_\nu x^\nu$ et $\phi(x) = \sum_{\nu=o}^{t} c_\nu x^\nu$, l'équation transformée de $(*)$ par $\mathcal{F}$ s'écrit:

$$\xi \sum_{\nu=o}^{t} (-1)^\nu c_\nu \hat{u}^{(\nu)} + \sum_{\nu=o}^{p} (-1)^\nu a_\nu \hat{u}^{(\nu)} = 0 \ , \qquad \xi = 2i\pi z$$

ou l'indice de dérivation indique la dérivée par rapport à ξ . Toute série formelle solution est admissible si elle fournit une forme régulière.

d) Chaque polynôme p_n , $n \geq 1$ vérifie une équation différentielle du second ordre:

$$J(x;n)p''_{n+1}(x) + K(x;n)p'_{n+1}(x) + L(x;n)p_{n+1}(x) = 0 \ , \qquad n \geq 0$$

où J,K et L sont des polynômes dont les degrés ne dépendent pas de n .

L'implication d) $\Longrightarrow$ c) est due a W. Hahn [5] . Voir les articles de A.M. Krall et LL. Littlejohn pour de nombreux exemples.

Introduisons la transformée de Stieltjes formelle de u :

$$S(u)(\frac{1}{z}) = -z\ F(u)(z)$$

e) La série formelle $S(z) = S(u)(z)$ vérifie l'équation suivante:

$$A(z)S'(z) = C(z)S(z)+D(z)$$

où A,C et D sont des polynômes.

c) $\Longrightarrow$ e) et on a: $A(z) = \phi(z)$; $C(z) = -\psi(z) - \phi'(z)$

$$D(z) = -u\theta_o\psi(z) - u\theta_o\phi'(z) - (Du)\theta_o\phi(z)$$

$$D(z) = -\ u\theta_o\psi(z) - D(u\theta_o\phi)(z)$$

Pour démontrer l'implication e) $\Longrightarrow$ c) , considérons les suites orthogonales de Laguerre-Hahn, c'est à dire celles pour lesquelles $S(u)(z)$ vérifie l'équation de Riccati:

$$A(z)S(z) = B(z)S^2(z) + C(z)S(z) + D(z)$$

où A,B,C, et D sont des polynômes.
Alors la forme u vérifie l'équation:

$$-D(Au) + (A'+C)u = x^{-1}(Bu^2) + (\theta_oB(1))\delta$$

avec

$$D(z) = -\ D(u\theta_oA)(z) + u\theta_o(A'+C)(z) - u^2\theta_o^2B(z)$$

Lorsque $B = 0$, on a bien le fait que u vérifie l'équation (*) et

$$\phi(z) = A(z) \quad , \quad \psi(z) = -A'(z) - C(z)$$

<u>Définition</u> [3] [4] . La suite $\{B_n\}_{n\geq o}$ est dite faiblement orthogonale d'index (p,q) parrapport à $v \in \mathcal{P}'$, s'il existe un couple d'entiers $p,q \geq 1$ tels que :

$$\langle v,B_{p-1}\rangle \neq 0 \quad , \quad \langle v,B_n\rangle = 0 \quad , \quad n \geq p$$

$$\langle xv,B_{q-1}\rangle \neq 0 \quad , \quad \langle xv,B_n\rangle = 0 \quad , \quad n \geq q$$

Notant H les sous-espace vectoriel de $\mathcal{P}'$ engendré par la suite duale de $\{B_n\}_{n\geq o}$, cela équivaut à dire qu'il existe $v \in \mathcal{P}'$ non nulle telle que v et $xv \in H$.

f) Il existe una forme linéaire $\tilde{u}$ telle que la suite $\{Q_n\}_{n\geq o}$ soit faiblement orthogonale d'index (p,q) par rapport à $\tilde{u}$.

La classe s de la suite $\{P_n\}_{n\geq o}$ est alors donnée par $1+s = \max(p,q-1)$. On a alors la conséquence suivante: notant H le sous-espace vectoriel engendré par la suite duale de $\{Q_n\}_{n\geq o}$, on peut dire que la suite orthogonale $\{P_n\}_{n\geq o}$ n'est pas semi-classique si et seulement si : quel que soit $v \in H - \{0\}$, $xv \notin H$.

Pour une généralisation des suites orthogonales semi-classiques, voir [6] et l'article de Al. Magnus dans ce volume.

REFERENCES

1. R.D. MORTON, A.M. KRALL, Distributional weight functions for orthogonal polynomials. SIAM, J. Math. Anal. 9 (1978), p. 604-626.

2. S.M. ROMAN, G.C. ROTA. The Umbral calculus. Adv. in Math. 27 (1978), p. 95-188.

3. P. MARONI. Une caractérisation des polynômes orthogonaux semi-classiques C.R. Acad. Sc. Paris, 301, série I, n°6 (1985), p 269-272.

4. P. MARONI. Prolégomènes a l'étude des polynômes orthogonaus semi-classiques. Publ. Labo. Anal. Num. Univ. P. et M. Curie, C.N.R.S. n° 85013 (1985) Paris.

5. W. HAHN. Uber Differentialgleichungen für Orthogonalpolynome. Monat. Math. 95, (1983), p. 269-274.

6. P. MARONI. Introduction à l'étude des δ-polynômes orthogonaux semi-classiques. Actas III Simposium Poli. Orto. y Apli. Segovia, Juin 1985. Edit. F. Marcellán, Dep. Mat., José Gutiérrez Abascal, 2, 28006 Madrid.

7. F. TREVES, Topological vector spaces, distributions and Kernels. Acad. Press (1967).

8. P. MARONI. Sur quelques espaces de distributions qui sont des formes linéaires sur l'espace vectoriel des polynômes. Symposium Laguerre, Bar-le-Duc (1984). Lecture Notes 1171 (1985).

THE L_p MINIMALITY AND NEAR-MINIMALITY OF ORTHOGONAL POLYNOMIAL APPROXIMATION AND INTEGRATION METHODS

J.C. Mason

Computational Mathematics Group
Royal Military College of Science
Shrivenham, Swindon, Wiltshire, England

ABSTRACT

It is known that the Chebyshev polynomials of the first and second kinds are minimal in L_p on $[-1,1]$ with respect to appropriate weight functions, namely certain powers of $1-x^2$, for $1 \leq p \leq \infty$. These properties are here exploited in two applications. First, convergence and optimality properties are established for a "complete" Chebyshev polynomial expansion method for the determination of indefinite integrals. Second, conjectures are derived concerning the near-minimality of the Laguerre polynomials $L_n^{\mp 1/2}(2\beta x)$ for $\beta \simeq 1$ with respect to appropriate exponentially weighted L_p norms on $[0,\infty)$.

1. INTRODUCTION

This paper discusses two distinct ways of exploiting minimal L_p properties of Chebyshev polynomials $T_k(x)$ and $U_k(x)$ of the first and second kinds, where k is the polynomial degree. Such minimal properties, together with a number of results concerning Chebyshev series, are discussed in full by Mason [1] and the two key properties are that, amongst all suitably normalised polynomials $Q_k(x)$ of degree k, for all $1 \leq p \leq \infty$

$$T_k(x) \quad \text{minimises} \quad \left[\int_{-1}^{1} (1-x^2)^{-1/2} |Q_k(x)|^p dx \right]^{1/p} \tag{1}$$

$$\text{and} \quad U_k(x) \quad \text{minimises} \quad \left[\int_{-1}^{1} (1-x^2)^{(p-1)/2} |Q_k(x)|^p dx \right]^{1/p} \tag{2}$$

In the first application, a function is integrated after first being expanded in a complete first and second kind Chebyshev series. Four minimal L_p properties (namely (1), (2) for $p = 1, \infty$) are then used to establish the optimality of the chosen method in certain canonical cases. We also establish L_∞ convergence for the integral and L_1 convergence for the integrand in the method. The present discussion extends and broadens the author's earlier treatment of integration methods in [1].

The second application is in the determination of orthogonal polynomial systems which have nearly minimal L_p norms on $[0,\infty)$, subject to weight functions closely related to e^{-x}. Two new conjectures are obtained, which extend to L_p norms some earlier results of the author in [2] for L_∞. These conjectures have already been tested and found to be valid for polynomials up to degree 10 in L_∞, and they are trivially valid for all polynomial degrees in L^2.

2. INDEFINITE INTEGRATION

2.1. The Chebyshev Method

Suppose that we require the value of the indefinite integral

$$h(x) = \int_{-1}^{x} f(x)\,dx \qquad \text{in} \qquad -1 \leq x \leq 1 , \tag{3}$$

and that $f(x)$ takes the form

$$f(x) = f^A(x) + (1-x^2)^{-1/2} f^B(x) , \tag{4}$$

where f^A and f^B are given continuous functions. This means that we are integrating functions which have $x^{-1/2}$ singularities at end points and a complementary smooth behaviour. (The analysis is actually valid if f^A and f^B are at most L_2-integrable, although the methods can be of limited accuracy in such cases). Now let us approximate f^A and f^B by the partial sums f^A_{n-1} and f^B_n of their expansions in $\{U_k\}$ and $\{T_k\}$ respectively, namely

$$f^A(x) \simeq f^A_{n-1}(x) = \sum_{n-1}^{n} a_k\, U_{k-1}(x) \tag{5}$$

$$f^B(x) \simeq f^B_n(x) = \sum_{k=0}^{n} b_k\, T_k(x) \tag{6}$$

where $a_k = \frac{2}{\pi} \int_{-1}^{1} (1-x^2)^{1/2} f^A(x)\, U_{k-1}(x)\,dx$

and $b_k = \frac{2}{\pi} \int_{-1}^{1} (1-x^2)^{-1/2} f^B(x)\, T_k(x)\,dx$.

On integrating (4) between -1 and x and using the approximations (5), (6), we obtain an indefinite integral in the form

$$h(x) \simeq h^A_n(x) + H^B_{n-1}(x) , \tag{7}$$

where $$h_n^A(x) = \int_{-1}^{x} f_{n-1}^A(x)\,dx = \sum_{k=1}^{n} \frac{a_k}{k} \left[T_k(x) - T_k(-1)\right] \tag{8}$$

and

$$H_{n-1}^B(x) = \int_{-1}^{x} f_n^B(x) = \frac{1}{2} b_o(\pi - \cos^{-1}x) - \sum_{k=1}^{n} \frac{b_k}{k} (1-x^2)^{1/2} U_{k-1}(x) \tag{9}$$

The integration is here greatly simplified by the formulae

$$\frac{d}{dx}\left[T_k(x)\right] = k\, U_{k-1}(x) \ ,$$

$$\frac{d}{dx}\left[(1-x^2)^{1/2} U_{k-1}(x)\right] = -k(1-x^2)^{-1/2} T_k(x)$$

The above method is essentially a generalisation to complete T_k and U_k expansions of a method of Filippi [3] , which was originally based on a U_k expansion. For practical implementation, however, the partial sums (5), (6) should normally be replaced by the (virtually indistinguishable) polynomials obtained by collocation at the respective Chebyshev zeros. This very much simplifies the calculation and, indeed, if the discrete orthogonality properties of T_k and U_k are exploited, then only $0(n)$ arithmetic operations are required in the method. (See [1]). However, it is more difficult to analyse the collocation method in the context of approximation theory, and that is why we have used the expansion method as our theoretical model here.

In the context of definite integration, the method can be viewed as a product integration rule with certain abscissae and weights, and then convergence can be studied from this viewpoint (See [4]).

2.2. The L_1 Convergence of the Integrand

Let us first analyse the error E_n in the approximation of the integrand using (5), (6) , namely

$$E_n(x) = f(x) - f_{n-1}^A(x) - (1-x^2)^{1/2} f_n^B(x)$$

On setting $x = \cos\theta$ (for $0 \le \theta \le \pi$) and multiplying through by $\sin\theta$, we obtain

$$\sin\theta\, E_n(\cos\theta) = \sin\theta\, f(\cos\theta) - \sum_{k=o}^{n} a_k T_k(\cos\theta) -$$

$$- \sum_{k=1}^{n} b_k \, U_{k-1}(\cos\theta) =$$

$$= \sin\theta \, f(\cos\theta) - \sum_{k=0}^{n} (a_k \cos k\theta + b_k \sin k\theta)$$

The right hand side is the error in the partial sum of a Fourier series expansion of a continuous function, namely $\sin\theta \, f(\cos\theta)$, and so by classical theory it converges to zero in L_2 as $n \to \infty$. Convergence to zero immediately follows in the weaker L_1 norm (in θ). Hence

$$||E_n(x)||_1 = \int_{-1}^{1} |E_n(x)|\,dx = \int_{0}^{\pi} |E_n(\cos\theta)|\sin\theta \, d\theta$$

$$= ||\sin\theta \, E_n(\cos\theta)||_1 \longrightarrow 0 \qquad (10)$$

This establishes the L_1 convergence (in x) of the integrand to $f(x)$ as $n \to \infty$.

2.3. The Uniform Convergence of the Integral

Turning now to the indefinite integral (which is obtained from the approximate integrand), the error ε_n in this is given by

$$\varepsilon_n(x) = h(x) - h_n^A(x) - H_{n-1}^B(x) = \int_{-1}^{x} E_n(x)\,dx$$

Now $$|\varepsilon_n(x)| = \left|\int_{-1}^{x} E_n(x)\,dx\right| \leq \int_{-1}^{x} |E_n(x)|\,dx ,$$

and hence

$$||\varepsilon_n(x)||_\infty = \max_{-1 \leq x \leq 1} |\varepsilon_n(x)| \leq \int_{-1}^{1} |E_n(x)|\,dx = ||E_n(x)||_1 . \qquad (11)$$

From (10) we immediately deduce the uniform convergence of ε_n to zero as $n \to \infty$.

The bound (11) is extremely conservative, since the modulus of the integral of an oscillatory function has been bounded by the integral of the modulus. Nevertheless, we show in §2.4 below the remarkable fact that, in two canonical cases, the method optimises both $||\varepsilon_n||_\infty$ and $||E_n||_1$ simultaneously.

2.4. The Optimality of the Method

The tacit assumption was made above that the expansions (5) (6) were particularly appropriate ones to adopt, and indeed their use certainly ensured a very simple integration procedure. But it is not clear that it might not, for example, be better to adopt a $\{T_k\}$ expansion to $f^A(x)$ in (5), and indeed this is the approach used in the original Chebyshev integration method of Clenshaw and Curtis [5]. However, even though a wide variety of orthogonal polynomial expansion methods would probably give reasonably comparable results, the respective choices of U_{k-1} and T_k in (5) (6) are optimal, in the sense that $||\varepsilon_n||_\infty$ and $||E_n||_1$ are minimised in two canonical cases, provided an appropriate small error is introduced (through a constant of integration).

2.4.1. Polynomials of Degree n for f(x)

Consider first the function

$$f(x) = f^A(x) = x^n \qquad (\text{with } f^B(x) = 0) , \tag{12}$$

which is representative of all n^{th} degree polynomials (of one degree higher than the approximation f^A_{n-1}). Now, in this case, the partial expansion of x^n up to degree n in $\{U_k\}$ is exact, and moreover the U_n term must have a unit coefficient of x^n. Hence

$$f^A_{n-1}(x) = x^n - 2^{-n}U_n(x) \tag{13}$$

Now integrating up to x,

$$\varepsilon_n(x) = \int^x E_n(x) + C = 2^{-n}(n+1)^{-1}\, T_{n+1}(x) + C$$

and, on setting $C = 0$,

$$\varepsilon_n(x) = 2^{-n}(n+1)^{-1}T_{n+1}(x) \tag{14}$$

By the minimality properties (1), (2) of $U_n(x)$, $T_{n+1}(x)$ for $p = 1, \infty$ applied to (13), (14), respectively, we deduce that both $||E_n||_1$ and $||\varepsilon_n||_\infty$ have been minimised (over all possible expansions of $f^A(x)$).

Note, however, that an error has been introduced at -1, namely

$$\varepsilon_n(-1) = 2^{-n}(n+1)^{-1}T_{n+1}(-1) = (-1)^{n+1}2^{-n}(n+1)^{-1} \tag{15}$$

However (15) tends to zero as $n \to \infty$, and hence $\varepsilon_n(x)$ still converges uniformly to zero.

2.4.2. Weighted Polynomials of Odd Degree $n+1$ for $f(x)$

Next consider the function

$$f(x) = (1-x^2)^{-1/2} f^B(x) = (1-x^2)^{-1/2} x^{n+1} \qquad \text{(with } f^A = 0\text{)} \tag{16}$$

where n is even, which is representative of all polynomials of degree $n+1$ (of one degree higher than f_n^B) weighted by $(1-x^2)^{-1/2}$. The oddness of f^B ensures that b_o vanishes in (9), so that the term in $\cos^{-1} x$ is not present. The partial expansion of x^{n+1} up to degree $n+1$ in $\{T_k\}$ is exact and the T_{n+1} term must have a unit coefficient of x^{n+1}. Hence

$$f_n^B(x) = x^{n+1} - 2^{-n} T_{n+1}(x) \tag{17}$$

Integrating from -1 to x,

$$\varepsilon_n(x) = \int_{-1}^{x} E_n(x)\,dx = -2^{-n}(n+1)^{-1}(1-x^2)^{-1/2} U_n(x) \tag{18}$$

By the minimal properties (1), (2) of $(1-x^2)^{1/2} T_{n+1}(x)$, $(1-x^2)^{-1/2} U_n(x)$ for $p=1,\infty$ applied to (17), (18), respectively, we deduce that both $||E_n||_1$ and $||\varepsilon_n||_\infty$ have been minimised (over all expansions of $f^B(x)$).

No error has been introduced at $x = -1$ in this case.

3. NEAR-MINIMALITY IN L_p WITH EXPONENTIAL WEIGHTS

Consider the following L_p norms of an appropriately normalised polynomial $P_k(x)$ of degree k, weighted by e^{-x} and $x^{1/2}e^{-x}$, respectively:

$$F_1(P_k) = \left[\int_0^\infty x^{-1/2} |e^{-x} P_k(x)|^p dx\right]^{1/p} \qquad (1 \le p \le \infty) \tag{19}$$

$$F_2(P_k) = \left[\int_0^\infty x^{-1/2} |x^{1/2} e^{-x} P_k(x)|^p dx\right]^{1/p} \qquad (1 \le p \le \infty) \tag{20}$$

Specific cases include

$$p = \infty: \quad F_1(P_k) = ||e^{-x} P_k||_\infty \;, \quad F_2(P_k) = ||x^{1/2} e^{-x} P||_\infty \tag{21}$$

$$p = 2 : \quad F_1(P_k) = \left[\int_o^\infty x^{-1/2} e^{-2x} P_k^2 \, dx \right]^{1/2} ,$$

$$F_2(P_k) = \left[\int_o^\infty x^{1/2} e^{-2x} P_k^2 \, dx \right]^{1/2} \qquad (22)$$

$$p = 1 : \quad F_1(P_k) = \int_o^\infty x^{-1/2} e^{-x} |P_k| dx , \quad F_2(P_k) = \int_o^\infty e^{-x} |P_k| dx \qquad (23)$$

Note that weights e^{-x} occur in F_1 for $p = \infty$ and F_2 for $p = 1$, $x^{-1/2}e^{-\alpha x}$ occur in F_1 for $p = 1,2$, and $x^{1/2}e^{-\alpha x}$ occur in F_2 for $p = 2,\infty$.

Now F_1, F_2 are functionals of P_k , which we desire to minimise . Explicit solutions are only known for $p = 2$, in which case

$$P_k = L_k^{-1/2} (2\beta x) , \quad L_k^{1/2}(2\beta x) \qquad (24)$$

where $\beta = 1$. For other values of p we aim instead to "nearly minimise" F_1, F_2 . The appropriate specification for the term "nearly minimise" is a matter of personal taste. However, we suggest that in practice P_k be accepted for $k \le 10$ if its functional is relatively within 20% of the minimum possible value.

This requirement can in fact be comfortably satisfied by (24) for $p = \infty$ and $\beta \simeq 1$ (see [2]) . For example, for n=10 , β =.975, F_1 and F_2 are within 10% of their minima. Let us now propose two conjectures which extend the above deductions for $p = 2,\infty$ and which have therefore already been confirmed in these 2 cases.

<u>Conjecture 3.1</u>. For any p $(1 \le p \le \infty)$, $F_1(P_k)$ is nearly minimised by $P_k = L_k^{-1/2}(2\beta_1 x)$ for some $\beta_1 = \beta_1(p)$ close to 1 .

<u>Conjecture 3.2</u>. For any p $(1 \le p \le \infty)$, $F_2(P_k)$ is nearly minimised by $P_k = L_k^{1/2}(2\beta_2 x)$ for some $\beta_2 = \beta_2(p)$ close to 1 .

Although we have no rigorous proofs, the following discussions give substantial support to the Conjectures, and are based on the application of bilinear transformations to (1), (2) which take $[-1,1]$ into $[0,\infty)$.

3.1. <u>Discussion of Conjecture 3.1</u>

$$\left[\int_{-1}^{1} (1-t^2)^{-1/2} |Q_k(t)|^p \, dt\right]^{1/p} \qquad (1 \leq p \leq \infty) \qquad (25)$$

is minimised over normalised polynomials Q_k of degree k (by (1)) when $Q_k = T_k$ and hence when

$$\int_{-1}^{1} (1-t^2)^{-1/2} \, Q_j(t) Q_k(t) \, dt = 0 \qquad (j < k) \qquad (26)$$

Setting $t = \frac{Ax-1}{Ax+1}$, $A = r^{-1}$, $r = k+p^{-1}$, so that

$$1-t^2 = \frac{2Ax}{(Ax+1)^2} \quad \text{and} \quad dt = \frac{2\,dx}{(Ax+1)^2} ,$$

it follows from (25) (26) that

$$\left[\int_{0}^{\infty} x^{-1/2} \, |(1+x/r)^{-r} P_k(x)|^p \, dx\right]^{1/p} \qquad (27)$$

is minimised over polynomials P_k of degree k when

$$\int_{0}^{\infty} x^{-1/2} (1+x/r)^{-(j+k+1)} P_j(x) P_k(x) \, dx = 0 \qquad (j<k) \qquad (28)$$

Replacing $(1+x/r)^{-r}$ by the comparable weight e^{-x} in both (27) and (28) , $F_1(P_k)$ is nearly minimised when

$$\int_{0}^{\infty} x^{-1/2} e^{-2\beta_{jk} x} P_j(x) \, P_k(x) \, dx = 0 \qquad (j < k) \qquad (29)$$

where $\beta_{jk} = 1/2 \ (j+k+1) r^{-1}$.

Now $\beta_{jk} = k/(k+p^{-1}) \doteq 1$ for $j = k-1$ (the key value of j) , and hence on replacing β_{jk} by a constant $\beta_1 \simeq 1$ in (29) we obtain Conjecture 3.1.

3.2. <u>Discussion of Conjecture 3.2</u>

$$\left[\int_{-1}^{1} (1-t^2)^{(p-1)/2} |Q_k(t)|^p dt\right]^{1/p} \qquad (1 \leq p \leq \infty) \qquad (30)$$

is minimised over normalised polynomials Q_k of degree k (by (2)), when $Q_k = U_k$ and hence when

$$\int_{-1}^{1} (1-t^2)^{1/2}\; Q_j(t)\; Q_k(t)\; dt = 0 \qquad (j < k) \qquad (31)$$

Setting $t = \frac{Ax-1}{Ax+1}$, $A = r^{-1}$, $r = k+1+p^{-1}$,

it follows from (30), (31) that

$$\left[\int_0^\infty x^{(p-1)/2}\;(1+Ax)^{-(p+1)}\,|(1+Ax)^{-k}P_k(x)|\,(1+Ax)^{-2}dx\right]^{1/p} =$$

$$=\left[\int_0^\infty x^{-1/2}\;|x^{1/2}(1+x/r)^{-r}\;P_k(x)|^p dx\right]^{1/p} \qquad (32)$$

is minimised over normalised polynomials P_k of degree k when

$$\int_0^\infty x^{1/2}(1+x/r)^{-(j+k+3)}P_j(x)P_k(x)\,dx = 0 \qquad (j < k) \qquad (33)$$

Replacing $(1+x/r)^{-r}$ by the comparable weight e^{-x} in both (32) and (33) , $F_2(P_k)$ is nearly minimised when

$$\int_0^\infty x^{1/2}e^{-2\beta_{jk}x}\;P_j(x)\;P_k(x)\,dx = 0 \qquad (j < k) \qquad (34)$$

where $\beta_{jk} = \frac{1}{2}(j+k+3)r^{-1}$

Now $\beta_{jk} = (k+1)/(k+1+p^{-1}) \simeq 1$ for $j = k-1$ (the key value of j), and hence on replacing β_{jk} by a constant $\beta_2 \simeq 1$ in (34) we obtain Conjecture 3.2.

REFERENCES

1. J.C. MASON, Some properties and applications of Chebyshev polynomial and rational approximation. In: "Rational Approximation and Interpolation". P. Graves-Morris, E.B. Saff, and R.S. Varga (Eds.), Springer-Verlag, Berlin, 1984, pp. 27-48.

2. J.C. MASON, Near-minimax approximation and telescoping procedures based on Laguerre and Hermite polynomials. In: "Polynômes Orthogonaux et Applications", C. Brezinski. A. Draux, A.P. Magnus, P. Maroni et A. Ronveaux (Eds.). Springer-Verlag, Berlin, 1985, pp. 419-425.

3. S. FILIPPI, Angenäherte Tschebyscheff-Approximation einer Stammfunktion- eine Modifikation des Verfahrens von Clenshaw und Curtis. Numer. Mathematik 6 (1964), 320-328.

4. I. SLOAN and W.E. SMITH, Properties to interpolating product integration rules. SIAM J. Numer. Anal. 19 (1982), 427-442.

5. C.W. CLENSHAW and A.R. CURTIS, A method for numerical integration on an authomatic computer. Numer. Mathematik 2 (1960), 197-205.

ORTHOGONAL RATIONAL FUNCTIONS WITH POLES IN A FINITE SUBSET OF R

Olav Njåstad

Department of Mathematics
University of Trondheim-NTH
N-7034 Trondheim
Norway

1. INTRODUCTION

R-functions are rational functions with no poles in the extended complex plane outside a given set $\{a_1,\ldots,a_p\}$ of points on the real axis. A positive linear functional Φ on the space of R-functions gives rise to an inner product, and thereby to orthogonal systems of R-functions. Orthogonal R-functions were studied in [7] , [8] , [9] . They have many properties analogous to those of orthogonal polynomials and orthogonal Laurent polynomials. Their zeros have certain characteristic properties, they give rise to Gaussian quadrature formulas, they are connected with Padé approximation problems. In [7] we used methods from the theory of orthogonal R-functions to solve an extended Hamburger moment problem. In [8] we studied the uniqueness question for this problem. It was shown that under certain conditions on the orthogonal system of R-functions associated with the problem, uniqueness of solution is equivalent with a limit point situation.

In this paper we consider the situation that Φ is not only positive, but positive on a certain interval $[\alpha,\beta]$ called a Stieltjes interval for the particular set $\{a_1,\ldots,a_p\}$. In this case the situation becomes simpler, and stronger and more complete results can be obtained. We shall in particular study properties of the zeros of the orthogonal R-functions in this case, and show that there exists at least one solution of the extended Hamburger moment problem with all its points of increase in $[\alpha,\beta]$. Unique solvability of the problem is in this case, without any extra conditions, equivalent to the limit point situation mentioned above.

For methods and results in the classical theory of orthogonal polynomials, see e.g. [1], [2], [3], [10]. For the analogous theory of orthogonal Laurent polynomials, see e.g. [4] , [5] , [6] .

2. ORTHOGONAL R-FUNCTIONS

Let $a_1,\dots,a_p$ be given (distinct) real numbers, ordered by size. Let $\mathcal{R}$ denote the linear space consisting of all functions of the form

$$(2.1)\qquad R(t) = \alpha_o + \sum_{i=1}^{p}\sum_{j=1}^{N_i} \frac{\alpha_{ij}}{(t-a_i)^j}, \qquad \alpha_o, \alpha_{ij} \in \mathbb{C} .$$

Elements of $\mathcal{R}$ are called R-functions. We denote by $\mathcal{R}_R$ the real space of all R-functions with real coefficients.

A function R belongs to $\mathcal{R}$ iff it can be written in the form $R(t) = \frac{P(t)}{Q(t)}$, where Q is a polynomial with all its zeros among the the points $a_1,\dots,a_p$, and where P is a polynomial with $\deg P \leq \deg Q$. We call R <u>degenerate</u> if $\deg P < \deg Q$. Note that $\mathcal{R}$ and $\mathcal{R}_R$ are closed under multiplication.

Let Φ denote a given linear functional on $\mathcal{R}$. This functional gives rise to a bilinear form $<,>$ on $\mathcal{R} \times \mathcal{R}$, defined by $<A,B> = \Phi(A.B)$. The functional Φ is said to be positive on an interval $[\alpha,\beta]$ (here $(-\infty,\beta]$, $[\alpha,\infty)$, $(-\infty,\infty)$ are allowed) if $\Phi(R) > 0$ for every $R \in \mathcal{R}$ with the property that $R(t) \geq 0$ for $t \in [\alpha,\beta] - \{a_1,\dots,a_p\}$, $R(t) \not\equiv 0$ for $t \in [\alpha,\beta] - \{a_1,\dots,a_p\}$. We shall call an interval $[\alpha,\beta]$ a <u>Stieltjes interval</u> if $\beta \leq a_1$, or $\alpha \geq a_p$, or $[\alpha,\beta] \subset [a_{s-1},a_s]$ for some s, $\mathcal{R}$ $(s = 2,\dots p)$.

We shall in the following mainly be interested in the situation that Φ is positive on a given Stieltjes interval $[\alpha,\beta]$. Note that when Φ is positive on some interval $[\alpha,\beta]$, then $<,>$ is an inner product on $\mathcal{R}_R \times \mathcal{R}_R$.

By a <u>distribution function</u> we shall mean a real-valued, bounded, non-decreasing function with a infinite number of points of increase. Let ϕ be a distribution function with all its points of increase in $[\alpha,\beta]$, and assume that every $R \in \mathcal{R}$ is Lebesgue-Stieltjes integrable with respect to ϕ. Then the linear functional Φ given by

$$(2.2)\qquad \Phi(R) = \int_{-\infty}^{\infty} R(t)\,d\phi(t)$$

is positive on $[\alpha,\beta]$. A linear functional Φ on $\mathcal{R}$ is said to be represented by a distribution function if formula (2.3) is valid for every $R \in \mathcal{R}$. We shall later point out that every linear functional on $\mathcal{R}$ which is positive on a Stieltjes interval $[\alpha,\beta]$ can be represented by at least one distribution function with all its points of increase in $[\alpha,\beta]$.

Again let Φ be an arbitrary linear functional on $\mathcal{R}$ wich is positive on $[\alpha,\beta]$. By applying the Gram-Schmidt orthogonalization process with respect to the inner product determined by Φ to the sequence

$$\{1 , \frac{1}{t-a_1} , \ldots , \frac{1}{t-a_p} , \frac{1}{(t-a_1)^2} , \ldots\}$$

we obtain an <u>orthogonal</u> sequence $\{Q_n : n=0,1,2,\ldots\}$ of functions in $\mathcal{R}_{\mathcal{R}}$.

Every natural number n has a unique decomposition $n = p.q_n+r_n$, $1 \leq r_n \leq p$. We write r for r_n , q for q_n when there is no danger of confusion. The meaning of statements about a_{r+1} when $r = p$ and a_{r-1} when $r = 1$ will be clear from the context.

We observe that the orthogonal R-function Q_n may be written in the following ways:

$$Q_n(t) = \beta_o^{(n)} + \frac{\beta_1^{(n)}}{(t-a_1)} + \ldots + \frac{\beta_{n-1}^{(n)}}{(t-a_{r-1})^{q+1}} + \frac{\beta_n^{(n)}}{(t-a_r)^{q+1}} , \qquad \beta_n^{(n)} \neq 0 , \tag{2.3}$$

$$Q_n(t) = \frac{B_n(t)}{(t-a_1)^{q+1}\ldots(t-a_r)^{q+1}(t-a_{r+1})^q\ldots(t-a_p)^p} , \tag{2.4}$$

where B_n is a polynomial of degree at most n .

3. ZEROS OF ORTHOGONAL R-FUNCTIONS

In [7] it was shown that B_n has at least $n-1$ zeros, all of which are real and simple. This result can be strengthened in the case that Φ is positive on a Stieltjes interval.

<u>Theorem 3.1. Let Φ be positive on a Stieltjes interval</u> $[\alpha,\beta]$. <u>Then B_n has n simple zeros in</u> (α,β) . <u>The function Q_n has the same zeros as</u> B_n .

<u>Proof:</u> We obtain the result by refining the argument in the proof of [7 , Proposition 2.4], taking into account the fact that all the factors $(t-a_i)$ have fixed signs in (α,β) .

#

The argument can be modified to show that if Φ is positive on an arbitrary interval $[\alpha,\beta]$, then B_n has at least n-1 zeros in (α,β), and Q_n has at least n-1-s zeros in (α,β), where s is the number of points in $(\alpha,\beta) \cap \{a_1,\ldots,a_p\}$ (cf. the proof of [7, Prop. 2.4]).

We call Q_n <u>regular</u> if a_{r_n-1} is not zero of B_n. Note that Q_n is regular iff $\beta_{n-1}^{(n)} \neq 0$ in (2.3). It follows from Theorem 3.1 that if Φ is positive on a Stieltjes interval $[\alpha,\beta]$, then all Q_n are regular and non-degenerate.

We define the function P_n associated with Q_n by the formula

$$(3.1)\qquad P_n(z) = \Phi_t\left(\frac{Q_n(t)-Q_n(z)}{t-z}\right), \qquad n = 0,1,2,\ldots$$

(here the notation Φ_t indicates that the functional Φ is operating on its argument as a function of t).

<u>Theorem 3.2.</u> <u>If Q_n is regular, in particular if Φ is positive on a Stieltjes interval $[\alpha,\beta]$, then the following hold:</u>

A. Q_n <u>and</u> Q_{n-1} <u>have no common zeros.</u>
B. P_n <u>and</u> P_{n-1} <u>have no common zeros.</u>
C. Q_n <u>and</u> P_n <u>have no common zeros.</u>

<u>Proof</u>: In [8, Theorem 3.4] the following formula analogous to the Liouville-Ostrogradski formula was proved:

$$(3.2)\qquad Q_{n-1}(z)P_n(z)-Q_n(z)P_{n-1}(z) = \frac{\beta_{n-1}^{(n)}(a_r-a_{r-1})}{\beta_{n-1}^{(n-1)}(z-a_r)(z-a_{r-1})}.$$

Since Q_n is regular $\beta_{n-1}^{(n)} \neq 0$, and the results immediately follow.

#

<u>Theorem 4.5.</u> <u>If Φ is positive on a Stieltjes interval $[\alpha,\beta]$, then the following hold:</u>

A. <u>Between two consecutive zeros of</u> Q_{n-1} <u>there is a zero of</u> Q_n, <u>and between two consecutive zeros of</u> Q_n <u>there is a zero of</u> Q_{n-1}.

B. <u>Between two consecutive zeros of</u> Q_n <u>there is a zero of</u> P_n, <u>and between two consecutive zeros of</u> P_n <u>there is a zero of</u> Q_n.

C. <u>Between two consecutive zeros of</u> P_{n-1} <u>there is a zero of</u> P_n, <u>and between two consecutive zeros of</u> P_n <u>there is a zero of</u> P_{n-1}.

Proof: From [8 , Theorems 3.5 and 3.6] we obtain by a limiting process the following Christoffel-Darboux type formula

$$(3.3)\qquad (z-a_{r-1})Q_{n-1}(z)\frac{d}{dz}[(z-a_r)Q_n(z)] - $$
$$- (z-a_r)Q_n(z)\frac{d}{dz}[(z-a_{r-1})Q_{n-1}(z)] =$$
$$= \frac{\beta_{n-1}^{(n)}}{\beta_{n-1}^{(n-1)}}(a_r-a_{r-1})\sum_{k=o}^{n-1} Q_k(z)^2 ,$$

and a similar formula involving the functions P_n . By the aid of these formulas standard arguments give separation properties for the functions of form $R_n(z) = (z-a_r)Q_n(z)$ and $S_n(z) = (z-a_r)P_n(z)$, hence for the functions of form $Q_n(z)$ and $P_n(z)$, since these have the same signs as $R_n(z)$ and $S_n(z)$ in (α,β) . (Note that $P_n(z)$ has only $n-1$ zeros , so they are all in (α,β) by the first part of B.)

#

4. THE EXTENDED HAMBURGER MOMENT PROBLEM

We set $c_o = \Phi(1)$, $c_j^{(i)} = \Phi(\frac{1}{(t-a_i)^j})$, $i = 1,\ldots,p$, $j=1,2,\ldots$. The problem of representing Φ by a distribution function ϕ is then equivalent to finding a distribution function ϕ such that $\int_{-\infty}^{\infty} d\phi(t) = c_o$, $\int_{-\infty}^{\infty} \frac{d\phi(t)}{(t-a_i)^j} = c_j^{(i)}$, $i=1,\ldots,p$, $j=1,2,\ldots$. We call this problem the <u>extended Hamburger moment problem</u> (EHMP) for Φ .

The zeros of the orthogonal R-functions give rise to Gaussian quadrature formulas. We shall here consider only the case that Φ is positive on a Stieltjes interval $[\alpha,\beta]$. Recall that $Q_n(t)$ has n simple zeros $t_1^{(n)},\ldots,t_n^{(n)}$, all in (α,β) .

<u>Theorem 4.1.</u> <u>Let</u> Φ <u>be positive on a Stieltjes interval</u> $[\alpha,\beta]$. <u>Then there exist positive weights</u> $\lambda_{n,1},\ldots,\lambda_{n,n}$ <u>such that the quadrature formula</u>

$$(4.1)\qquad \Phi(R) = \sum_{k=1}^{n} \lambda_{n,k} R(t_k^{(n)})$$

<u>is valid for every</u> $R(t)$ <u>of form</u> (2.1) , <u>where</u> $N_i \leq 2q+2$ <u>for</u> $i < r$, $N_i \leq 2q$ <u>for</u> $i > r$, $N_r \leq 2q+1$.

Proof: Follows from [7 , Proposition 4.4 and 4.7] , [8 , Theorem 2.6] and the fact that Q_n is non-degenerate.

Theorem 4.2. Let ϕ be positive on a Stieltjes interval $[\alpha,\beta]$. Then there exists at least one solution ϕ of the EHMP for Φ with all points of increase in $[\alpha,\beta]$.

Proof: We use a standard procedure, defining the non-decreasing, uniformly bounded functions $\phi_n(t)$ by

$$(4.2) \qquad \phi_n(t) = \sum \{\lambda_{n,k} : t_k^{(n)} \leq t\} .$$

For sufficiently large n we may by Theorem 4.1 write

$$c_j^{(i)} = \Phi\left(\frac{1}{(t-a_i)^j}\right) = \sum_{k=1}^{n} \lambda_{n,k} \frac{1}{(t_k^{(n)}-a_i)^j} = \int_{-\infty}^{\infty} \frac{d\phi_n(t)}{(t-a_i)^j} .$$

Application of Helly's selection and convergence theorems shows that the sequence $\{\phi_n(t)\}$ contains at least one subsequence converging to a distribution function ϕ , and the limit function ϕ of every such convergent subsequence satisfies $\int_{-\infty}^{\infty} d\phi(t) = c_0$, $\int_{-\infty}^{\infty} \frac{d\phi(t)}{(t-a_i)^j} = c_j^{(i)}$, $i=1,\ldots p$, $j=1,2,\ldots$ (For details , see [7 , Section 5].) Furthermore the limit function ϕ of every convergent subsequence has all its points of increase in $[\alpha,\beta]$, since all the points $t_1^{(n)},\ldots,t_n^{(n)}$ lie in (α,β) .

#

We recall that the Stieltjes transform $\hat{\phi}$ of a distribution function ϕ is given by

$$(4.3) \qquad \hat{\phi}(z) = \int_{-\infty}^{\infty} \frac{d\phi(t)}{t-z} , \qquad \operatorname{Im} z \neq 0 .$$

Let $\Sigma(z)$ denote the set of all points of the form $w = \hat{\phi}(z)$ for all solutions ϕ of the EHMP for ϕ .

For every complex number z outside the real axis and every complex number ζ we define

$$(4.4) \qquad f_n(z,\zeta) = - \frac{P_n(z) - \zeta \frac{z-a_{r-1}}{z-a_r} P_{n-1}(z)}{Q_n(z) - \zeta \frac{z-a_{r-1}}{z-a_r} Q_{n-1}(z)}$$

The mapping $\zeta \to f_n(z,\zeta)$ is a linear fractional transformation, with determinant $\frac{(z-a_{r-1})}{(z-a_r)}[P_n(z)Q_{n-1}(z)-P_{n-1}(z)Q_n(z)]$. The mapping is non-singular if Q_n is regular. The real axis is then mapped onto a circle $\Gamma_n(z)$ bounding a closed disc $\Delta_n(z)$. If both Q_n and Q_m are regular and $m > n$, then $\Delta_m(z) \subset \Delta_n(z)$ (see [10 , Theorem 4.2]). There are infinitely many regular Q_n (see [10 , Theorem 2.5]). Since the discs $\Delta_n(z)$ form a nested sequence, the intersection $\Delta(z)$ is either a single point or a closed disc. It can be shown that $\Delta(z)$ is a single point for every z or a closed disc for every z (See [8 , Theorem 4.2].) In view of this we may use the terms limit point case and limit circle case without reference to any particular point z.

It can be shown that the inclusion $\sum(z) \subset \Delta(z)$ is true in general. The inclusion $\Delta(z) \subset \sum(z)$ can be proved if there exists infinitely many regular Q_n such that not both Q_n and Q_{n-1} are degenerate. This is the case when $[\alpha,\beta]$ is a Stieltjes interval, all Q_n then being regular and non-degenerate. For more details on these connections, see [8] .

Taking into account the essential uniqueness of the inverse Stieltjes transform , we see that the above remarks lead to the following result:

Theorem 4.3. Let Φ be positive on a Stieltjes interval $[\alpha,\beta]$. Then $\sum(z) = \Delta(z)$, and the EHMP has a unique solution iff the limit point case occurs.

REFERENCES

1. AKIEZER, N.I., The classical moment problem and some related questions in analysis. Hafner Publishing Company, New York, (1965).

2. BREZINSKI, C., Padé-type approximation and general orthogonal polynomials, Birkhauser Verlag, Basel-Boston-Stuttgart, (1980).

3. CHIHARA, T.S., Introduction to orthogonal polynomials, Mathematics and its applications series, Gordon; (1978).

4. JONES, WILLIAM B., NJÅSTAD, OLAV and THRON, W.J., "Orthogonal Laurent polynomials and the strong Hamburger moment problem", J. Math. Anal. Appl. 98, (1984), 528-554.

5. NJÅSTAD, OLAV and THRON, W.J., "The theory of sequences of orthogonal L-polynomials", Padé approximants and continued fractions (Eds., Haakon Waadeland and Hans Wallin), Det Kongelige Norske Videnskabers Selskab, Skrifter (1983), No 1, 54-91.

6. NJÅSTAD, OLAV and THRON, W.J., "Unique solvability of the strong Hamburger moment problem", J. Austral. Math. Soc., (Series A) 40 (1986), 5-19.

7. NJÅSTAD, OLAV, "An extended Hamburger moment problem", Proc. Edinb. Math. Soc., 28 (1985), 167-183.

8. NJÅSTAD, OLAV, "Unique solvability of an extended Hamburger moment problem", J. Math. Anal. Appl., to appear.

9. NJÅSTAD, OLAV, "A multi-point Padé approximation problem", Analytic Theory of Continued Fractions, Proceedings, (Ed. W.J. Thron), Pitlochry, Scotland, 1985, Lecture Notes in Mathematics, No. 1199, Springer Verlag, Berlin (1986).

0. SHOHAT, J.A. and TAMARKIN, J.D., The problem of moments, Mathematical surveys No 1, Amer. Math. Soc., Providence, R.I., (1943).

N^{th}-ROOT ASYMPTOTICS OF ORTHONORMAL POLYNOMIALS AND NON-DIAGONAL PADE APPROXIMANTS

Herbert Stahl
Technische Fachhochschule Berlin/FB 2
Luxemburger Strasse 10
D-1000 BERLIN 65
Fed.Rep.Germany

Abstract

Connections between n-th root asymptotics of orthonormal polynomials associated with measures of first and second kind are investigated. This problem arises in the study of convergence and divergence of essentially non-diagonal sequences of Padé approximants to Markov functions.

1. Introduction

In the present talk we consider a problem that belongs to the area of n-th root asymptotics of orthonormal polynomials and arises in connection with investigations of the convergence and divergence of essentially non-diagonal sequences of Padé approximants to Markov functions, i.e. to functions of the form

$$f(z) = \int \frac{d\mu(x)}{x - z} \tag{1.1}$$

where μ is a positive measure with compact support $S(\mu) \subseteq \mathbb{R}$.

For functions of this type, there exists a rather complete convergence theory for non-diagonal sequences of Padé approximants in the upper half or triangle of the Padé table, i.e. for Padé approximants with numerator degree $m \geq$ denominator degree n (cf. [GrM], [Sa], [St1], [St2]). The basic idea of the proofs of convergence for such sequences is the following: The inverse denominator polynomials of the Padé approximants $[m/n]$ are orthogonal with respect to the measures $x^{m-n}d\mu(x)$. This orthogonality is used to derive asymptotics for the denominator polynomials. From them, one gets asymptotic estimates for the error function of the Padé approximants, which then allow to specify the convergence and divergence domains. In Section 2, we assemble all results relevant for our discussion.

In the lower triangle of the Padé table, a similar procedure cannot work, since there does not exist an orthogonality of the inverse denomi nator polynomials comparable with that in the upper triangle. However, it is possible to transfer convergence results from the upper triangle to the lower triangle of the Padé table by passing to the reciprocal of the function f and simultaneously to reciprocals of the Padé approximants $[m/n]$. We shall describe the technique in more detail in Section 3.

The reciprocal of a Markov function is, up to two leading terms, again a Markov function, but the new function is defined by a different measure, which we shall denote by ν. It is the so-called measure of the second kind associated with μ. In this connection, the original measure μ is called measure of the first kind. We remark that passing to the reciprocal is the first step of a continued fraction development of the function f, which explains the names first and second kind.

After these introductorial remarks, we come to core part of the talk: The connections between measures of first and second kind with respect to asymptotics of the associated orthonormal polynomials. The interest in this topic arises from the following circumstances: In order to get exact asymptotic estimates for the error function in case of sequences in the upper triangle of the Padé table, it is necessary to assume that the measure μ satisfies some condition, which guarantees regular n-th root asymptotic behaviour of the inverse denominator polynomials. The key question in this paper is wether this property carries over to the measure ν of the second kind. The question will be investigated in Section 4, and we shall get an affirmative answer.

In Section 5, these results are then used to extend the convergence and divergence theorem of Section 2 to sequences in the lower triangle of the Padé table.

2. Notations and Results in the Upper Triangle of the Padé Table.

Let Π_n denote the set of polynomials of degree at most $n \in \mathbb{N}$. A rational function $[m/n](z) = p_{mn}(1/z) / q_{mn}(1/z)$ is called m,n-Padé approximant to a function (1.1), developed at infinity, if the two polynomials $p_{mn} \in \Pi_m$ and $q_{mn} \in \Pi_n$, q_{mn} not identically zero, satisfy

$$f(z)\, q_{mn}(\tfrac{1}{z}) - p_{mn}(\tfrac{1}{z}) = O(z^{-m-n-1}) \quad \text{for} \quad z \to \infty , \qquad (2.1)$$

where $O(.)$ denotes Landau's symbol. Weassume that the polynomials p_{mn} and q_{mn} are normalized in such a way that the so-called inverse denominator polynomial $Q_{mn}(z) = z^n q_{mn}(1/z)$ is monic. It is well known (cf. [Pe], §67) that for $m \geq n$ these polynomials are uniquely determined by the orthogonality

$$\int x^j Q_{mn}(x) x^{m-n} d\mu(x) = 0 \quad \text{for } j=0,\ldots,n-1 \tag{2.2}$$

and for the error function of the Padé approximant, we have the explicit formula (cf. [St1;Lemma 3.4])

$$f(z) - [m/n](z) = \frac{1}{Q^2_{mn}(z)\, z^{m-n}} \int \frac{Q^2_{mn}(x)\, x^{m-n}}{z - x} d\mu(x). \tag{2.3}$$

In order to avoid non-normality phenomena, we assume in the sequel that

$$m - n \in 2\mathbb{Z} \tag{2.4}$$

We will consider the convergence of so-called ray-sequences $\{[m/n];\ (m,n) \in N(\lambda) \subseteq \mathbb{N}^2\}$, which are characterized by the fact that the sequence of indices $N = N(\lambda)$ satisfies

$$\lim_{m+n\to\infty,\,(m,n)\in N} \frac{m}{n} = \lim_{N} \frac{m}{n} = \lambda \in (0,\infty). \tag{2.5}$$

The number λ describes the angle of the ray, along which the sequence tends to infinity within the Padé table. If the sequence is contained in the upper triangle of the Padé table, then we have $\lambda \geq 1$, and if it is contained in the lower triangle, $\lambda \leq 1$. In (2.5), we have used two types of notation for limits of ray sequences, a complete and a short one. In the sequel we shall use only the second one.

We need a special logarithmic potential, which we introduce now : Let $P(S)$ denote the set of all probability measures on $S = S(\mu)$. For θ with $0 \leq \theta < 1$ and $\psi \in P(S)$, we consider the logarithmic potentials

$$p_\theta(\psi;z) = (1-\theta) \int \log|z-x|^{-1}\, d\psi(x) + \theta \log|z|^{-1} \tag{2.6}$$

There uniquely exists an equilibrium distribution $\psi_\theta \in P(S)$, which has minimal energy

$$I_\theta(\psi_\theta) = \inf\{I_\theta(\psi);\ \psi \in P(S)\},$$

where the definition of energy

$$I_\theta(\psi) = (1-\theta)^2 \iint \log|x-y|^{-1} d\psi(x)\, d\psi(y) + 2\theta(1-\theta)\int \log|x|^{-1} d\psi(x)$$

corresponds to the special structure of (2.6) (cf. [St1;Lemma 3.1], [Sa; Section 2]).

Lemma 2.1 ([St1; Lemma 3.1]): The potential $p_\theta(\psi_\theta;.)$ is equal to a constant c_θ on $S(\psi_\theta)$ with possible exceptions on a set of capacity zero, if $\theta > 0$ and $\mathrm{cap}(S(\mu)) > 0$, then the potential is greater than

c_θ in a domain surrounding $z = 0$, and it is smaller than c_θ elsewhere. The set $S(\psi_\theta)$ is the intersection of $S(\mu)$ with one or the union of two closed intervals. We have $\mathrm{cap}(S(\psi_\theta)) > 0$, where cap(.) denotes the (logarithmic) capacity.#

The function

$$G_\theta(z) = \exp\{p_\theta(\psi_\theta;z) - c_\theta\}, \quad 0 \leq \theta < 1, \tag{2.7}$$

the so-called <u>divergence domain</u>

$$D_\theta = \{z \in \mathbb{C};\ G_\theta(z) > 1\}, \tag{2.8}$$

and the so-called <u>convergence domain</u>

$$C_\theta = \overline{\mathbb{C}} \setminus (\overline{D}_\theta \cup J \cup \{0\}) \tag{2.9}$$

play a fundamental role in the results stated in the present section. The set J in (2.9) is the smallest interval containing $S(\mu)$. The subindex θ is related to the parameter λ in the definition of ray sequences $N = N(\lambda)$ by

$$\theta = \theta(\lambda) = |1-\lambda|/|1+\lambda| \ . \tag{2.10}$$

We have $D_\theta = \emptyset$, if and only if $\theta = 0$ or $\mathrm{cap}(S(\mu)) = 0$. The domain D_θ as well as the function G_θ increase monotonically and continuously with θ. Examples with analytic representations of the so-called <u>extremal potential</u> $p_\theta(\psi_\theta;.)$ for special sets $S = S(\mu)$ can be found in [St1], and for the case $S(\mu) = [0,1]$, also in [Sa] and [GrM].

Using orthogonality (2.2), the following asymptotics have been proved for ray sequences of inverse denominator polynomials Q_{mn} in the upper triangle:

<u>Theorem 2.2</u> ([St1;Lemma 3.5 and 3.6], [Sa; Section 2]):

Let μ be a positive measure with compact support $S(\mu) \subsetneq \mathbb{R}$, $N = N(\lambda)$ a ray sequence with $\lambda \geq 1$, and J the smallest interval containing $S(\mu)$. If the measure μ satisfies Ullman's or Erdös' condition on the set $S(\psi_\theta)$, then we have

$$\lim_N \left| \frac{Q_{mn}^2(z)\ z^{m-n}}{\|Q_{mn}^2\|_{L^2(x^{m-n}d\mu)}} \right|^{-1/(m+n)} = G_\theta(z) \tag{2.11}$$

locally uniformly for $z \in \mathbb{C} \setminus J$. The function G_θ has been defined in (2.7) and the parameters θ and λ are connected by (2.10).#

Remark 1. Ullman's and Erdös' conditions have been designed to ensure regular n-th root asymptotic behaviour of orthonormal polynomials. We will formulate them here for compact subsets of the set $S(\mu)$. The stricter condition of both is

Ullman's Condition (cf. [Ul]): Let $S \subseteq S(\mu)$ be a compact set, and $\Gamma(\mu)$ be the set of all carriers of μ (i.e. the set of all Borel sets $B \subseteq \mathbb{R}$ with $\mu(\mathbb{R}\setminus B) = 0$), and $\underline{c}(\mu,S) :=$ $= \inf \{\mathrm{cap}(B \cap S);\ B \in \Gamma(\mu)\}$. The measure μ is said to satisfy Ullman's condition on S, if $\underline{c}(\mu,S) = \mathrm{cap}\,(S(\mu) \cap S)$. ($\underline{c}(\mu)$ is called minimal carrier capacity of μ on S).

Simpler to handle and more classical is

Erdös' Condition ([ErTu]): Let μ be the Radon-Nikodym derivative of μ (with respect to the linear Lebesgue measure). Erdös' condition is said to be satisfied on a compact set $S \subseteq S(\mu)$, if $\mu' > 0$ almost everywhere on $S \cap S(\mu)$.

Remark 2: If neither Ullman's, nor Erdös', nor any other condition for regular asymptotic behavior is satisfied, then instead of (2.11) only

$$\limsup_{N} \left| \frac{Q_{mn}^2(z)\, z^{m-n}}{\|Q_{mn}^2\|_{L^2(x^{m-n}d\mu)}} \right|^{-1/m+n} \leq G_\theta(z) \tag{2.12}$$

holds true locally uniformly for $z \in \mathbb{C}\setminus J$ (cf. [St1; Lemma 3.5]). This shows that the additional condition in Theorem 2.2 is necessary in order to ensure exact asymptotics in (2.11).

From Theorem 2.2 together with the remainder formula (2.3), we rather immediately get the next theorem.

Theorem 2.3: Let the measure μ, the sequence $N = N(\lambda)$, $\lambda \geq 1$, and the parameter θ be the same as those in Theorem 2.2, and let the measure μ satisfy Ullman's or Erdös' condition on the set $S(\psi_\theta)$. Then for the error function of a ray sequence $\{[m/n];\ (m,n) \in N(\lambda)\}$ of Padé approximants to the Markov function (1.1), we have

$$\lim_{N} |f(z) - [m/n](z)|^{1/m+n} = G_\theta(z) \tag{2.13}$$

locally uniformly for $z \in \mathbb{C}\setminus(J \cup \{0\})$.

Corollary: Under the same assumptions as in Theorem 2.2 or 2.3, the ray sequence $\{[m/n];\ (m,n) \in N(\lambda)\}$ converges locally uniformly to f in the domain C_θ, and diverges in D_θ. Both domains have been defined in (2.8) and (2.9) .

Remark 1: The divergence domain D_θ is non-empty if and only if $\lambda > 1$ and $\mathrm{cap}\ (S(\mu)) > 0$. The case $\lambda = 1$ includes diagonal Padé approximants $[n/n]$, $n \in \mathbb{N}$, for which the corollary is known as Markov's Theorem [Ma].

Remark 2: If neither Ullman's, nor Erdös', nor any other condition for regular asymptotic behaviour is satisfied, then instead of (2.11) only inequality

$$\limsup_{N} \ |f(z) - [m/n](z)|^{1/m+n} = G_\theta(Z) \qquad (2.14)$$

can be proved. Thus, the convergence result stated in the corollary remains true, but it is no longer possible to determine the divergence domain.

3. Connections between the Upper on Lower Triangle of the Padé Table.

In order to transfer results from the upper to the lower triangle of the Padé table, we simultaneously consider reciprocals of the function f and of the Padé approximants $[m/n]$. It is easy to verify that

$$\frac{1}{f(z)} = -a_1 z + a_o - \int \frac{d\nu(x)}{x-z} = -a_1 z + a_o - g(z) \qquad (3.1)$$

where ν is a positive measure with compact support $S(\nu) \subseteq \mathbb{R}$. Hence, the function g on the right-hand side of (3.1) is again of Markov type, and ν is the measure of the <u>second kind</u> associated with μ . The two supports $S(\mu)$ and $S(\nu)$ are identical up to at most denumerably many isolated points, which implies that

$$\mathrm{cap}\ (S(\mu)) = \mathrm{cap}\ (S(\nu)) \ . \qquad (3.2)$$

The smallest interval containing $S(\nu)$ is contained in J , the smallest interval containing $S(\mu)$. Further, we have

$$a_1 = \frac{1}{||\mu||}, \quad a_o = \frac{1}{||\mu||^2}\int x d\mu \ , \quad \text{and}$$

$$||\nu|| = \frac{1}{||\mu||^2} \int \left(x - \frac{1}{||\mu||}\int x\, d\mu\right) d\mu \qquad (3.3)$$

Analoguous to (3.1), we get for the Padé approximants the identities

$$\frac{1}{[m/n]_f(z)} = -a_1 z + a_o - [n-1/m-1]_g(z) \tag{3.4}$$

$$(f(z)-[m/n]_f(z)) = f(z)[m/n]_f(z)(g(z)-[m-1/n-1]_g(z)) \tag{3.5}$$

where $[m/n]_f$ and $[n-1/m-1]_g$ denote approximants to the functions f and g, respectively.

From (3.2) it follows that we arrive at the same function G_θ and the same measure ψ_θ, if in Section 2 we start from ν and $S(\nu)$ instead of μ and $S(\mu)$.

Identity (3.5) transfers results related to the function g from the upper triangle to corresponding results in the lower triangle of the Padé table, but now related to the function f, and vice versa. Since g is a Markov function, we can expect that Theorem 2.2 also holds true for inverse denominator polynomials• of Padé approximants $[m/n]_g$ .Whether this is the case, will be investigated in the next section.

By $Q_{mn}(\mu;.)$, and $Q_{mn}(\nu;.) \in \Pi_n$, $m,n \in \mathbb{N}$, we denote the monic polynomials that satisfy orthogonality (2.2) with the measure μ and ν, respectively. Hence, they are the inverse denominator polynomials of the Padé approximants $[m/n]_f$ and $[m/n]_g$, respectively. The polynomial $Q_{mn}(\mu;.)$ is identical with the polynomial Q_{mn} which has been investigated extensively in the last section.

4. Connections between Measures of the First and Second Kind.

In order to get exact n-th root asymptotics for the polynomials $Q_{mn}(\mu;z)$, $m \geq n$, we had to assume in Theorem 2.2 that the measure μ satisfies Ullman's or Erdös' condition on $S(\psi_\theta)$. In the present section we shall investigate,whether the consequences of this assumption carry over to ν and to polynomials associated with the measure ν.

The basic instrument in our investigation is a theorem about equivalent descriptions of locally regular n-th root asymptotic behaviour of orthonormal polynomials, which has been proved in [St3], and will be published elsewhere:

Theorem 4.1: Let μ be a positive measure with compact support $S(\mu) \subseteq \mathbb{R}$, $K \subseteq \mathbb{R}$ a compact set satisfying

$$\text{cap}\,(K \cap S(\mu)) > 0 \quad \text{and} \tag{4.1}$$

$$\text{cap}\,(K \cap S(\mu)) = \text{cap}\,(\mathring{K} \cap S(\mu)) \tag{4.2}$$

μ_K the restriction of the measure μ to the set K, J_K the smallest interval containing $K \cap S(\mu)$, $g_{K\cap S(\mu)}(z,\infty)$ Green's function of the domain $\mathbb{C}\setminus(K\cap S(\mu))$, and $P_n(\mu_K;\cdot) \in \Pi_n$, $n\in\mathbb{N}$, the orthonormal polynomial associated with μ_K. Then the following five assertions are equivalent:

(a) Locally uniformly for $z \in \mathbb{C} \setminus I_K$, we have

$$\lim_{n\to\infty} \left|P_n(\mu_K;z)\right|^{1/n} = e^{g_{K\cap S(\mu)}(z,\infty)}. \tag{4.3}$$

(b) Locally uniformly for $z \in \mathbb{C} \setminus I$, we have

$$\limsup_{n\to\infty} \left|P_n(\mu;z)\right|^{1/n} \leq e^{g_{K\cap S(\mu)}(z,\infty)}. \tag{4.4}$$

(c) For all $z \in K\cap S(\mu)$, with possible exceptions on a set of capacity zero, we have

$$\limsup_{n\to\infty} |P_n(\mu;z)|^{1/n} = 1 . \tag{4.5}$$

(d) For any infinite sequence of polynomials $U_n \in \Pi_n$, $n \in N\subseteq\mathbb{N}$, with $\|U_n\|_{L^2(\mu_K)} \neq 0$ for all $n \in N$, we have

$$\limsup_{n\to\infty,\, n\in N} \left|\frac{U_n(z)}{\|U_n\|_{L^2(\mu_K)}}\right|^{1/n} \leq e^{g_{K\cap S(\mu)}(z,\infty)}. \tag{4.6}$$

locally uniformly for $z \in \mathbb{C}$.

(e) For any infinite sequence of polynomials as in assertion (d), we have

$$\limsup_{n\to\infty,\, n\in N} \left|\frac{U_n(z)}{\|U_n\|_{L^2(\mu_K)}}\right|^{1/n} \leq 1. \tag{4.7}$$

for all $z \in K\cap S(\mu)$, with possible exceptions on a set of capacity zero. #

With the help of Theorem 4.1, we can prove the next theorem, which is an extension of theorem 2.2, and the main result of the present section.

Theorem 4.2: Let the measure μ and the ray sequence $N = N(\lambda)$, $\lambda \geq 1$, be the same as those in Theorem 2.2, and let λ and θ be connected by (2.2). If the measure μ satisfies Ullman's or Erdös' condition on the set $S(\psi_\theta)$, then not only (2.11), but also

$$\lim_{N} \left| \frac{Q_{mn}(\nu;z)\,z^{m-n}}{\|Q_{mn}(\nu;.)^2\|_{L^2(x^{m-n}d\nu)}} \right|^{-1/m+n} = G_\theta(z) \tag{4.8}$$

holds true locally uniformly for $z \in \mathbb{C} \setminus J$. The function G_θ has been defined in (2.7).

Remark: Of course, Remark 2 of Theorem 2.2 is also valid for the polynomials $\{Q_{mn}(\nu;.)\}$.

Proof: Let us consider the diagonal sequence $\{[n/n]_f(z) = p_{nn}(1/z)/q_{nn}(1/z)\ ;\ n \in \mathbb{N}\}$ of Padé approximants. From (2.2) it follows that

$$P_n(\mu;z) = \frac{Q_{nn}(\mu;z)}{\|Q_{nn}(\mu;z)\|_{L^2(\mu)}}, \quad n \in \mathbb{N} \tag{4.9}$$

Since $p_{nn}(0) = 0$, identity (3.4) together with (3.3) implies that

$$Q_{n-1,n-1}(\nu;z) = \frac{-1}{\|\mu\|} z^n\, p_{nn}\left(\frac{1}{z}\right) \in \pi_n \tag{4.10}$$

is the inverse denominator polynomial of the Padé approximant $[n-1/n-1]_g$. Using upper and lower estimates for the integral on the right-hand side of representation (2.3) (with $z \notin J$ fixed), one immediately gets

$$|P_n(\mu;z)| = \frac{|Q_{nn}(\mu;z)|}{\|Q_{nn}(\mu;.)\|_{L^2(\mu)}} = \frac{c_n(z)}{\sqrt{|f(z)-[n/n]_f(z)|}} \tag{4.10}$$

for $n \in \mathbb{N}$ and $z \in \mathbb{C} \setminus J$, where the real function c_n has upper and lower bounds

$$0 < a(z) \leq c_n(z) \leq b(z) < \infty \tag{4.11}$$

that are independent of $n \in \mathbb{N}$. For the function g and the measure ν, there exists an identity analoguous to (4.10). If we divide both identities, then from (3.5) and the bounds (4.11), it follows that

$$\lim_{n\to\infty} \left| \frac{P_n(\mu;z)}{P_{n-1}(\nu;z)} \right|^{1/n} = 1 \tag{4.12}$$

locally uniformly for $z \in \mathbb{C} \setminus J$. Hence, the orthogonal polynomials $P_n(\mu;.)$ and $P_{n-1}(\nu;.)$ have identical n-th root asymptotics.

After these preparations, we now use Theorem 4.1: Let μ_θ and ν_s denote the restrictions of the measures μ and ν to the set $S(\psi_\theta)$, respectively. By Lemma 2.1, $S(\psi_\theta)$ is the intersection of $S(\mu)$ with one or the union of two closed intervals, and $\mathrm{cap}\,(S(\psi_\theta)) > 0$. Hence, both assumptions (2.1) and (2.2) of Theorem 4.1 are satisfied for the set $K = S(\psi_\theta)$.

Since μ satisfies Ullman's or Erdös's condition on $S(\psi_\theta)$, it follows that assertion (a) of Theorem 4.1 holds true for the sequence $\{P_n(\mu|_K;.)\}$ (cf. [Ul]). From assertion (b) in Theorem 4.1 and (4.12), we then get

$$\limsup_{n\to\infty} |P_n(\nu;z)|^{1/n} \leq e^{g_{S(\psi_\theta)}(z;\infty)} \tag{4.13}$$

and further with assertion (e) of Theorem 4.1

$$\limsup_{N} \left| \frac{Q^2_{mn}(\nu;z)\, z^{m-n}}{\|Q^2_{mn}(\nu;.)\|_{L^2(x^{m-n}d\nu)}} \right|^{1/m+n} \leq 1 \tag{4.14}$$

for $z \in S(\psi_\theta)$, with possible exceptions on a set of capacity zero. In [St1; Lemma 3.5 and 3.6], it has been shown that the opposite inequality in (4.14) holds true for all $z \in \mathbb{C}$ as a consequence of the orthogonality (2.2), and further that equality holds true in (4.14), with possible exceptions on a set of capacity zero, if and only if (4.8) holds true. Thus, by (4.14) we have proved (4.8).

Q.E.D.

5. Divergence and Convergence Results for the Whole Padé Table.

In this last section, we use Theorem 4.2 to extend the results of Section 2 to the whole Padé table, which means that now also ray sequences $N(\lambda)$ with $0 < \lambda \leq 1$ may be considered.

From Theorem 4.2, Theorem 2.3, and identity (3.5), we rather immediately get the next theorem. It is only necessary to regard that $f(z) \neq 0$ and $[m/n]_f(z) \neq 0$ for $z \notin J$ and $m \geq n$ sufficiently large.

Theorem 5.1: Let the measure μ and the ray sequence $N = N(\lambda)$, $\lambda \geq 1$, be the same as those in Theorem 2.2, but now without any restriction on $\lambda \in (0,\infty)$, and let λ and θ be connected by (2.2). If the measure μ satisfies Ullman's or Erdös' condition on the set $S(\psi_\theta)$, then the following three assertions hold true:

(i) The ray sequence $\{[m/n]\ ;\ (m,n) \in N(\lambda)\}$ converges to the function (1.1) locally uniformly in the convergence domain C_θ, defined in (2.9), and we have

$$\lim_{N} |f(z) - |m/n|(z)|^{1/m+n} = G_\theta(z) \qquad (5.1)$$

locally uniformly for $z \in C_\theta$.

(ii) If $\lambda > 1$, then the sequences $\{[m/n]; (m,n) \in N(\lambda)\}$ diverges to ∞ for $n+m \to \infty$, $(n,m) \in N(\lambda)$, in the divergence domain D_θ, defined in (2.8).

(iii) If $\lambda < 1$, then the sequences $\{[m/n]\ ;\ (m,n) \in N(\lambda)\}$ converges to 0 for $n+m \to \infty$, $(n,m) \in N(\lambda)$, in the divergence domain D_θ. We note that $f(z) \neq 0$ for all $z \in \mathbb{C} \setminus J$, and therefore the convergence to 0 is equivalent to divergence.

Remark 1: The divergence behavior in D_θ corresponds to that of section of power series (case (ii)), and of reciprocals of sections of power series (case (iii)).

Remark 2: The function g, which we use in identity (3.5) to prove Theorem 5.1, has only a technical status of the approximation problem. It does even not appear in the formulation of the approximation problem. The same is true with respect to the measure ν. Hence, it would therefore be rather inappropriate to make technical assumptions with respect to the measure ν, as for instance that it should satisfy Ullman's or Erdös' condition, in order to get Theorem 4.2 without the deeper results of Theorem 4.1.

References

[ErTu] P. Erdös and P. Turan: On Interpolation III; Ann. of Math 41 (1940) pp. 510-553.

[GrM] P.R. Graves-Morris: The Convergence of Ray Sequences of Padé Approximants of Stieltjes Functions; J. Comp. and Appl. Math. 7, (1981), pp. 191-201.

[Ma] A. Markov: Deux demonstrations de la convergence de certaines fractions continues; Act. Math. 19, (1895), pp. 93-104.

[Pe] O. Perron: Die Lehre von den Kettenbruchen; Chelsea, New York 1929.

[Sa] E.B. Saff: Incomplete and Orthogonal Polynomials; in: Approximation Theory IV, Eds.: C.K. Chui et al., Academic Press, New York 1983, pp. 219-255.

[St1] H. Stahl: Beiträge zum Problem der Konvergenz von Padéapproximierenden, Dissertation, TU-Berlin 1976.

[St2] H. Stahl: Non-Diagonal Padé Approximants to Markov Functions; in: Approximation Theory V, Eds.: C.K. Chui et al., Academic Press, New York 1986, pp. 571-574.

[St3] H Stahl: N^{th}-Root Asymptotics of Orthogonal Polynomials Associated with General Measures; Manuscript, TFH-Berlin, 1987.

[Ul] J.L. Ullman: A Survey of Exterior Asymptotics for Orthogonal Polynomials Associated with a Finite Interval and a Study of the Case of General Weight Measures; in: Approximation Theory and Spline Functions. Eds.: S.P. Singh et al., D. Reidel, Dodrecht 1984, pp. 467-478.

ZEROS OF ORTHOGONAL POLYNOMIALS ON HARMONIC ALGEBRAIC CURVES

Torrano, E. and Guadalupe, R.*
Dept. Matemáticas para Químicos. Fac.C.Químicas. Universidad Complutense. 28040 Madrid. Spain.
(*) Facultad de Informática. Universidad Politécnicas. Madrid. Spain

SUMMARY

Starting with an infinite, complex, positive-definite, hermitian matrix, we build a triangular +1 matrix $\hat{D}_n$ generalising the tridiagonal one of the Hankel case. We apply this construction to moments matrices arising from distributions on curves $Im(A(z)) = 0$, $A(z) \in C[z]$. We study the matrix $A(\hat{D}_n)$ and its relationship with the truncated matrix of the corresponding symmetric operator and we give conditions for the zeros to be on the curve.

Proposition 1:

Let $M = (c_{ij})_{i,j=1}^{\infty}$ be an infinite, hermitian, quasi-defined matrix. Let us call $M_n = (c_{ij})_{i,j=1}^{n}$ and $M'_n = (c_{ij+1})_{i,j=1}^{n}$ being $\{\tilde{P}_n(z)\}_{n=0}^{\infty}$ the S.O.P.M. corresponding to M. Let $M_n = L_n R_n$ be the LR decomposition of M_n such that L_n and R_n are lower and upper triangular, respectively, and $R_n = (r_{ij})_{i,j=1}^{n}$ which $r_{ij} = 1$, $1 \leq i \leq n$. Let us call $E_n = (e_i \delta_{ij})_{i,j=1}^{n}$ which $e_i = |M_i|/|M_{i-1}|$ and $|M_0| = 1$, we have then:

i) $\tilde{D}_n = L_n^{-1} M'_n R_n^{-1}$ is an upper Hessenberg matrix, and if $\tilde{D}_n = (\tilde{d}_{ij})_{i,j=1}^{n}$ we have $\tilde{d}_{i,i-1} = 1$, $2 \leq i \leq n$ which $n \geq 2$.

ii) $\tilde{D}_n$ is a principal submatrix of $\tilde{D}_{n+1}$.

iii) $\tilde{P}_n(z) = |zI_n - \tilde{D}_n|$, $n = 1, 2, \ldots$

iv) If, furthermore, M is positive definite and $\hat{D}_n$ is the matrix of (5), then:

$$\tilde{D}_n = (\sqrt{E_n}^{-1} \hat{D}_n \sqrt{E_n}), \quad n = 1, 2, \ldots$$

Proof.

i) We have (see 5 page 1) $M_n^{-1} M'_n = F_n$ where F_n is the Frobenius matrix associated to $P_n(z)$, since $M_n = L_n R_n$ we have $R_n^{-1} L_n^{-1} M'_n = F_n$ and therefore $\tilde{D}_n = L_n^{-1} M'_n R_n^{-1} = R_n F_n R_n^{-1}$. In the second equality R_n y R_n^{-1} are upper triangular with $r_{ii} = 1$, $i = 1, \ldots, n$ and F_n is Frobenius, it is immediate than $\tilde{D}_n$ is upper Hessenberg and furthermore that $d_{i,i-1} = 1$ $i = 1, 2, \ldots, n$.

ii) M'_n is a principal submatriz of M'_{n+1}, L_n and R_n are principal submatrices of L_{n+1} and R_{n+1}, respectively, and, being triangular, L_n^{-1} and R_n^{-1} are principal submatrices of L_{n+1}^{-1} and R_{n+1}^{-1}, resp., and therefore lower and upper, resp. $\tilde{D}_n = L_n^{-1}M_n'R_n^{-1}$ is a principal submatrix of $\tilde{D}_{n+1}$ which is $\tilde{D}_{n+1} = L_{n+1}^{-1}M_{n+1}'R_{n+1}^{-1}$.

iii) We have seen that $\tilde{D}_n$ and F_n are similar, since $\tilde{D}_n = R_nF_nR_n^{-1}$. Therefore $|zI_n - \tilde{D}_n| = |zI_n - F_n| = \tilde{P}_n(z)$.

iv) Since M is positive-definite, M_n admits the Choleski (see [4] page 172), the LR [1] page 35) and the Gauss-Banachiewicz (see [1] page 38) decompositions: $M_n = T_nT_n^H = L_nR_n = V_nE_nV_n^H$. Since M_n is positive-definite, $|M_n| > 0$ $i = 1, \ldots, n$ $e_i > 0$ $\forall i$ and we may speak of the positive root of E_n, it is clear, therefore, that $T_n = V_n \sqrt{E_n}$, $L_n = V_nE_n$, $R_n = V_n^H$ whence $\tilde{D}_n = L_n^{-1}M_n'R_n^{-1} = E_n^{-1}V_n^{-1}M_n'V_n^{-H}$ and (see [5] page,2) $D_n' = T_n^{-1}M_n'T_n^{-}$ substituting $(\sqrt{E_n})^{-1}M_n'V_n^{-H}(\sqrt{E_n})^{-H}$ and since E_n is real diagonal, we have:

$$\tilde{D}_n = (\sqrt{E_n})^{-1}\ \hat{D}_n(\sqrt{E_n}) \quad n \in N \qquad \text{q.e.d.}$$

Notation

The matrix M defines in $C[z]$ the scalar product $<z^i, z^j> = c_{j+1,i+1}$ we shall call Π the pre-hilbert space $(C[z], < >)$ and $\{\hat{P}_k(z)\}_{k=0}^{\infty}$ will be the S.N.O.P. and $\{P_k(z)\}_{k=0}^{\infty}$ the S.M.O.P. corresponding to M.

Given a matrix M_n of order n, $(M_n)_k$ $k \in N$ y $k \leq n$, will be the matrix resulting from eliminating in M_n the last n-k rows and columns.

Proposition 2a

The matrix $\tilde{D}_n$ of the previous proposition is the matrix resulting from truncating the matrix expression of the operator U: $\Pi\to\Pi$ (shift-right), with respect to the S.M.O.P. similarly, $\hat{D}_n$ plays the same rôle with respect to the S.N.O.P.

Proof

Let us consider the matrix of the change of basis from the canomical one to the S.M.O.P.

$$\begin{pmatrix} P_0(z) \\ P_1(z) \\ \cdot \\ \cdot \\ P_n(z) \end{pmatrix} = A_{n+1} \begin{pmatrix} 1 \\ z \\ \cdot \\ \cdot \\ z^n \end{pmatrix}$$

A_{n+1} is lower triangular and $a_{ii} = 1$, $1 \leq i \leq n+1$, the orthogonality relation of the S.M.O.P. translates into matrix terms as $A_{n+1}M_{n+1}A^H_{n+1} = E_{n+1}$ since we know, with our notation, that $||P_k(z)||^2 = e_{k+1}$ since A_{n+1} is invertible we have $M_{n+1} = A^{-1}_{n+1}E_{n+1}A^{-H}_{n+1}$. The uniqueness of the decomposition $M_{n+1} = V_{n+1}E_{n+1}V^H_{n+1}$ allows us to conclude that $V_{n+1} = A^{-1}_{n+1}$, from proposition 1, we know that $\tilde{D}_{n+1} = R_{n+1}F_{n+1}R^{-1}_{n+1}$ and, since $M_{n+1} = L_{n+1}R_{n+1} = (V_{n+1}E_{n+1})V^H_{n+1}$ it is clear that $R_{n+1} = V^H_{n+1} = A^{-H}_{n+1}$ and therefore $\tilde{D}_{n+1} = A^{-H}_{n+1}F_{n+1}A^H_{n+1}$.

If we call U_{n+1} the matrix resulting from taking the first n+1 rows and columns in the expression of U with respect to the canonical basis, since F_{n+1} is Frobenius and A^H_{n+1} and A^{-H}_{n+1} are upper triangular, it is clear that the matrices $A^{-H}_{n+1}U_{n+1}A^H_{n+1}$ and $A^{-H}_{n+1}F_{n+1}A^H_{n+1}$ differ only in the last column, in other words $(A^{-H}_{n+1}U_{n+1}A^H_{n+1})_n = (A^{-H}_{n+1}F_{n+1}A^H_{n+1})_n = (\tilde{D}_{n+1})_n$. Since by ii) of proposition 1 $(\tilde{D}_{n+1})_n = \tilde{D}_n$ it results in general $\forall k \geq n+1$ $(A^{-H}_k U_k A^H_k)_n = \tilde{D}_n$. The second assertion is immediate making use of iv) of proposition 1.

Notation

We shall say, for brevity, that a distribution $\sigma: \gamma \to R$, where $\gamma \subset \mathbb{C}$ is a curve formed by the union of Jordan arcs, verifies (I) if the following conditions are fulfilled:

1º. $\sigma(z)$ is non-decreasing when γ is traced out positively and it has an infinite set of points of effective growth.

2º. $\left| \int_\gamma z^i \bar{z}^j \, d\sigma(z) \right| < +\infty \quad i, j = 0, 1, 2, \ldots$

It what follows we shall assume to be an algebraic harmonic curve, and therefore of equation $\operatorname{Im}(A(z)) = 0$ with $A(z) = a_p z^p + a_{p-1}z^{p-q} + \ldots + a_0$ $a_k \in \mathbb{C}$ $k = 0, 1, \ldots, p$. We shall always denote the degree of $A(z)$ with the letter p.

Corollary

Let $\Pi : \gamma \to R$ be a distribution on γ verifying (I) and let $M = (c_{ij})^\infty_{i,j=1}$ the corresponding moments matrix. Let $\mathcal{A} : \Pi \to \Pi$ be the operator such that $\mathcal{A}(q(z)) = A(z)q(z)$ and let $\hat{H}_n$ be the truncated matrix of order n of the matrix expression of $\mathcal{A}$ with respect to the S.N.O.P. then:

$$\hat{H}_n = (A(\hat{D}_{n+p+i}))_n \quad -1 \leq i \quad e \quad i \in Z$$

Proof

The study of the operator multiplication by a polynominal associated to an algebraic harmonic curve γ, its symmetry and the analysis of its matrix expression with respect to the S.N.O.P. has been undertaken by

J. Vinuesa (see [6] pages 33-43).

Let us point out that, using Halmos' notation (see [2] (167)), $\hat{D}_n$ is triangular +1. Therefore $\hat{D}_n^k$ is triangular +k, on the other hand, a nested sequence $\{\hat{D}_k\}_{k=1}^{\infty}$ of this type of matrices clearly verifies:

$$(\hat{D}_n^2)_{n-1} = (\hat{D}_{n+1}^2)_{n-1} = (\hat{D}_{n+2}^2)_{n-1} = \ldots \quad n \geq 2$$

similarly:

$$(\hat{D}_n^k)_{n-k+1} = (\hat{D}_{n+1}^2)_{n-k+1} = (\hat{D}_{n+2}^k)_{n-k+1} = \ldots \quad n \geq k$$

hence, is $\hat{D}_n$ is the truncated matrix of order n of the operator multiplication by z with respect the S.N.O.P., it is immediate that $(\hat{D}_{n+p-1}^p)_n$ will be the truncated matrix of order n of the operator multiplication by z^p $p \in N$, with respect to the S.N.O.P. we have implicitly made use of an obvious fact, namely, that any natural power k of an infinite triangular +1 matrix is a perfectly defined infinite triangular +k matrix, since its elements are always finite sums and no problem of convergence arises.

Hance we can conclude that the truncated matrix of order n of the operator multiplication by A(z), a polynominal of degree p with complex coefficients, is $(A(\hat{D}_{n+p-1}))_n$ or, if preferred, $(A(\hat{D}_{n+p+i}))_n$ with $i \geq -1$ and $i \in Z$, this is true in particular for $i = 0$. The previous matrix is precisely that one which results from truncating, with order n, the Jacobi's α-matrix appearing in [6], corresponding to the operator multiplication by A(z) with respect to the S.N.O.P. and which we called $\hat{H}_n$, i.e.:

$$\hat{H}_n = (A(\hat{D}_{n+p+i}))_n \quad i \geq -1 \ , \ i \in Z \qquad \text{q.e.d.}$$

Observation

Let $n > p$ and consider the matrix $A(\hat{D}_n)$. Let us note the following:

1º. $A(\hat{D}_n)$ is triangular +p.

2º. $\hat{H}_{n-p+1} = (A(\hat{D}_n))_{n-p+1}$ is Jacobi with 2p+1 diagonals and hermitian, however $A(\hat{D}_n)$ is not, in general, hermitian.

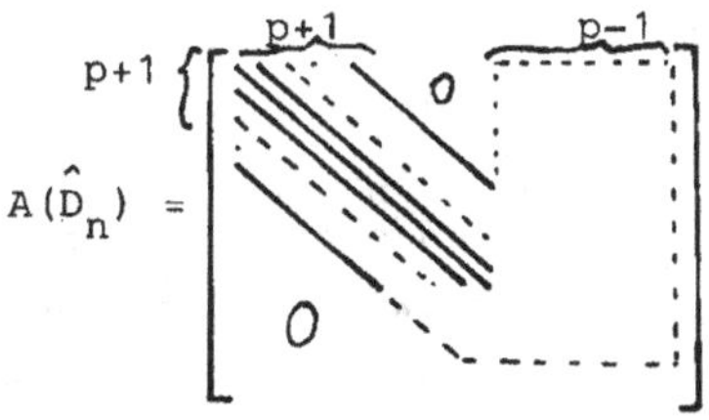

Let us see what happens when $A(\hat{D}_n)$ is hermitian.

Proposition 3a

Let $M = (c_{ij})_{i,j=1}^{\infty}$ be the moments matrix corresponding to a distribution $\sigma : \gamma \to R$ verifying (I), with equation $\text{Im}(A(z)) = 0$.

If $A(\hat{D}_n)$ is hermitian, with $n > p$, then all the zeros of $P_n(z) \in$ S.O.P. besides being simple they are on the curve.

The absolutely exceptional case is excluded when all the multiple roots of $P_n(z) = 0$ are multiple roots of $A(z) = 0$ and they are so with the same or greater multiplicity.

Proof

We shall first prove by reduction to absurdity that $P_n(z)$ has simple roots.

$\tilde{P}_n(z) = |I_n z - \hat{D}_n|$ let $\{z_{ni}\}_{i=1}^{n} = \sigma(\hat{D}_n)$. Let us suppose, without loss of generality, that $P_n(z)$ has only one multiple root z_{nj} of multiplicity m since $\hat{D}_n = T_n^H F_n T_n^{-H}$, $\hat{D}_n$ was, being equivalent, the same Jordan form that F_n, this one is Frobenius. Hence (see [3] page 240), to each different eigenvalue of F_n there corresponds one and only one block in Jordan's decomposition. The block associated to the eigenvalue z_{nj} cannot be:

$$\begin{pmatrix} z_{nj} & & & & & & \\ & z_{nj} & & & & & \\ & & z_{nj} & & & & \\ & & & \cdot & & & \\ & & & & \cdot & & \\ & & & & & z_{nj} & \\ & & & & & & z_{nj} \end{pmatrix} \quad \text{but} \quad \begin{pmatrix} z_{nj} & 1 & & & & & \\ & z_{nj} & 1 & & & & \\ & & z_{nj} & 1 & & & \\ & & & \cdot & & & \\ & & & & \cdot & 1 & \\ & & & & & z_{nj} & 1 \\ & & & & & & z_{nj} \end{pmatrix}$$

since the first one would be not one but m blocks. If we call J_n the Jordan form $\hat{D}_n$, we will have $\hat{D}_n = Q_n J_n Q_n^{-1}$, whence (see [1] pages 97-100) $A(\hat{D}_n) = Q_n A(J_n) Q_n^{-1}$, where $A(J_n)$ is the matrix written down in the following page.

This matrix, except when $A'(z_{nj}) = A''(z_{nj}) = \ldots = A^{(m-1)}(z_{nj}) = 0$, and this we assumed not to happen, is not diagonal. Since it is triangular, its eigenvalues are $\{A(z_{ni})\}_{i=1}^{n}$ a simple calculation shows that it cannot be diagonalised $\dim(\text{Ker}(A(J_n) - I_n A(z_{nj}))) = 1 \neq m$, therefore $A(\hat{D}_n) = Q_n A(J_n) Q_n^{-1}$ cannot be diagonalised either, against the hypothesis that $A(\hat{D}_n)$ is hermitian, which, as we know (see [1] page 274) implies diagonalisability. Thus $\hat{D}_n$ does not possess multiple eigenvalues.

$$A(J_n) = \begin{bmatrix} A(z_{n1}) & & & & & & & \\ & \ddots & & & & & & \\ & & A(z_{n,j-1}) & & & & & \\ & & & A(z_{nj}) & A'(z_{nj})/1! \ldots & A^{(m-1)}(z_{jn})/(m-1)! & & \\ & & & & A(z_{nj}) \ldots\ldots & A^{(m-2)}(z_{nj})/(m-2)! & & \\ & & & & \ddots & \vdots & & \\ & & & & & A(z_{nj}) & & \\ & 0 & & & & & A(z_{n,j+m}) & \\ & & & & & & \ddots & \\ & & & & & & & A(z_{nn}) \end{bmatrix}$$

Let us see, in the second place, that the zeros of $P_n(z) = 0$ are located on the curve. Since $\hat{D}_n$ does not have multiple eigenvalue it is equivalent to a diagonal $\hat{D}_n = Q_n Z_n Q_n^{-1}$ with $Z_n = (z_{ni}\delta_{ij})_{i,j=1}^n$. It is clear that $A(\hat{D}_n) = Q_n A(Z_n) Q_n^{-1} . A(\hat{D}_n)$ is equivalent to the diagonal $(A(z_{ni}) . \delta_{ij})_{i,j=1}^n$ which is its Jordan form. Since it is hermitian, it is unitarily equivalent to a real diagonal. Because of the uniqueness of the Jordan form, we have $A(z_{ni}) \in R$, $I_m(A(z_{ni})) = 0$ hence $z_{ni} \in \gamma$ $1 \leq i \leq n$ q.e.d.

Observation

We have not yet been able to prove the reciprocal of the previous proposition. We have proved, however, a restricted form of the reciprocal. We know that for algebraic curves in general (see [7] page 3), if M_{n+1} is a γ-typical extension of M_n (as we know, typical and relative to γ), then the roots of $P_n(z)$ are on the curve and are simple. We shall prove, adding this hypothesis, that in that case $A(\hat{D}_n)$ is hermitian.

Proposition 4a

With the same assumptions as in the previous proposition and adding, furthermore, the condition that, for a certain $n \in N$, M_{n+1} is a γ-typical extension of M_n, $A(\hat{D}_n)$ is hermitian.

Proof

We shall start with the conditions that the zeros of $P_n(z) = 0$ are simple and located on the curve and we shall see what can we conclude from these assumptions. Afterwards we shall add our further hypothesis.

If the zeros of $P_n(z) = 0$ $\{z_{ni}\}_{i=1}^{n}$ are simple, the Vandermonde matrix:

$$V_n = \begin{pmatrix} 1 & z_{n1} & z_{n2}^2 & z_{n1}^3 & \cdots & z_{n1}^{n-1} \\ 1 & z_{n2} & z_{n2}^2 & z_{n2}^3 & \cdots & z_{n2}^{n-1} \\ \cdot & & & & & \\ \cdot & & & & & \\ 1 & z_{nn} & z_{nn}^2 & z_{nn}^3 & \cdots & z_{nn}^{n-1} \end{pmatrix}$$

is invertible. Let us call $Z_n = (z_{ni}\ \delta_{ij})_{i,j=1}^{n}$ and F_n the Frobenius matrix asociated to $P_n(z)$, we know (see [1] page 69) that $V_nF_n = Z_nV_n$ and hence $F_n = V_n^{-1}Z_nV_n$. Since $\hat{D}_n = T_n^HF_nT_n^{-H}$ we have $\hat{D}_n = T_n^HV_n^{-1}\ Z_nV_nT_n^{-H}$ and therefore $A(\hat{D}_n) = T_n^HV_n^{-1}\ A(Z_n)\ V_nT_n^{-H}$. By assumption $z_{ni} \in \gamma$, $1 \leq i \leq n$. and hence $A(z_{ni}) \in R$ with $1 \leq i \leq n$. Without restricting the reciprocal $A(\hat{D}_n)$ is equivalent to a real diagonal. Let us see that with the added hypothesis, it is furthermore unitarily so.

M_{n+1} is a γ-typical extension of M_n and, if $\{z_{ni}\}_{i=1}^{n}$ are the roots of $P_n(z) = 0$, we know that they are simple and that there exist constants (see [7] page 2) $\{p_i\}_{i=1}^{n}$ $p_i > 0$ $1 \leq i \leq n$, and $e_n > 0$ such that:

$$c_{ij} = \sum_{k=1}^{n} z_{nk}^{j-1}\ \bar{z}_{nk}^{i-1}\ p_k \quad \text{with} \quad 1 \leq i,j \leq n$$

except for c_{nn} which is:

$$c_{nn} = \sum_{k=1}^{n} z_{nk}^{j-1}\ \bar{z}_{nk}^{i-1}\ p_k + e_n$$

we may write $M_n = V_n^HL_nV_n$ where $L_n = (p_i\ \delta_{ij})_{i,j=1}^{n}$ since M_n is positive-definite and hermitian, it admits the Choleski decomposition and we shall have $M_n = T_nT_n^H = V_n^HL_nV_n$, whence $(T_n^{-1}V_n^H\ \sqrt{L_n})\ (\sqrt{L_n}V_nT_n^{-H}) = I_n$, i.e., the matrix $\sqrt{L_n}V_nT_n^{-H}$ is unitary, since L_n is diagonal with positive real elements. It is clear that $A(Z_n) = (\sqrt{L_n})^{-1}A(Z_n)\ \sqrt{L_n}$, and since $A(\hat{D}_n) = T_n^H\ V_n^{-1}\ A(Z_n)\ V_nT_n^{-H}$ we conclude that:

$$A(\hat{D}_n) = (T_n^H\ V_n^{-1}\ (\sqrt{L_n})^{-1})\ A(Z_n)(\sqrt{L_n}\ V_n\ T_n^{-H}) \qquad \text{q.e.d.}$$

References

[1] Gantmacher, F.R. "The theory of matrices", Chelsea Publishing Company, New York, N.Y. (1974).

[2] Halmos, Paul R."A Hilbert Space Problem Book", Springer-Verlag New York (Second Edition) (1982).

[3] Lancaster, P. & Tismenetsky M: "The theory of matrices", Computer Science and Applied Mathematics, Academic Press (1985).

[4] Stoer, J. & Bulirsch, R. "Introduction to Numerical Analysis",

Springer-Verlag, New York (1976).

[5] Torrano, E. & Vinuesa, J. "Ceros de polinomios ortogonales en el campo complejo", VII Congresso do Grupo de Matematicos de Expressao Latina, Coimbra (1985).

[6] Vinuesa, J. "Polinomios Ortogonales relativos a curvas algebraicas armónicas", Departamento de Teoría de Funciones de la Universidad de Zaragoza (1973).

[7] Vinuesa, J. "Demostración de la conjetura de Vigil", Reunión de la Agrupación de Matemáticos de Expresión Latina, Palma de Mallorca (1977).

OPEN PROBLEMS

1. ALLEN, P.B. (Communicated by M.L. Mehta).

Define $P_n(x) = x^n - \dots$ a monic polynomial,

$$\int_{-\infty}^{\infty} \frac{P_n(x)P_m(x)}{\mathrm{Ch}^2\,\pi x}\,dx = h_n\,\delta_{nm}\,,\quad \text{orthogonal with weight } (\mathrm{ch}\ \pi x)^{-2}.$$

i.e.

$$P_n(x) = \frac{n!^3}{(2n)!}\ (-i)^n\ {}_3F_2\left(-n,n+1,\tfrac{1}{2}+ix\ ;\ 1,1;1\right)\,,$$

and

$$h_n = \frac{2}{\pi}\ \frac{n!^6}{(2n)!\,(2n+1)!}$$

or, $P_{n+1}(x) = xP_n(x) - R_nP_{n-1}(x)\,,\qquad R_n = \dfrac{n^4}{4(2n-1)(2n+1)}$

Let

$$I^{\pm}_{mn}(y) = \frac{1}{4}\ \left\langle \left\{P_m\left(\tfrac{x+y}{2}\right) \pm P_m\left(\tfrac{x-y}{2}\right)\right\}\left\{p_n\left(\tfrac{x+y}{2}\right) \pm p_n\left(\tfrac{x-y}{2}\right)\right\} \right\rangle\,,$$

where by definition

$$\langle f(x)\rangle \equiv \frac{\mathrm{sh}\ \pi y}{2y}\int_{-\infty}^{\infty} f(x)\ \frac{dx}{\mathrm{Ch}\ \pi x + \mathrm{Ch}\ \pi y}\ ;$$

in particular

$$\langle x^n\rangle \equiv \frac{\mathrm{Sh}\ \pi y}{2y}\int_{-\infty}^{\infty} x^n\ \frac{dx}{\mathrm{Ch}\ \pi x + \mathrm{Ch}\ \pi y} = \frac{1}{y}\left(\frac{d}{dz}\right)^n \left(\frac{\mathrm{Sh}\ yz}{\sin\ z}\right)\Bigg|_{z=0}$$

= an even polynomial in y.

Question. Can one give a general expression for $I^{\pm}_{mn}(y)$, or a recursive relation?

Remarks.

(i) $I^{\pm}_{mn}(y) = 0$ if $m+n$ is odd
= an even polynomial of order $m+n$ if $m+n$ is even.

(ii) $I^{-}_{mo} = 0\ ,\quad I^{+}_{oo} = 1\ ,\quad I^{+}_{mn}(0) = \frac{\pi}{2}\,h_m\delta_{mn} = \dfrac{n!^6}{(2n)!\,(2n+1)!}\,\delta_{nm}$.

(iii) $I^{\pm}_{mn}(i) = \frac{1}{4}\ \{1\pm(-)^m\}\ \{1\pm(-)^n\}\ P_m(\tfrac{i}{2})P_n(\tfrac{i}{2}) =$

$= \frac{1}{4}\ \{1\pm(-)^n\}\ \{1\pm(-)^n\}.(-i)^{m+n}.\ \dfrac{(m!\ n!)^3}{(2m)!\ (2n)!}$

$I^{+}_{mn}(y)$ (resp. $I^{-}_{mn}(y)$) has y^2+1 as a factor if m is odd (resp. even).

(iv) For small m,n, one can compute $I^{\pm}_{mn}$. Thus

$$I^{+}_{o2} = \frac{1}{3} y^2 , \qquad I^{+}_{22} = \frac{1}{180} (21 y^4 + 5y^2 + 4) ,$$

$$I^{-}_{22} = \frac{1}{12} y^2 (y^2+1)$$

(v) $A_n(y) = \langle x^{2n} \rangle$ = an even polynomial in y with all its coefficients positive.

It is easier to deal with $A_n(y)$ than with $I^{\pm}_{mn}(y)$.

2. DEHESA, J.S. Orthogonal Polynomials with an Asymptotic Distribution of Zeros of Gaussian Type.

Let $P_n(x)$ a system of orthogonal polynomials characterised by the three-term recurrence relation

$$P_{n+1}(x) = (x-a_n)P_n(x) - b^2_{n-1}P_{n-1}(x)$$

How are the coefficients a_n, b_n so that the asymptotic density of zeros (i.e. the number of zeros per unit interval in the limit of infinite n) of the polynomial $P_n(x)$ be a Gaussian function? .

This problem is of interest in nuclear physics, the reason being connected to the fact that the Hamiltonian of a nucleus in a subspace of the Hilbert space of the system has a Gaussian density of energy levels.

3. ISMAIL, M.E.H. and LETESSIER, J. Monotonicity of zeros of ultraspherical polynomials.

Let $X(\lambda)$ be a positive zero of an ultraspherical (Gegenbauer) polynomial $C^{\lambda}_n(x)$. Markov's Theorem, Szegö [2, §6.21] and a quadratic transformation for a hypergeometric function can be used to prove that $X(\lambda)$ is a decreasing function of λ if $\lambda > 0$. Laforgia [1] conjectured that $\lambda X(\lambda)$ increases with λ when $\lambda > 0$. Laforgia mentioned in [1] that numerical evidence supports his conjecture. We believe the following stronger conjecture is true.

Conjecture. If $X(\lambda)$ is a positive zero of an ultraspherical polynomial $C_n^\lambda(x)$ then $\lambda^c X(\lambda)$ is an increasing function of λ for all positive λ when $c \geq 1/2$.

The above conjecture is false without the assumption $c \geq 1/2$ since $C_2^\lambda(x) = \lambda[2(\lambda+1)x^2-1]$ so $X(\lambda)$ is $[2(\lambda+1)]^{-1/2}$ and $\lambda^c X(\lambda)$ is not an increasing function of λ on $(0,\infty)$ for any $c < 1/2$. Clearly it suffices to prove the conjecture for $c = 1/2$.

Our conjecture is based on numerical evidence and one can easily prove the conjecture for $n = 2,3,4$. We tabulated the values of $\lambda^{1/2}X(\lambda)$ for $6 \leq n \leq 18$, $\lambda = j/10$, $j = 1,2,\ldots,199$ and the numerical evidence strongly supports our conjecture.

We now show how to settle the conjecture when $X(\lambda)$ is the largest positive zero of $C_n\lambda(x)$. Let $f_n(x) = \lambda^{-n/2}C_n\lambda(x\lambda^{-1/2})$. The f_n's satisfy

$$xf_n(x) = \frac{\lambda(n+1)}{2(n+\lambda)} f_{n+1}(x) + \frac{n+2\lambda}{2(n+\lambda)} f_{n-1}(x) .$$

Therefore the zeros of $f_n(x)$ are eigenvalues of a tridiagonal matrix A whose i,j entry $a_{i,j}$, $0 \leq i,j < n$, satisfies

$$2(\lambda+i)a_{i,j} = \lambda(i+1)\delta_{i+1,j} + (2\lambda+i)\delta_{i,j+1}$$

The entries of A are nonnegative and increase with λ when $\lambda > 0$. The result now follows from the Perron-Frobenius theorem. [3, chapter 3].

References.

1. A. Laforgia, Monotonicity properties for the zeros of orthogonal polynomials and Bessel functions, in "Polynômes orthogonaux et Applications", Edited by C. Brezinski, A. Draux, A. Magnus, P. Maroni and A. Ronveaux, Lecture Notes in Mathematics, Springer-Verlag, New York, 1985, pp. 267-277.

2. G. Szegö, Orthogonal Polynomials, fourth edition, Colloquium Publication, Volume 23, American Mathematical Society, Providence, 1975.

3. R. Varga, Matrix Iterative Methods, Prentice Hall, Englewood Cliffs, New Jersey, 1962.

4. MAGNUS, A.P., Stability of M(a,b) with respect to masspoints.

M(a,b) is the set of measures $d\alpha$ such that the recurrence coefficients of the related orthonormal polynomials $a_{n+1}p_{n+1} =$ $= (x-b_n)p_n - a_n p_{n-1}$ satisfy $a_n \to a$ and $b_n \to b$ when $n \to \infty$ [P. Nevai, Memoirs AMS 213 (1979)] . The essential support of a $d\alpha \in M(a,b)$ is [b-2a,b+2a] [p. 29 in A. Maté, P. Nevai & V. Totik : Twisted difference operators and perturbed Chebyshev polynomials, preprint 94 (June 1986) U.C. Louvain] . If we add to such a $d\alpha$ a series of masspoints accumulating at $b \pm 2a$ and of finite total mass, is the result still in $M(a,b)$?

5. MAGNUS, A.P., Pure point measures in $M(a,b)$.

Give an explicit example of a pure point measure in $M(a,b)$ (dense set of masspoints in [b-2a, b+2a]). A solution in terms of random functions is in F. Delyon, B. Simon and B. Souillard, Ann. Inst. Poincaré Phys. théorique 42 (1985) 283-309 [kindly communicated to me by W. Van Assche] .

6. MAGNUS, A.P., Rakmanov theorem of several intervals.

Let $S = [\alpha_1,\beta_1] \cup [\alpha_2,\beta_2] \cup \dots \cup [\alpha_m,\beta_m]$, $-\infty < \alpha_1 < \beta_2 < \dots < \beta_m < \infty$ $n < \infty$. A Chebyshev measure $d\tau(x;r_1,\dots,r_{m-1};\varepsilon_1,\dots,\varepsilon_{m-1})$ of essential support S has the form $\pi^{-1}|\prod_1^{m-1}(x-r_k)|^{-1} \prod_1^m |(x-\alpha_k)(x-\beta_k)|^{1/2}$, $x \in S$, plus possible masspoints at each r_i with weight

$$|\prod_{k\neq i}(r_i-r_k)|^{-1} \prod_i^m |(r_i-\alpha_k)(r_i-\beta_k)|^{1/2}(1+\varepsilon_i) , \quad \beta_i \leq r_i \leq \alpha_{i+1} ,$$

$\varepsilon_i = +1$ or -1 . The corresponding recurrence coefficients have a predictable law $a_n(r,\varepsilon)$ and $b_n(r,\varepsilon)$ [Ahiezer, Sov. Math. 1 (1960) 989-992 , 2 (1961) 687-690 ; A.P. Magnus, Lecture Notes Math 765, 150-171; P. Turchi et al., J. Phys. C: Solid State Phys. 15 (1982) 2891-2924,...] . What is the condition on a measure $d\alpha$ on S ensucing the existance of $r_1,\dots,r_{m-1}$ and $\varepsilon_1,\dots,\varepsilon_{m-1}$ such that $a_n - a_n(r,\varepsilon)$ and $b_n - b_n(r,\varepsilon) \to 0$ when $n \to \infty$? [This is also Nex's problem in Bar-le-Duc proceedings] . Szegö's condition

$$\int_S \log \alpha'(x) \prod_1^m |(x-\alpha_k)(x-\beta_k)|^{-1/2} dx > -\infty$$ has been shown to be sufficient [A.I. Aptekarev, Math. USSR Sb. 53 (1986) 233-260] . What about Rahmanov's condition $\alpha'(x) > 0$ a.e. on S ?

7. MEHTA, M.L.

Let ω be the nth roof of unity , $\omega = \exp\ (2\pi i/n)$, $i = \sqrt{-1}$; and $\varphi_k(x)$ be the harmonic oscilator function,

$$\varphi_k(x) = e^{x^2/2} \left(-\frac{d}{dx}\right)^k e^{-x^2} = e^{-x^2/2} H_k(x) ,$$

i.e. the k-th Hermite polynomial multiplied by a Gaussian. Consider a square matrix A and a rectangular matrix F with elements

$$A = \left[\frac{1}{\sqrt{n}} \omega^{jk}\right]_{j,k=0,1,\ldots,n-1} ,$$

$$F = \left[\sum_{p=-\infty}^{\infty} \varphi_k\left(\sqrt{\frac{2\pi}{n}}\,(pn+j)\right)\right]_{\substack{j=0,1,\ldots,n-1\\ k=0,1,2,\ldots}}$$

One verifies that

$$(AF)_{jk} = i^k F_{jk}$$

i.e. the columns of F are the eigenvectors of A . Now A being a finite $n \times n$ matrix, at most n of the columns of F are linearly independent. The question is to choose n such columns and express every other column of F as a linear combination of them preferably in an "elegant" way .

8. VAN DOORN, E.A.

In [1] an inequality is proven which amounts to

$$\frac{c_n(-N-1,a)}{c_n(-N,a)} \leq c_n(-1,a+N) , \qquad (*)$$

where $a > 0$, n and N are non-negative integers and $c_n(x,\alpha)$ is a Charlier polynomial, recurrently defined by

$$c_o(x,\alpha) = 1 , \qquad \alpha c_1(x,\alpha) = \alpha - x$$

$$\alpha c_{n+1}(x,\alpha) = (n+\alpha-x)c_n(x,\alpha) - nc_{n-1}(x,\alpha) , \qquad n = 1,2,\ldots$$

Q_1 . The proof in [1] is rather cumbersome. Does there exist a short proof for (*) ?

Q_2 . Can (*) be generalized, e.g., to

$$\frac{c_n(-x-1,a)}{c_n(-x,a)} \le \frac{c_n(-x,a+1)}{c_n(-x+1,a+1)} \quad ?$$

1 F. Le Gall and J. Bernussou, An estimate for the time congestion of overflow processes. IEEE Trans. Commun., vol. COM-31 (1983), 1202-1203.

Solution by M.L. Mehta:

Put $P_n(x,\alpha) = \alpha^n C_n(-x,\alpha)$, (1)

then

$$P_o(x,\alpha) = 1 , \qquad P_1(x,\alpha) = x+\alpha \tag{2}$$

and

$$P_{n+1}(x,\alpha) = (x+n+\alpha)P_n(x,\alpha)-n\alpha P_{n-1}(x,\alpha) , \qquad n = 1,2,... \tag{3}$$

One can directly verify (by induction for example), that

$$P_n(x,\alpha) = \sum_{r=o}^{n} \binom{n}{r} \alpha^r \frac{\Gamma(x+n-r)}{\Gamma(x)} \tag{4}$$

satisfies the recurrence relation (3) and the initial conditions (2) . From (4) we deduce

$$P_n(x,\alpha) \geqq 0 \quad \text{for} \quad x \geqq 0 \quad , \quad \alpha \geqq 0 \tag{5}$$

$$P_n(x+1,\alpha)-P_n(x,\alpha) = n\, P_{n-1}(x+1,\alpha) , \tag{6}$$

$$P_n(x,\alpha)-xP_{n-1}(x+1,\alpha) = \alpha P_{n-1}(x,\alpha) . \tag{7}$$

We will prove that for $x \geqq 1$, $\alpha \geqq 0$, one has

$$\frac{P_n(x+1,\alpha)}{P_n(x,\alpha)} \le \frac{P_n(x,\alpha+1)}{P_n(x-1,\alpha+1)} , \qquad n \geqq 0 \tag{8}$$

$n = 0$ is trivial, so suppose $n \geqq 1$.

Subtracting 1 from each side and using (6) , inequality (8) is equivalente to

$$\frac{P_{n-1}(x+1,\alpha)}{P_n(x,\alpha)} \le \frac{P_{n-1}(x,\alpha+1)}{P_n(x-1,\alpha+1)}$$

or (in view of (5)) to

$$\frac{P_n(x,\alpha)}{P_{n-1}(x+1,\alpha)} \geq \frac{P_n(x-1,\alpha+1)}{P_{n-1}(x,\alpha+1)}$$

Now subtract x from each side and use (7) and (6) to get an equivalent statement

$$\frac{\alpha P_{n-1}(x,\alpha)}{P_{n-1}(x+1,\alpha)} \geq \frac{(\alpha+1)P_{n-1}(x-1,\alpha+1)}{P_{n-1}(x,\alpha+1)} - 1 =$$

$$= \frac{\alpha P_{n-1}(x-1,\alpha+1)}{P_{n-1}(x,\alpha+1)} - \frac{(n-1)P_{n-2}(x,\alpha+1)}{P_{n-1}(x,\alpha+1)}$$

In view of (5) , the last inequality is a consequence of

$$\frac{P_{n-1}(x,\alpha)}{P_{n-1}(x+1,\alpha)} \geq \frac{P_{n-1}(x-1,\alpha+1)}{P_{n-1}(x,\alpha+1)}$$

or of

$$\frac{P_{n-1}(x+1,\alpha)}{P_{n-1}(x,\alpha)} \leq \frac{P_{n-1}(x,\alpha+1)}{P_{n-1}(x-1,\alpha+1)} \qquad (9)$$

Inequality (9) is the same as (8) except that n is replaced by n-1 . Since (8) is true for n = 0 and n = 1 , it is true for every $n \geq 0$.

ol. 1173: H. Delfs, M. Knebusch, Locally Semialgebraic Spaces. XVI, 29 pages. 1985.

ol. 1174: Categories in Continuum Physics, Buffalo 1982. Seminar. dited by F.W. Lawvere and S.H. Schanuel. V, 126 pages. 1986.

'ol. 1175: K. Mathiak, Valuations of Skew Fields and Projective ljelmslev Spaces. VII, 116 pages. 1986.

'ol. 1176: R.R. Bruner, J.P. May, J.E. McClure, M. Steinberger, H_∞ Ring Spectra and their Applications. VII, 388 pages. 1986.

'ol. 1177: Representation Theory I. Finite Dimensional Algebras. roceedings, 1984. Edited by V. Dlab, P. Gabriel and G. Michler. XV, 40 pages. 1986.

ol. 1178: Representation Theory II. Groups and Orders. Proceedgs, 1984. Edited by V. Dlab, P. Gabriel and G. Michler. XV, 370 ages. 1986.

ol. 1179: Shi J.-Y. The Kazhdan-Lusztig Cells in Certain Affine Weyl roups. X, 307 pages. 1986.

ol. 1180: R. Carmona, H. Kesten, J.B. Walsh, École d'Été de obabilités de Saint-Flour XIV – 1984. Édité par P.L. Hennequin. X, 38 pages. 1986.

ol. 1181: Buildings and the Geometry of Diagrams, Como 1984. eminar. Edited by L. Rosati. VII, 277 pages. 1986.

ol. 1182: S. Shelah, Around Classification Theory of Models. VII, 279 ages. 1986.

ol. 1183: Algebra, Algebraic Topology and their Interactions. Proceengs, 1983. Edited by J.-E. Roos. XI, 396 pages. 1986.

ol. 1184: W. Arendt, A. Grabosch, G. Greiner, U. Groh, H.P. Lotz, Moustakas, R. Nagel, F. Neubrander, U. Schlotterbeck, Onearameter Semigroups of Positive Operators. Edited by R. Nagel. , 460 pages. 1986.

ol. 1185: Group Theory, Beijing 1984. Proceedings. Edited by Tuan .F. V, 403 pages. 1986.

ol. 1186: Lyapunov Exponents. Proceedings, 1984. Edited by L. nold and V. Wihstutz. VI, 374 pages. 1986.

ol. 1187: Y. Diers, Categories of Boolean Sheaves of Simple gebras. VI, 168 pages. 1986.

ol. 1188: Fonctions de Plusieurs Variables Complexes V. Séminaire, 79–85. Edité par François Norguet. VI, 306 pages. 1986.

ol. 1189: J. Lukeš, J. Malý, L. Zajíček, Fine Topology Methods in Real alysis and Potential Theory. X, 472 pages. 1986.

ol. 1190: Optimization and Related Fields. Proceedings, 1984. dited by R. Conti, E. De Giorgi and F. Giannessi. VIII, 419 pages. 86.

ol. 1191: A.R. Its, V.Yu. Novokshenov, The Isomonodromic Deforation Method in the Theory of Painlevé Equations. IV, 313 pages. 86.

l. 1192: Equadiff 6. Proceedings, 1985. Edited by J. Vosmansky and . Zlámal. XXIII, 404 pages. 1986.

l. 1193: Geometrical and Statistical Aspects of Probability in anach Spaces. Proceedings, 1985. Edited by X. Fernique, B. einkel, M.B. Marcus and P.A. Meyer. IV, 128 pages. 1986.

ol. 1194: Complex Analysis and Algebraic Geometry. Proceedings, 85. Edited by H. Grauert. VI, 235 pages. 1986.

ol.1195: J.M. Barbosa, A.G. Colares, Minimal Surfaces in $\mathbb{R}^3$. X, 124 ges. 1986.

ol. 1196: E. Casas-Alvero, S. Xambó-Descamps, The Enumerative eory of Conics after Halphen. IX, 130 pages. 1986.

l. 1197: Ring Theory. Proceedings, 1985. Edited by F.M.J. van staeyen. V, 231 pages. 1986.

l. 1198: Séminaire d'Analyse, P. Lelong – P. Dolbeault – H. Skoda. eminar 1983/84. X, 260 pages. 1986.

l. 1199: Analytic Theory of Continued Fractions II. Proceedings, 85. Edited by W.J. Thron. VI, 299 pages. 1986.

l. 1200: V.D. Milman, G. Schechtman, Asymptotic Theory of Finite mensional Normed Spaces. With an Appendix by M. Gromov. VIII, 6 pages. 1986.

Vol. 1201: Curvature and Topology of Riemannian Manifolds. Proceedings, 1985. Edited by K. Shiohama, T. Sakai and T. Sunada. VII, 336 pages. 1986.

Vol. 1202: A. Dür, Möbius Functions, Incidence Algebras and Power Series Representations. XI, 134 pages. 1986.

Vol. 1203: Stochastic Processes and Their Applications. Proceedings, 1985. Edited by K. Itô and T. Hida. VI, 222 pages. 1986.

Vol. 1204: Séminaire de Probabilités XX, 1984/85. Proceedings. Edité par J. Azéma et M. Yor. V, 639 pages. 1986.

Vol. 1205: B.Z. Moroz, Analytic Arithmetic in Algebraic Number Fields. VII, 177 pages. 1986.

Vol. 1206: Probability and Analysis, Varenna (Como) 1985. Seminar. Edited by G. Letta and M. Pratelli. VIII, 280 pages. 1986.

Vol. 1207: P.H. Bérard, Spectral Geometry: Direct and Inverse Problems. With an Appendix by G. Besson. XIII, 272 pages. 1986.

Vol. 1208: S. Kaijser, J.W. Pelletier, Interpolation Functors and Duality. IV, 167 pages. 1986.

Vol. 1209: Differential Geometry, Peñíscola 1985. Proceedings. Edited by A.M. Naveira, A. Ferrández and F. Mascaró. VIII, 306 pages. 1986.

Vol. 1210: Probability Measures on Groups VIII. Proceedings, 1985. Edited by H. Heyer. X, 386 pages. 1986.

Vol. 1211: M.B. Sevryuk, Reversible Systems. V, 319 pages. 1986.

Vol. 1212: Stochastic Spatial Processes. Proceedings, 1984. Edited by P. Tautu. VIII, 311 pages. 1986.

Vol. 1213: L.G. Lewis, Jr., J.P. May, M. Steinberger, Equivariant Stable Homotopy Theory. IX, 538 pages. 1986.

Vol. 1214: Global Analysis – Studies and Applications II. Edited by Yu.G. Borisovich and Yu.E. Gliklikh. V, 275 pages. 1986.

Vol. 1215: Lectures in Probability and Statistics. Edited by G. del Pino and R. Rebolledo. V, 491 pages. 1986.

Vol. 1216: J. Kogan, Bifurcation of Extremals in Optimal Control. VIII, 106 pages. 1986.

Vol. 1217: Transformation Groups. Proceedings, 1985. Edited by S. Jackowski and K. Pawalowski. X, 396 pages. 1986.

Vol. 1218: Schrödinger Operators, Aarhus 1985. Seminar. Edited by E. Balslev. V, 222 pages. 1986.

Vol. 1219: R. Weissauer, Stabile Modulformen und Eisensteinreihen. III, 147 Seiten. 1986.

Vol. 1220: Séminaire d'Algèbre Paul Dubreil et Marie-Paule Malliavin. Proceedings, 1985. Edité par M.-P. Malliavin. IV, 200 pages. 1986.

Vol. 1221: Probability and Banach Spaces. Proceedings, 1985. Edited by J. Bastero and M. San Miguel. XI, 222 pages. 1986.

Vol. 1222: A. Katok, J.-M. Strelcyn, with the collaboration of F. Ledrappier and F. Przytycki, Invariant Manifolds, Entropy and Billiards; Smooth Maps with Singularities. VIII, 283 pages. 1986.

Vol. 1223: Differential Equations in Banach Spaces. Proceedings, 1985. Edited by A. Favini and E. Obrecht. VIII, 299 pages. 1986.

Vol. 1224: Nonlinear Diffusion Problems, Montecatini Terme 1985. Seminar. Edited by A. Fasano and M. Primicerio. VIII, 188 pages. 1986.

Vol. 1225: Inverse Problems, Montecatini Terme 1986. Seminar. Edited by G. Talenti. VIII, 204 pages. 1986.

Vol. 1226: A. Buium, Differential Function Fields and Moduli of Algebraic Varieties. IX, 146 pages. 1986.

Vol. 1227: H. Helson, The Spectral Theorem. VI, 104 pages. 1986.

Vol. 1228: Multigrid Methods II. Proceedings, 1985. Edited by W. Hackbusch and U. Trottenberg. VI, 336 pages. 1986.

Vol. 1229: O. Bratteli, Derivations, Dissipations and Group Actions on C^*-algebras. IV, 277 pages. 1986.

Vol. 1230: Numerical Analysis. Proceedings, 1984. Edited by J.-P. Hennart. X, 234 pages. 1986.

Vol. 1231: E.-U. Gekeler, Drinfeld Modular Curves. XIV, 107 pages. 1986.

Vol. 1232: P.C. Schuur, Asymptotic Analysis of Soliton Problems. VIII, 180 pages. 1986.

Vol. 1233: Stability Problems for Stochastic Models. Proceedings, 1985. Edited by V.V. Kalashnikov, B. Penkov and V.M. Zolotarev. VI, 223 pages. 1986.

Vol. 1234: Combinatoire énumérative. Proceedings, 1985. Edité par G. Labelle et P. Leroux. XIV, 387 pages. 1986.

Vol. 1235: Séminaire de Théorie du Potentiel, Paris, No. 8. Directeurs: M. Brelot, G. Choquet et J. Deny. Rédacteurs: F. Hirsch et G. Mokobodzki. III, 209 pages. 1987.

Vol. 1236: Stochastic Partial Differential Equations and Applications. Proceedings, 1985. Edited by G. Da Prato and L. Tubaro. V, 257 pages. 1987.

Vol. 1237: Rational Approximation and its Applications in Mathematics and Physics. Proceedings, 1985. Edited by J. Gilewicz, M. Pindor and W. Siemaszko. XII, 350 pages. 1987.

Vol. 1238: M. Holz, K.-P. Podewski and K. Steffens, Injective Choice Functions. VI, 183 pages. 1987.

Vol. 1239: P. Vojta, Diophantine Approximations and Value Distribution Theory. X, 132 pages. 1987.

Vol. 1240: Number Theory, New York 1984–85. Seminar. Edited by D.V. Chudnovsky, G.V. Chudnovsky, H. Cohn and M.B. Nathanson. V, 324 pages. 1987.

Vol. 1241: L. Gårding, Singularities in Linear Wave Propagation. III, 125 pages. 1987.

Vol. 1242: Functional Analysis II, with Contributions by J. Hoffmann-Jørgensen et al. Edited by S. Kurepa, H. Kraljević and D. Butković. VII, 432 pages. 1987.

Vol. 1243: Non Commutative Harmonic Analysis and Lie Groups. Proceedings, 1985. Edited by J. Carmona, P. Delorme and M. Vergne. V, 309 pages. 1987.

Vol. 1244: W. Müller, Manifolds with Cusps of Rank One. XI, 158 pages. 1987.

Vol. 1245: S. Rallis, L-Functions and the Oscillator Representation. XVI, 239 pages. 1987.

Vol. 1246: Hodge Theory. Proceedings, 1985. Edited by E. Cattani, F. Guillén, A. Kaplan and F. Puerta. VII, 175 pages. 1987.

Vol. 1247: Séminaire de Probabilités XXI. Proceedings. Edité par J. Azéma, P.A. Meyer et M. Yor. IV, 579 pages. 1987.

Vol. 1248: Nonlinear Semigroups, Partial Differential Equations and Attractors. Proceedings, 1985. Edited by T.L. Gill and W.W. Zachary. IX, 185 pages. 1987.

Vol. 1249: I. van den Berg, Nonstandard Asymptotic Analysis. IX, 187 pages. 1987.

Vol. 1250: Stochastic Processes – Mathematics and Physics II. Proceedings 1985. Edited by S. Albeverio, Ph. Blanchard and L. Streit. VI, 359 pages. 1987.

Vol. 1251: Differential Geometric Methods in Mathematical Physics. Proceedings, 1985. Edited by P.L. García and A. Pérez-Rendón. VII, 300 pages. 1987.

Vol. 1252: T. Kaise, Représentations de Weil et GL_2 Algèbres de division et GL_n. VII, 203 pages. 1987.

Vol. 1253: J. Fischer, An Approach to the Selberg Trace Formula via the Selberg Zeta-Function. III, 184 pages. 1987.

Vol. 1254: S. Gelbart, I. Piatetski-Shapiro, S. Rallis. Explicit Constructions of Automorphic L-Functions. VI, 152 pages. 1987.

Vol. 1255: Differential Geometry and Differential Equations. Proceedings, 1985. Edited by C. Gu, M. Berger and R.L. Bryant. XII, 243 pages. 1987.

Vol. 1256: Pseudo-Differential Operators. Proceedings, 1986. Edited by H.O. Cordes, B. Gramsch and H. Widom. X, 479 pages. 1987.

Vol. 1257: X. Wang, On the C*-Algebras of Foliations in the Plane. V, 165 pages. 1987.

Vol. 1258: J. Weidmann, Spectral Theory of Ordinary Differential Operators. VI, 303 pages. 1987.

Vol. 1259: F. Cano Torres, Desingularization Strategies for Three Dimensional Vector Fields. IX, 189 pages. 1987.

Vol. 1260: N.H. Pavel, Nonlinear Evolution Operators and Semi groups. VI, 285 pages. 1987.

Vol. 1261: H. Abels, Finite Presentability of S-Arithmetic Groups. Compact Presentability of Solvable Groups. VI, 178 pages. 1987.

Vol. 1262: E. Hlawka (Hrsg.), Zahlentheoretische Analysis II. Semina 1984–86. V, 158 Seiten. 1987.

Vol. 1263: V.L. Hansen (Ed.), Differential Geometry. Proceedings 1985. XI, 288 pages. 1987.

Vol. 1264: Wu Wen-tsün, Rational Homotopy Type. VIII, 219 pages 1987.

Vol. 1265: W. Van Assche, Asymptotics for Orthogonal Polynomials VI, 201 pages. 1987.

Vol. 1266: F. Ghione, C. Peskine, E. Sernesi (Eds.), Space Curves Proceedings, 1985. VI, 272 pages. 1987.

Vol. 1267: J. Lindenstrauss, V.D. Milman (Eds.), Geometrical Aspec of Functional Analysis. Seminar. VII, 212 pages. 1987.

Vol. 1268: S.G. Krantz (Ed.), Complex Analysis. Seminar, 1986. VI 195 pages. 1987.

Vol. 1269: M. Shiota, Nash Manifolds. VI, 223 pages. 1987.

Vol. 1270: C. Carasso, P.-A. Raviart, D. Serre (Eds.), Nonline Hyperbolic Problems. Proceedings, 1986. XV, 341 pages. 1987.

Vol. 1271: A.M. Cohen, W.H. Hesselink, W.L.J. van der Kallen, J.F Strooker (Eds.), Algebraic Groups Utrecht 1986. Proceedings. X 284 pages. 1987.

Vol. 1272: M.S. Livšic, L.L. Waksman, Commuting Nonselfadjoi Operators in Hilbert Space. III, 115 pages. 1987.

Vol. 1273: G.-M. Greuel, G. Trautmann (Eds.), Singularities, Repr sentation of Algebras, and Vector Bundles. Proceedings, 1985. XI 383 pages. 1987.

Vol. 1274: N. C. Phillips, Equivariant K-Theory and Freeness of Grou Actions on C*-Algebras. VIII, 371 pages. 1987.

Vol. 1275: C.A. Berenstein (Ed.), Complex Analysis I. Proceeding 1985–86. XV, 331 pages. 1987.

Vol. 1276: C.A. Berenstein (Ed.), Complex Analysis II. Proceeding 1985–86. IX, 320 pages. 1987.

Vol. 1277: C.A. Berenstein (Ed.), Complex Analysis III. Proceeding 1985–86. X, 350 pages. 1987.

Vol. 1278: S.S. Koh (Ed.), Invariant Theory. Proceedings, 1985. 102 pages. 1987.

Vol. 1279: D. Ieşan, Saint-Venant's Problem. VIII, 162 Seiten. 1987.

Vol. 1280: E. Neher, Jordan Triple Systems by the Grid Approach. X 193 pages. 1987.

Vol. 1281: O.H. Kegel, F. Menegazzo, G. Zacher (Eds.), Gro Theory. Proceedings, 1986. VII, 179 pages. 1987.

Vol. 1282: D.E. Handelman, Positive Polynomials, Convex Integ Polytopes, and a Random Walk Problem. XI, 136 pages. 1987.

Vol. 1283: S. Mardešić, J. Segal (Eds.), Geometric Topology a Shape Theory. Proceedings, 1986. V, 261 pages. 1987.

Vol. 1284: B.H. Matzat, Konstruktive Galoistheorie. X, 286 page 1987.

Vol. 1285: I.W. Knowles, Y. Saitō (Eds.), Differential Equations a Mathematical Physics. Proceedings, 1986. XVI, 499 pages. 1987.

Vol. 1286: H.R. Miller, D.C. Ravenel (Eds.), Algebraic Topolog Proceedings, 1986. VII, 341 pages. 1987.

Vol. 1287: E.B. Saff (Ed.), Approximation Theory, Tampa. Procee ings, 1985–1986. V, 228 pages. 1987.

Vol. 1288: Yu. L. Rodin, Generalized Analytic Functions on Riema Surfaces. V, 128 pages, 1987.

Vol. 1289: Yu. I. Manin (Ed.), K-Theory, Arithmetic and Geomet Seminar, 1984–1986. V, 399 pages. 1987.